RESPONSES OF PLANTS TO AIR POLLUTION

PHYSIOLOGICAL ECOLOGY

A Series of Monographs, Texts, and Treatises

EDITED BY

T. T. KOZLOWSKI

University of Wisconsin
Madison, Wisconsin

T. T. Kozlowski. Growth and Development of Trees, Volumes I and II – 1971

Daniel Hillel. Soil and Water: Physical Principles and Processes, 1971

J. Levitt. Responses of Plants to Environmental Stresses, 1972

V. B. Youngner and C. M. McKell (Eds.). The Biology and Utilization of Grasses, 1972

T. T. Kozlowski (Ed.). Seed Biology, Volumes I, II, and III – 1972

Yoav Waisel. Biology of Halophytes, 1972

G. C. Marks and T. T. Kozlowski (Eds.). Ectomycorrhizae: Their Ecology and Physiology, 1973

T. T. Kozlowski (Ed.). Shedding of Plant Parts, 1973

Elroy L. Rice. Allelopathy, 1974

T. T. Kozlowski and C. E. Ahlgren (Eds.). Fire and Ecosystems, 1974

J. Brian Mudd and T. T. Kozlowski (Eds.). Responses of Plants to Air Pollution, 1975

Rexford Daubenmire, Plant Geography, 1978

RESPONSES OF PLANTS TO AIR POLLUTION

Edited by

J. Brian Mudd
Department of Biochemistry
University of California
Riverside, California

T. T. Kozlowski
Department of Forestry
University of Wisconsin
Madison, Wisconsin

ACADEMIC PRESS New York San Francisco London 1975

A Subsidiary of Harcourt Brace Jovanovich, Publishers

ACADEMIC PRESS, INC.
111 Fifth Avenue, New York, New York 10003

United Kingdom Edition published by
ACADEMIC PRESS, INC. (LONDON) LTD.
24/28 Oval Road, London NW1

Library of Congress Cataloging in Publication Data
Main entry under title:

Responses of plants to air pollution.

(Physiological ecology series)
Includes bibliographies and index.
1. Plants, Effect of air pollution on. I. Mudd, John Brian, (date) II. Kozlowski, Theodore Thomas, (date)
QK751.R47 581.2'4 75-3969
ISBN 0-12-509450-7

PRINTED IN THE UNITED STATES OF AMERICA

CONTENTS

3 Ozone

Robert L. Heath

4 Fluorides

Chong W. Chang

5 Peroxyacyl Nitrates

J. B. Mudd

6 Oxides of Nitrogen

O. C. Taylor, C. R. Thompson, D. T. Tingey, and R. A. Reinert

7 Particulates

Shimshon L. Lerman and Ellis F. Darley

13 Interaction of Air Pollutants and Plant Disease

Michael Treshow

14 Interactions of Air Pollution and Agricultural Practices

Saul Rich

LIST OF CONTRIBUTORS

Numbers in parentheses indicate the pages on which the authors' contributions begin.

JESSE H. BENNETT (273),* Department of Biology, University of Utah, Salt Lake City, Utah

CHONG W. CHANG (57), USDA, Agricultural Research Service, Western Cotton Research Laboratory, Phoenix, Arizona

ELLIS F. DARLEY (141), Statewide Air Pollution Research Center, University of California, Riverside, California

A. S. HEAGLE (159), Agricultural Research Service, Plant Pathology Department, North Carolina State University, Raleigh, North Carolina

ROBERT L. HEATH (23), Departments of Biology and Biochemistry, University of California, Riverside, California

W. W. HECK (159), Agricultural Research Service, Botany Department, North Carolina State University, Raleigh, North Carolina

A. CLYDE HILL (273), Department of Biology, University of Utah, Salt Lake City, Utah

T. T. KOZLOWSKI (1), Department of Forestry, University of Wisconsin, Madison, Wisconsin

FABIUS LEBLANC (237), Department of Biology, University of Ottawa, Ottawa, Ontario, Canada

* Present address: USDA, Agricultural Research Center, Beltsville, Maryland

SHIMSHON L. LERMAN (141), Statewide Air Pollution Research Center, University of California, Riverside, California

JOE R. MCBRIDE (195), Department of Forestry and Resource Management, College of Natural Resources, University of California, Berkeley, California

PAUL R. MILLER (195), Air Pollution Research Center, University of California, Riverside, California

J. B. MUDD (1, 9, 97), Department of Biochemistry, University of California, Riverside, California

DHRUVA N. RAO (237), Department of Botany, Banaras Hindu University, Vanarasi, India

R. A. REINERT (121, 159), USDA, Agricultural Research Service Department of Plant Pathology, North Carolina State University, Raleigh, North Carolina 27607

SAUL RICH (335), Department of Plant Pathology and Botany, The Connecticut Agricultural Experiment Station, New Haven, Connecticut

O. C. TAYLOR (121),* Statewide Air Pollution Research Center, University of California, Riverside, California

C. R. THOMPSON (121), Statewide Air Pollution Research Center, University of California, Riverside, California

WILLIAM W. THOMSON (179), Department of Biology, University of California, Riverside, California

D. T. TINGEY (121), Environmental Protection Agency, National Ecological Research Laboratory, Corvallis, Oregon

MICHAEL TRESHOW (307), Department of Biology, University of Utah, Salt Lake City, Utah

* Present address: EPA, National Environmental Research Center, Corvallis, Oregon

PREFACE

Unrelenting pollution of the earth resulting from rapidly growing population and industry is one of the most serious problems confronting mankind. We must now deal with a wide array of toxic substances, including gases, particulates, and radioactive materials, which affect our food supply, health, and economy. In particular, the yields of practically all crop plants and structures of natural ecosystems are adversely affected by atmospheric pollutants. The tragedy of all this is that the pollution problem is manmade, and has a strong base in years of well-demonstrated apathy. It is against such a background that this treatise examines the effects of air pollutants, individually and synergistically, on both higher and lower plants.

The book is comprehensive, authoritative, and well-documented. The opening chapter presents an overview of sources of air pollution, costs of air pollution, and mechanisms of pollution injury to plants. Separate chapters on sulfur dioxide, ozone, fluorides, peroxyacyl nitrates, oxides of nitrogen, and particulates follow. Subsequent chapters are devoted to plant responses to combinations of pollutants; to effects of pollutants on plant ultrastructure, on forests, and on lichens and bryophytes; to interactions of pollutants with canopies of vegetation; to interactions of pollutants and plant diseases; and to interactions of pollutants with agricultural practices.

We hope that this volume will be interesting and useful to scientists in many disciplines as well as to others who share our concern that we can no longer expect clean air to be the normal environment for plants or animals. The book will be particularly useful as a reference work or text for upper level undergraduate students, graduate students, researchers, and growers of plants. The subject matter overlaps into a wide variety of disciplines including agronomy, plant anatomy, biochemistry, cryptogamic botany, ecology, entomology, forestry, horticulture, landscape architecture, meteorology, microscopy, plant pathology, plant physiology, and soil science.

We are indebted to our distinguished collaborators for their enthusiastic responses to contribute to this joint effort, for sharing their expertise, and for their patience during the production phases. Mr. P. E. Marshall and Mr. T. L. Noland assisted with preparation of the Subject Index.

J. Brian Mudd
T. T. Kozlowski

RESPONSES OF PLANTS TO AIR POLLUTION

1

INTRODUCTION

T. T. Kozlowski and J. B. Mudd

Dealing with environmental pollution promises to be one of man's most urgent problems in the years to come. We can no longer expect clean air to be the normal environment for plant and animal growth and are faced with the fact that in relatively recent times the total amount and complexity of toxic pollutants in the environment have increased alarmingly. Whereas prior to 1940 the major recognized pollutants were particulate matter and sulfur dioxide in smoke, we now realize we must cope with a wide array of environmental pollutants including gases, particulates, agricultural chemicals, radioactive materials in the atmosphere, sewage and chemicals in water, and solid wastes on land. Individually and synergistically these toxic substances adversely affect man's food supply, health, and well-being. The polluting substances poison our air and waters. They enter food chains of higher animals, alter reproductive capacity of animals, cause or aggravate eye and respiratory diseases, corrode metals and building materials, and adversely affect industrial equipment.

Growing plants are particularly susceptible to pollution, with reduction in photosynthesis and growth often occurring before visible symptoms of injury are noted. Yields of virtually all important crop plants have been greatly depressed by air pollution. Because of the high concentration of pollutants around large cities it has become increasingly difficult to grow truck crops in many areas, especially leafy ones such as lettuce. In some heavily polluted areas, as in the Los Angeles basin, it has become necessary to abandon many citrus groves and truck farms.

Accumulation of toxic substances in the biosphere is causing serious changes in the structure and function of natural ecosystems. In forested areas trees are eliminated first by low dosages of pollutants. As duration of exposure increases, tall shrubs are eliminated, followed in order by lower shrubs, herbs, mosses, and lichens (Woodwell, 1970). In addition to affecting ecosystems directly, atmospheric pollutants undoubtedly will have some long–term effects on plants by influencing CO_2 content, light intensity, temperature, and precipitation. With our present state of knowledge it is difficult to predict specifically what those changes will be (Wenger *et al.*, 1971).

I. Sources of Pollution

For the most part environmental pollution is a man-made problem. As may be seen in Table I the motor vehicle is the most important source of atmospheric pollution (about 100 million tons released in 1970). Industrial sources are a distant second (26 million tons in 1970), emitting about one-fourth as much as transportation does. Generation of steam and electric power produces slightly less (22 million tons in 1970) than industry. Space heating emits slightly less than 10% (9 million tons in 1970) as much as transportation. The composition of pollutants from various sources differs greatly, with industry emitting the most diversified pollutants. Whereas carbon monoxide is the major component of pollution by motor vehicles, sulfur oxides are primary pollutants of industry, power generation, and space heating.

A number of agricultural chemicals and biocides adversely affect growth and development of plants. In addition to excess nutrients from fertilizers, which wash into lakes and streams and cause undesirable changes in aquatic environments, such compounds include some insecticides, fungicides, herbicides, and antitranspirants. Commonly some of these chemicals inhibit photosynthesis by clogging stomata or causing changes in optical properties of leaves, heat balance of leaves, leaf metabolism, leaf anatomy, or by combinations of these. These chemicals also reduce photosynthesis by causing lesions on leaves, chlorosis, or abscission (Kozlowski and Keller,

TABLE I
SOURCE OF ESTIMATED AIR POLLUTION IN 1968 BY WEIGHT[a]

Source	% Sulfur oxides	% Nitrogen oxides	% Carbon monoxide	% Hydro-carbons	% Particu-lates	Total
Transportation	0.7	4.2	46.5	8.5	0.7	60.6
Industry	6.3	1.4	1.4	2.8	4.2	16.1
Power generation	8.5	2.2	0.7	0.7	2.1	14.2
Space heating	2.1	0.7	1.4	0.7	0.7	5.6
Refuse burning	0.7	0.7	0.7	0.7	0.7	3.5
	18.3	9.2	50.7	13.4	8.4	100.0

[a] From Hesketh (1973).

1966). The influence of agricultural chemicals on plants varies greatly with the specific chemical, rate, method, time, number of applications, species, soil type, weather, and other factors (Kozlowski, 1971).

Plants themselves have sometimes been considered to contribute to pollution by giving off chemical substances which inhibit growth of other plants (Rice, 1974). Such allelopathic chemicals may be released by plants to soil by leaching, volatilization, excretion, exudation, and decay either directly or by activity of microorganisms. Among the naturally occurring compounds which have inhibitory effects on growth of neighboring plants are organic acids, lactones, fatty acids, quinones, terpenoids, steroids, phenols, benzoic acid, cinnamic acid, coumarins, flavonoids, tannins, amino acids, polypeptides, alkaloids, cyanohydrins, sulfides, mustard oil glycosides, purines, and nucleosides. Allelopathic chemicals are ecologically important because they influence succession, dominance, vegetation dynamics, species diversity, structure of plant communities, and productivity (Whittaker, 1970).

Several investigators have reported inhibitory effects of specific plants on seed germination and growth of adjacent plants. A few examples will be given. Perhaps the best known allelopathic chemical is juglone in *Juglans*. It is washed into the soil from leaves and fruits and inhibits growth of adjacent plants (Brooks, 1951). The toxic leachates of leaves of *Artemisia absinthium* (Funke, 1943), *Encelia farinosa* (Gray and Bonner, 1948), and *Ailanthus* (Mergen, 1959) have also been reported. In California the annual vegetation adjacent to naturalized stands of *Eucalyptus camaldulensis* is greatly inhibited. Where *Eucalyptus* litter accumulates annual herbs rarely survive to maturity (del Moral and Muller, 1970), indicating allelopathic influences.

II. Cost of Air Pollution

Some idea of the magnitude of the national cost of air pollution in 1968 in terms of additional costs of health care, reduction in property value, degradation of materials, and damage to vegetation (including agricultural productivity) may be gained from Table II. The total annual cost was in excess of $16 billion, and in the following order: health > residential property > materials > vegetation. According to Barrett and Waddell (1973), particulates and sulfur oxides were responsible for most health costs of air pollution. Damage to residential property was traced primarily to sulfur oxides and particulates, acting individually and synergistically. Several air pollutants caused damage to materials. Corrosion of steel was attributed entirely to sulfur oxides, and fading of dyes to oxidants and nitrogen oxides. About 90% of direct losses of plant crops was traceable to effects of oxidants ($100 million out of a total loss of $210 million). Waddell (1970) indicated that sulfur oxides accounted for direct damage of $13 million, with the rest of the damage to crops caused by particulates, such as dusts of fluorides, lead, and other pollutants.

III. Mechanisms of Pollution Injury to Plants

In addition to killing plants, atmospheric pollutants adversely affect plants in various ways. Pollution injury is most commonly classed as acute, chronic (chlorotic), or hidden. In acute injury collapsed marginal or intercostal leaf areas are noted which at first have a water-soaked appearance. Later these dry and bleach to an ivory color in most species, but in some they become brown or brownish red. These lesions are caused by absorption of enough gas to kill the tissue. Chronic injury involves leaf yellowing which may progress slowly through stages of bleaching until most of the

TABLE II

COST OF POLLUTION DAMAGE, BY POLLUTANTS IN 1968 (BILLIONS OF DOLLARS)[a]

Loss category	Sulfur oxides	Particulates	Oxidants	Nitrogen oxides	Total
Residential property	2.808	2.392	—	—	5.200
Materials	2.202	0.691	1.127	0.732	4.752
Health	3.272	2.788	—	—	6.060
Vegetation	0.013	0.007	0.060	0.040	0.120
Total	8.295	5.878	1.187	0.772	16.132

[a] From Barrett and Waddell (1973).

chlorophyll and carotenoids are destroyed and interveinal portions of the leaf are nearly white. Chronic injury is caused by absorption of an amount of gas that is somewhat insufficient to cause acute injury or it may be caused by absorption over a long period of time of sublethal amounts of gas (Thomas, 1951, 1961). Sometimes the effects of different pollutants are separable and at other times indistinguishable. Solberg and Adams (1956) found histological responses of dicotyledonous plants to hydrogen fluoride and sulfur dioxide to be indistinguishable. Costonis (1970) did not find any histological differences of diagnostic significance between SO_2 and ozone injury to *Pinus strobus*.

The histological changes most commonly noted in pollution-injured leaves include plasmolysis, granulation or disorganization of cell contents, cell collapse or disintegration, and pigmentation of affected tissue (Darley and Middleton, 1966). Sometimes the physiological activity of affected plants is impaired before any external symptoms are visible. For this reason many investigators referred to "hidden," "invisible," or "physiological" injury of pollutants. Early criteria for such invisible injury included the following: (1) it involved a disturbance of the life of the plant that was eventually expressed as an effect on growth, (2) the disturbance was not evident to the naked eye, and (3) it was present where plants underwent prolonged exposure to concentrations of pollutants that did not produce visible markings. Haselhoff and Lindau (1903), however, believed that "invisible injury" was not a useful term because anatomical changes in response to pollution were visible with a microscope. They believed that pollutant effects could be readily described as either acute or chronic. They considered that chlorotic or necrotic lesions were evidence of decreased photosynthetic tissue and that both visible symptoms and decreased growth were due to disruption of cell structure and activity. Many other investigators did not find evidence of hidden injury following experimental fumigation or field observations and contended that there was no visible damage (Hill and Thomas, 1933; Johnson, 1932; Katz, 1949; Thomas, 1951, 1956; Thomas *et al.*, 1943). On the other hand, some investigators noted that pollution produced a complex of physiological effects such as reduced water uptake, stunted growth, and stomatal closure (Hull and Went, 1952; Koritz and Went, 1953). Inasmuch as these effects were not immediately obvious and were not accompanied by foliar lesions, they were described as "hidden damage." Working with hydrogen fluoride Thomas (1956, 1961) and Thomas and Hendricks (1956) redefined "invisible" injury as the reduction of photosynthesis below the level expected from the amount of leaf destruction. Hidden injury was considered to have two main characteristics: (1) it did not occur unless a threshold concentration of hydrogen fluoride was exceeded, and (2) it had a certain magnitude and a limited duration once the exposure to the pollutant ceased. These characteristics were determined by plant species and variety. The existence of some form of hidden injury was also recognized by Brewer *et al.* (1960), Reckendorfer (1952), and Applegate and Adams (1960). One of the problems in pollution research is that no standardized definition of hidden injury has been widely accepted. Rather a wide variety of terms and concepts has been used.

A complication in evaluating the physiological mechanism of pollution injury is that such factors as light, water, temperature, and mineral nutrition affect the response of plants to pollutants (Rich, 1964; Darley and Middleton, 1966). Still another difficulty is that more than one pollutant often is responsible for injury in the field. For example, both ozone and SO_2 were responsible for the physiogenic needle disorder of *Pinus strobus* known as "chlorotic dwarf" (Dochinger and Heck, 1969). Even tolerable levels of a single pollutant may injure plants when present with another pollutant at an equally low level.

Air pollutants appear to be relatively nonspecific agents which have many sites of action. They inhibit many enzyme systems and metabolic processes. The effect of a pollutant depends on its concentration in a cell as well as metabolic patterns in cells. Most air pollutants decrease photosynthesis directly or indirectly by causing loss of photosynthetic tissues (e.g., leaf abscission, chlorosis, necrosis) and by affecting stomatal aperture.

The following chapters will examine in greater detail the impact of individual atmospheric pollutants (gases and particulates) and synergisms between them, on both lower and higher plants. Attention will be given to uptake of pollutants, plant susceptibility and symptoms, and especially to biochemical responses of plants to air pollutants.

References

Applegate, H. G., and Adams, D. F. (1960). "Invisible injury" of bush beans by atmospheric and aqueous fluorides. *Int. J. Air Pollut.* **3,** 231–248.

Barrett, L. B., and Waddell, T. E. (1973). "Cost of Air Pollution Damage: A Status Report," Publ. AP-85. Environmental Protection Agency, Washington, D.C.

Brewer, R. F., Sutherland, F. H., Guillemet, F. B., and Crevcling, R. K. (1960). Some effects of hydrogen fluoride gas on bearing navel orange trees. *Proc. Amer. Soc. Hort. Sci.* **76,** 208–214.

Brooks, M. G. (1951). Effect of black walnut trees and their products on other vegetation. *W. Va., Agr. Exp. Sta., Bull.* **347.**

Costonis, A. C. (1970). Acute foliar injury of eastern white pine induced by sulfur dioxide and ozone. *Phytopathology* **60,** 994–999.

Darley, E. F., and Middleton, J. T. (1966). Problems of air pollution in plant pathology. *Annu. Rev. Plant Pathol.* **4,** 103–118.

del Moral, R., and Muller, C. H. (1970). The allelopathic effects of *Eucalyptus camaldulensis*. *Amer. Midl. Natur.* **83,** 254–282.

Dochinger, L. S., and Heck, W. W. (1969). An ozone–sulfur dioxide synergism produces symptoms of chlorotic dwarf of eastern white pine. *Phytopathology* **59,** 399.

Funke, G. L. (1943). The influence of *Artemisia absinthium* on neighboring plants. *Blumea* **5,** 281–293.

Gray, R., and Bonner, J. (1948). An inhibitor of plant growth from the leaves of *Encelia farinosa*. *Amer. J. Bot.* **34,** 52–57.

Haselhoff, E. and Lindau, G. (1903). Die Beschädigung der Vegetation durch Rauch. Gebrüder Borntraeger, Leipzig.

Hesketh, H. E. (1973). "Understanding and Controlling Air Pollution." Ann Arbor Sci. Publ., Ann Arbor, Michigan.

Hill, G. R., and Thomas, M. D. (1933). Influence of leaf destruction by sulfur dioxide and by clipping on the yield of alfalfa. *Plant Physiol.* **8,** 223–245.

Hull, H. M., and Went, F. W. (1952). Life processes of plants as affected by air pollution. *Proc. Nat. Air Pollut. Symp., 2nd., 1952,* pp. 122–128.

Johnson, A. B. (1932). A study of the action of sulfur dioxide at low concentrations on the wheat plant, with particular reference to the question of "invisible injury." Ph.D. Thesis, Stanford University, Palo Alto, California.

Katz, M. (1949). Sulfur dioxide in the atmosphere and its relation to plant life. *Ind. Eng. Chem.* **42,** 2450–2465.

Koritz, H. G., and Went, F. W. (1953). The physiological action of smog on plants. I. Initial growth and transpiration studies. *Plant Physiol.* **28,** 50–62.

Kozlowski, T. T. (1971). "Growth and Development of Trees," Vol. I. Academic Press, New York.

Kozlowski, T. T., and Keller, T. (1966). Food relations of woody plants. *Bot. Rev.* **32,** 293–382.

Mergen, F. (1959). A toxic principle in the leaves of *Ailanthus*. *Bot. Gaz. (Chicago)* **121,** 32–36.

Reckendorfer, P. (1952). Ein Beitrag zur Microchemie des Rauchschadens durch Fluor. Die Wanderung des Fluors im pflanzlichen Gewebe. I. Die unsichtbaren Schaden. *Pflanzenschutzberichte* **9,** 33–55.

Rice, E. (1974). "Allelopathy." Academic Press, New York.

Rich, S. (1964). Ozone damage to plants. *Annu. Rev. Phytopathol.* **2,** 253–266.

Solberg, R. A., and Adams, D. F. (1956). Histological responses of some plant leaves to hydrogen fluoride and sulfur dioxide. *Amer. J. Bot.* **43,** 755–760.

Thomas, M. D. (1951). Gas damage to plants. *Annu. Rev. Plant Physiol.* **2,** 293–322.

Thomas, M. D. (1956). The invisible injury theory of plant damage. *J. Air Pollut. Contr. Ass.* **5,** 205–208.

Thomas, M. D. (1961). Effects of air pollution on plants. *In* "Air Pollution," Monogr. No. 46, pp. 233–278. World Health Organ., Geneva.

Thomas, M. D., and Hendricks, R. H. (1956). Effects of air pollutants on plants. *In* "Air Pollution Handbook" (P. L. Magill, F. R. Holden, and C. Ackley, eds.), Sect. 9, pp. 9–41.

Thomas, M. D., Hendricks, R. H., Collier, T. R., and Hill, G. R. (1943). The utilization of sulphate and sulphur dioxide for the sulphur nutrition of alfalfa. *Plant Physiol.* **18,** 345–371.

Waddell, T. E. (1970). "Economic Effects of Sulfur Oxide Air Pollution from Point Sources on Vegetation and Environment." U.S.D.H.E.W., P.H.S., E.H.S., Nat. Air Pollut. Contr. Admin., Raleigh, North Carolina.

Wenger, K. F., Ostrom, C. E., Larson, P. R., and Rudolph, T. D. (1971). Potential effects of global atmospheric conditions on forest ecosystems. *In* "Man's Impact on Terrestrial and Ocean Ecosystems" (W. M. Matthews, F. E. Smith, and E. D. Goldberg, eds.), pp. 192–202. MIT Press, Cambridge, Massachusetts.

Whittaker, R. H. (1970). The biochemical ecology of higher plants. *In* "Chemical Ecology" (E. Sondheimer and J. B. Simeone, eds.), Chapter 3, pp. 43–70. Academic Press, New York.

Woodwell, G. M. (1970). Effects of pollution on the structure and physiology of ecosystems. *Science* **168,** 429–433.

2

SULFUR DIOXIDE

J. B. Mudd

I. Introduction

Sulfur dioxide has a more venerable history as an air pollutant than any other chemical. The most common source of sulfur dioxide in the atmosphere is the combustion of fossil fuels, coal, and oil. This source can be minimized by using fuels low in sulfur content. If sulfur-containing fuels are combusted, various control techniques are available (National Air Pollution Control Administration, 1969). The contributors of sulfur dioxide to the air of the world are coal combustion 51 metric tons sulfur/year, petroleum refining 3 metric tons sulfur/year, petroleum combustion 11 metric tons sulfur/year, smelting 8 metric tons sulfur/year (data of 1965).

The toxicity of sulfur dioxide to plants is considered in various aspects in several chapters in this volume. This chapter will cover some of the physiological and biochemical effects of sulfur dioxide.

II. Chemical Considerations

A. Laboratory Studies

Some of the chemical properties of sulfur dioxide are discussed in the book by Schroeter (1966). The solubility of SO_2 in water is 228 g/liter at 0°C. This is equivalent to dissolving 80 volumes of SO_2 in one volume of water. SO_2 is also quite soluble in some organic solvents, for example the partition coefficient H_2O/chloroform is 0.72, and for H_2O/benzene it is 0.44.

Hydration of SO_2 is very rapid:

$$SO_2 + H_2O \underset{k_{-1}}{\overset{k_1}{\rightleftharpoons}} HSO_3^- + H^+$$

$$k_1 = 3.4 \times 10^6 \text{ sec}^{-1}$$

$$k_{-1} = 2 \times 10^8 \ M^{-1} \text{ sec}^{-1}$$

at 20°C and 0.1 M ionic strength. It is notable that in the presence of surfactants the rate of absorption of SO_2 in water is decreased because SO_2 is taken up in the surfactant layer.

There is considerable doubt that H_2SO_3 exists in solution, thus the un-ionized sulfur species is SO_2 dissolved in water. Nevertheless, ionization constants have been determined.

$$H_2SO_3(H_2O + SO_2) \rightleftharpoons HSO_3^- + H^+ \qquad K = 1.72 \times 10^{-2}$$

$$HSO_3^- \rightleftharpoons SO_3^{2-} + H^+ \qquad K = 6.24 \times 10^{-8}$$

B. Atmospheric Studies

SO_2 is the main sulfur compound emitted into the atmosphere. It can be oxidized to SO_3 which forms H_2SO_4 when hydrated:

$$SO_2 \rightarrow SO_3$$

$$SO_3 + H_2O \rightarrow H_2SO_4 \rightarrow \text{aerosol}$$

The relationship of SO_2 to sulfate has been shown from collection of monitoring data in the United States (Altshuller, 1973). The relationship between SO_2 and sulfate is nonlinear: at lower SO_2 concentrations the sulfate increases in proportion to SO_2, but at higher SO_2 concentrations there is no further increase in sulfate. In the specific case of the city of Philadelphia, the SO_2 sulfate ratio was twice as high in central Philadelphia as compared to suburban sites. Thus some of the sulfate may be derived from SO_2 oxidation.

Cox and Penkett (1971) have reported the photooxidation of SO_2 in the absence and presence of NO and 2-pentene. The oxidation of SO_2 to H_2SO_4 aerosol was greatly enhanced in the presence of NO and 2-pentene. This enhancement is undoubtedly an offshoot of the mechanism for the production of "photochemical smog," but the actual oxidant involved is unknown. Ozone does not seem likely because it does not react readily with SO_2 at low concentrations. Cox and Penkett (1971) suggest that the oxidant may be peroxy radicals or peroxyacyl nitrates.

The reaction of SO_2 or sulfate with plants depends on activity in solution: Thomas *et al.* (1943) estimate sulfite to be 30 times more toxic than sulfate, so the degree of oxidation of SO_2 to sulfate in the atmosphere has important consequences as far as the plant is concerned.

III. Susceptibility

A. Symptoms

Excellent photographs of SO_2 damage to plants can be found in the book of Jacobson and Hill (1970). The most common type of acute injury is interveinal chlorosis (Fig. 1). Interveinal areas become bleached in most cases, but sometimes the destroyed tissue is brown. Sensitive species include spinach (*Spinacea oleracea*), cucumber (*Cucumis sativus*), and oats (*Avena sativa*), and resistance is found in maize (*Zea mays*), celery (*Apium graveolens*), and citrus species.

It is generally accepted that SO_2 injury depends on entry through the stomata, and so conditions which favor open stomata at the time of exposure predispose the plants to injury. For example, water stress tends to close the stomata and protect the plant from injury. The young but fully expanded leaves are the first to show injury, whereas those which are still expanding are least affected.

B. Dosage

Susceptible species may be damaged by 0.05–0.5 ppm SO_2 for 8 hours. For shorter term exposures (30 minutes) the concentration must be raised

FIG. 1. Interveinal chlorosis caused by sulfur dioxide. Top (left to right): cucumber, tomato. Bottom (left to right): rose, strawberry. (Photo courtesy Statewide Air Pollution Research Center, Riverside, California.)

to 1–4 ppm. Resistant species will require 2 ppm for 8 hours or 10 ppm for 30 minutes. These values come from the use of SO_2 as a single pollutant, and the effect may be greater if SO_2 is administered in the presence of another pollutant.

There is much information in the literature concerning a "threshold" for SO_2 damage to plants. The essence of this concept is that below a certain concentration of SO_2, damage does not take place, presumably because the plant is able to metabolize the dissolved SO_2 to nontoxic products. Once this metabolic capability is exceeded, the toxic compounds accumulate and the threshold has been crossed. The conception of "threshold" was embodied in the equation O'Gara (1922) used to describe the conditions for development of SO_2 damage.

$$(C - C_R)t = K$$

C = concentration of SO_2
C_R = threshold concentrations (0.33 ppm in O'Gara's case)
t = time in hours required to initiate damage
K = constant; the threshold dose

This equation has been modified and substitutes have been suggested, but the variables that can affect the development of symptoms are so numerous that further sophistication does not seem profitable. The essential points in O'Gara's equation are (1) that below a certain concentration no damage will take place no matter how long the fumigation, and that (2) above that certain concentration damage can be elicited by combinations of concentration and times of exposure.

IV. Physiological Effects

A. Stomatal Opening

Unsworth *et al.* (1972) reported that concentrations of SO_2 in the range 0.1–0.5 ppm stimulated stomatal opening. This was especially true in water stressed plants where stomata tended to be partially closed. They reported that SO_2 could even cause partial opening of stomata in the dark. This stimulation of stomatal opening could be particularly relevant in cases of synergism of other pollutants with SO_2: SO_2 keeps stomata open and facilitates entry of the other pollutant. In the synergism of SO_2 and O_3 reported by Menser and Heggestad (1966) the symptoms were indeed those of O_3, although the increase in stomatal aperture with $O_3 + SO_2$ was not dramatic in this case and was most noticeable on the upper leaf surface.

But the effects of SO_2 on stomata do not seem to be simple. Majernik and Mansfield (1972) have reported a series of studies on the effect of SO_2 and other variables on stomatal opening. First they reported that SO_2 stimulated stomatal opening at relative humidities greater than 80% and later further qualifications were made. They summarized their work as follows:

a. In moist air (water vapor pressure deficit less than 7 mm Hg) containing 320–330 ppm CO_2, the stomata normally open wider in the presence of SO_2 at concentrations of 0.25 ppm and above (Majernik and Mansfield, 1970, 1971)

b. When relative humidity is low (water vapor pressure deficit greater than 7 mm Hg) the stomata close in response to the same range of SO_2 concentrations (Mansfield and Majernik, 1970).

c. The opening reaction to SO_2 in moist air can be counteracted by raised ambient CO_2 levels.

d. The very wide stomatal opening in moist CO_2 free air is only slightly enhanced by SO_2.

The last two points are not important as far as natural vegetation is concerned, but the first two are, and help explain the observation of earlier workers on the interaction of SO_2 and humidity. Majernik and Mansfield (1972) emphasize that their findings were made entirely with *Vicia faba* which may not be typical of other species. Indeed, the conditions used by Menser and Heggestad (1966) for SO_2 fumigations of tobacco would have caused opening of *Vicia faba* stomata but caused the closure of the tobacco stomata.

B. Synergism

The synergism of SO_2 and O_3 in causing plant damage to tobacco was first reported by Menser and Heggestad (1966). Concentrations of the gases were in the order of 0.03 ppm O_3 and 0.25 ppm SO_2 fumigated in the light at 24–26°C and 80–100% relative humidity for 2 or 4 hours. Neither gas alone produced symptoms. These results were confirmed by MacDowall and Cole (1971). The latter authors determined threshold doses for the two gases and gave the figures for ozone as 20 pphm/hour (e.g., 20 pphm for 1 hour) and 300 pphm/hour for SO_2. The synergism was most pronounced at the threshold doses for the two gases.

The topic of synergism is more completely discussed in Chapter 8 by Reinert, Haegle and Heck.

C. Amino Acid Composition

Arndt (1970) measured the effect of SO_2 fumigation (0.24 ppm) on the free amino acids of *Phaseolus vulgaris, Trifolium repens,* and a mixture of pasture grasses. There was a tendency for the content of free amino acids to be increased in the fumigated plants, but there was no indication that the increase could generally be accounted for by one or two amino acids. The method of analysis was quantitative paper chromatography, and one wonders if more subtle changes could be detected by automated amino acid analysis using ion exchange columns.

Jäger and Pahlich (1972) have concentrated on the effects of SO_2 on glutamic acid and glutamine of the amino acids of *Pisum sativum* seedlings, and have also measured effects on glutamate dehydrogenase and glutamine synthetase (Jäger *et al.,* 1972; Pahlich *et al.,* 1972). The effect of 1.3 ppm SO_2 for 24 hours was to lower the amount of glutamate (40–20% of control) and increase the amount of glutamine (130–240% of control). Glutamate dehydrogenase was activated in the direction of reductive amination and inactivated in the direction of oxidative deamination:

$$H_2O + \text{glutamate} + NAD^+ \underset{\text{activated}}{\overset{\text{inactivated}}{\rightleftharpoons}} \alpha\text{-ketoglutarate} + NH_3 + H^+ + NADH$$

There was no apparent effect on glutamine synthetase. In general, the effect of SO_2 fumigation is to lower the protein content of the fumigated leaves. Pahlich (1973) has also determined the activity of glutamate-oxalacetate transaminase in SO_2 fumigated plants and found that the mitochondrial form of the enzyme is more susceptible than the cytoplasmic form.

In common with many physiological observations on the effect of air pollutants, it is difficult to propose a specific point of attack for SO_2 as a result of the above observations. The reactions of SO_2 with biochemicals are detailed in Section V, and any one of these, or others yet to be discovered, may be involved in changing the amino acid and protein content of the leaves.

D. Photosynthesis and Chlorophyll

Brinckmann *et al.* (1971) have summarized previous work which shows that various aspects of photosynthesis are inhibited by the addition of compounds which form from sulfite and aldehydes or ketones. It is somewhat less certain that these addition compounds are really formed in the leaves of plants exposed to SO_2 concentrations found in polluted air. Jiracek *et al.* (1972) have isolated such addition compounds of glyceraldehyde, α-ketoglutarate, and pyruvate + oxalacetate from various parts of *Pisum sativum* seedlings exposed to 1% SO_2 for 1–4 days. The products are undoubtedly formed, but the high concentrations of SO_2 used cast some doubt on whether the α-hydroxysulfonates are produced in the ppm range of SO_2 concentrations.

Nevertheless, the α-hydroxysulfonates must be taken into consideration as a potential source of the toxicity of SO_2. The α-hydroxysulfonates were first recognized as inhibitors of glycolic acid oxidase and of stomatal opening: for example, 0.8 m*M* α-hydroxydecanesulfonate caused 50% inhibition of stomatal opening (Zelitch, 1971). But these inhibitors have more recently been recognized as capable of more general inhibitions. Lüttge *et al.* (1972) reported that the glyoxal bisulfite compound at a concentration of 0.5 m*M* inhibited CO_2 fixation and light dependent changes in pH. Both these authors and Murray and Bradbeer (1971) interpreted the effects of the α-hydroxysulfonates as being compatible with effects on membrane and transport properties.

There is ample evidence concerning the effect of SO_2 on chlorophyll. Rao and Le Blanc (1966) showed that exposure of the lichen, *Xanthoria fallax,* to SO_2 caused the conversion of chlorophyll to pheophytin. It is well known that lowering the pH will cause the loss of Mg^{2+} from the chlorophyll to form pheophytin and so the effect of SO_2 may be entirely explained in this case by acidification. Grill and Haertl (1972) have concluded that SO_2 resistance of spruce needles depends on buffer capacity. Older needles have less buffering capacity in the acid range and are more susceptible to injury by SO_2. Syratt and Wanstall (1969) have measured the destruction of chlorophyll in bryophytes. They found that chlorophyll destruction was greater at higher SO_2 concentrations and at higher humidities, and suggested that SO_2 was acting as sulfurous acid. The same conclusion was reached by Gilbert (1969) as a result of his studies on lichens and bryophytes in North Eastern England.

E. Metabolism of SO_2

Sulfur is an essential nutrient for plants. Usually plants take up sulfate and use this source for formation of essential compounds, such as cysteine and methionine in proteins, glutathione, coenzyme A, lipoic acid, biotin, sulfur-containing bases in nucleic acids, and sulfoquinovose (in the sulfolipid). In the vast majority of cases, the sulfur is required in a reduced state, and much study has been devoted to the enzymes of sulfate reduction in plants. It is frequently presumed that sulfite is an intermediate in this process and that exogenous sulfite should fit into the scheme for sulfate reduction. Although the possibility has not been eliminated that the form of the intermediates may be such that while sulfate can be activated, sulfite cannot. If the latter were the case, sulfite would have to be converted to sulfate before it could be metabolized. Nevertheless, cell-free preparations from plant tissues are perfectly capable of reducing sulfite to sulfide. Schiff and Hodson (1973) have suggested that sulfite and sulfite reductase may have no role in the normal reduction of sulfate.

When radioactive sulfate is supplied to plant tissue, the products include radioactive sulfite, sulfide, cysteine, and methionine. Exposure of plants to $^{35}SO_2$ showed the products to include sulfate, cysteine, glutathione, and at least one uncharacterized compound (Weigal and Ziegler, 1962). The data presented by these authors showed sulfate to be the major product. This observation is commonly made: sulfate accumulation takes place in plants exposed to SO_2. De Cormis (1969) has also exposed plants to $^{35}SO_2$ and has found that soon after treatment, 98% of the ^{35}S in the cells is in the form of sulfate. After 15 days one finds 2.5% in free amino acids, 5% in protein, and 92.5% as sulfate. It was also reported that exposure to $^{35}SO_2$ resulted in the subsequent evolution of $H_2{}^{35}S$, especially if the plants were illuminated. These results again show that oxidation of the SO_2 taken up is the predominant reaction. There is no indication whether the reductive process uses dissolved SO_2 directly or passes through the sulfate intermediate.

Guderian (1970) has made interesting comparisons of SO_2 damage, sulfur accumulation, and photosynthetic capability in (alder) *Alnus glutinosa* leaves of different ages. In all cases the peak of photosynthetic activity coincided with the peak of accumulation of sulfur (the distribution of sulfur in different compounds was not determined). However, the peak of damage shifted from the older to the younger leaves as the SO_2 concentration was shifted from 1 ppm to 2 ppm. In the latter case, maximum damage was observed when photosynthesis (and sulfur accumulation) was maximal. These results show that the conclusion that conversion of sulfite to sulfate as a mechanism of protection may be too simple. Further complications are apparent in Guderian's (1971) findings that nutritional status plays an important role in determining damage by SO_2. Increased nutrient levels of Ca^{2+} and K^+ tend to lower the damage, while increased phosphate levels increase the damage.

Sulfite oxidation has not been studied extensively in plants. Syratt and Wanstall (1969) found that respiration was stimulated by SO_2 in some mosses, but sulfate accumulation was not correlated with respiration. Sulfite oxidation in the bacterium *Thiobacillus novellus* depends on cytochrome c as a cofactor (Charles and Suzuki, 1966), but it remains to be seen whether a similar system is present in higher plants. The hepatic sulfite oxidase (Cohen *et al.*, 1972) has been given a role in SO_2 detoxification in animals. It may be relevant that in his studies of ethylene formation in plants, Yang (1970) discovered that during nonenzymic oxidation of methionine to methionine sulfoxide, sulfite was oxidized to sulfate:

$$CH_3SR + 2SO_3^{2-} + 1.5O_2 \xrightarrow{Mn^{2+}} CH_3S(O)R + 2SO_4^{2-}$$

The reaction was inhibited by superoxide dismutase, indicating that superoxide was an intermediate in the oxidations.

Yang (1973) and Yang and Saleh (1973) have reported that sulfite is oxidized to sulfate in the presence of Mn^{2+}, oxygen, and either tryptophan or indoleacetic acid. During this process the indole compounds are also oxidized. These reactions may be involved in the oxidation of sulfite in plant cells and also have important consequences in terms of the indole compounds oxidized.

V. Biochemical Effects

A. Aldehydes and Ketones

The reaction of sulfite with aldehydes and ketones has been known for many years:

$$R_1(R_2)CO + HSO_3^- \rightleftharpoons R_1(R_2)C(OH)SO_3^-$$

The equilibrium position is in favor of the sulfonate, especially in the case of aldehydes. These reactions can affect biological systems in at least two ways: (1) trapping metabolic intermediates which have aldehyde or ketone groups; (2) forming α-hydroxysulfonates which are inhibitors of enzyme catalyzed reactions. The relevance of the reaction with aldehydes and ketones to biological materials depends on the rates of reaction of HSO_3^- with the keto compounds at the concentrations that HSO_3^- is likely to be found in the cell, and on the equilibrium position. These considerations have not yet been given sufficient attention.

B. Olefinic Compounds

The reaction of sulfite with olefinic compounds proceeds by a free radical mechanism and usually produces sulfonic acids. These reactions proceed more rapidly as the pH is increased, suggesting that the reactive species is SO_3^{2-}.

Sulfite will react with maleic or fumaric acids to produce sulfocarboxylic acids, and the reaction with *N*-ethyl maleimide has been used in the determination of sulfite (Mudd, 1968).

Lehmann and Benson (1964) have reported the addition of sulfite to methyl glucoseenide, producing methyl-6-sulfo-α-D-quinovoside in good yield, and the reaction of sulfite with phosphoenol pyruvate where the products are CO_2, sulfoacetic acid and phosphate.

The metabolism of $^{35}SO_2$ has not turned up any products that one would expect from the above reactions. The unknown compounds noted by Weigal and Ziegler (1962) may contain such products, but the biological significance of the reaction with olefinic compounds is unknown.

C. Disulfides

Sulfite reacts with disulfides such as cystine with the production of thiol and S-sulfonate:

$$RSSR + SO_3^{2-} \rightleftharpoons RS^- + RSSO_3^-$$

The reaction is reversible both with small molecular weight compounds such as cystine, and with proteins. In order to pull the reaction to the right, RS^- must be trapped, for example, by mercury ions. The reactions of sulfite with disulfide bonds of proteins has been used for analytical purposes, usually with high concentrations of sulfite. Once again we must doubt whether the reaction takes place in plants exposed to SO_2. It has been reported that the SH content of SO_2 damaged pine needles is 2–4 times as high as that in control leaves (Grill and Esterbauer, 1972), but one may wonder how direct this effect is. Perhaps the metabolism of the leaf is changed by SO_2 to produce more SH compounds.

D. Pyrimidines

Mukai *et al.* (1970) reported that mutants of *Escherichia coli* were formed in the presence of sodium bisulfite and that these mutants could be attributed to changes at C-G pairs. Similar results had been reported for λ phage by Hayatsu and Miura (1970). In these cases the concentrations of bisulfite used were 1 and 3 *M*. Bisulfite reacts with uracil or cytosine by adding at the 5–6 double bond forming stable 5,6-dihydro-6-sulfonate derivatives (Shapiro *et al.*, 1970; Hayatsu *et al.*, 1970). The reactions are reversible, being favored at slightly acid pH and reversed at slightly alkaline pH.

These results may not seem relevant to the effects of SO_2 pollution on plants, but Ma *et al.* (1973) have reported that 0.075 ppm SO_2 caused chromatid aberrations in the pollen tubes of *Tradescantia paludosa,* making further work on mutagenic effects of great interest.

E. Enzymes

In addition to the reaction of sulfite with disulfide bonds of proteins, there are other methods of enzyme inactivation. The mechanisms suggested depend on the similarity of sulfite with some other anion.

(1) Ziegler (1972, 1973) has concluded that the inhibition of ribulose-diphosphate carboxylase and phosphoenolpyruvate carboxylase can be attributed partially to competitive inhibition for the bicarbonate site. Mukerji and Yang (1974) have also studied the inhibitions of phosphoenolpyruvate carboxylase by sulfite and have also concluded that the inhibition is competitive with respect to bicarbonate. Mukerji and Yang (1974) made an interesting comparison of sulfite, glyoxal bisulfite, glyoxalate bisulfite, and α-hydroxypyridine-methanesulfonic acid. Only the latter was not an inhibitor: perhaps the inhibition depends on the charged group, but the pyridine moiety is too large or too hydrophobic to enter the active site.

(2) Kamogawa and Fukui (1973) studied the inhibition of potato and rabbit muscle phosphorylase by bisulfite. They found that the bisulfite inhibition was competitive with respect to phosphate, glucose 1-phosphate, and arsenate. The effect was very specific for bisulfite, there being no effect with sulfate, azide, cyanide, but there was some effect with bicarbonate. The effect was explained in terms of the structure of the anions involved (sulfite, arsenate, and carbonate), competing for the phosphate site on the enzyme.

VI. Conclusions

1. Symptomology and dose-response considerations for SO_2 as a single pollutant are essentially complete.
2. Interaction of SO_2 with other pollutants is not fully understood, either at the level of symptomology or physiology.
3. The mechanism by which SO_2 is converted to sulfate in plants is not understood.
4. The chemical and biochemical basis for physiological effects, e.g., on stomatal opening, needs further study.
5. The relevance of chemical reactions of SO_2 to biological systems deserves further development.

References

Altshuller, A. P. (1973). Atmospheric sulfur dioxide and sulfate. *Environ. Sci. Technol.* **7,** 709–712.

Arndt, U. (1970). Konzentrationsänderungen bei freien Aminosäuren in Pflanzen unter dem Einfluss von Fluorwasserstoff und Schwfefeldioxid. *Staub* **30,** 256–259.

Brinckmann, E., Lüttge, U., and Fischer, K. (1971). Die Wirkung von SO_2 und Bisulfitverbindungen auf Photosynthese, Ionentransport und Spaltöffnungs regulation bei Blättern von höheren Pflanzen. *Ber. Deut. Bot. Ges.* **84,** 523–524.

Charles, A. M., and Suzuki, I. (1966). Purification and properties of sulfite: Cytochrome c oxidoreductase from *Thiobacillus norellus. Biochim. Biophys. Acta* **128,** 522–534.

Cohen, J. J., Betcher-Lange, S., Kessler, D. L., and Rajagopalan, K. V. (1972). Hepatic sulfite oxidase. *J. Biol. Chem.* **247,** 7759–7766.

de Cormis, L. (1969). Quelques aspects de l'absorption du soufre par les plantes soumises à une atmosphère contenant du SO_2. *Proc. Eur. Congr. Air Pollut., 1st, 1968,* pp. 75–78.

Cox, R. A., and Penkett, S. A. (1971). Photo-oxidation of atmospheric SO_2. *Nature* (*London*) **229,** 486–488.

Gilbert, O. L. (1969). The effect of SO_2 on lichens and bryophytes around Newcastle-upon-Tyne. *Proc. Eur. Congr. Air Pollut. 1st, 1968,* pp. 223–235.

Grill, D., and Esterbauer, H. (1972). Water-soluble sulfhydryl compounds in intact and SO_2-damaged pine needles. *Soc. Ecol., Proc. Conf. Load Loadabil. Ecosyst., 1972,* Giessen, West Germany, pp. 155–156.

Grill, D., and Haertl, O. (1972). Cell physiological and biochemical studies SO_2-fumigated spruce needles, resistance and buffer capacity. *Mitt. Forst. Bundesvers.* **97,** 367–386.

Guderian, R. (1970). Untersuchungen uber quantitative Beziehungen zwischen dem Schwefelgehalt von Pflanzen und dem Schwefeldioxidgehalt der Luft. *Z. Pflanzenkr. Pflanzenschutz* **77,** 289–308.

Guderian, R. (1971). Einfluss der Nährstoffversorgung auf die Aufnahme von Schwefeldioxid aus der Luft und auf die Pflanzenanfälligkeit. *Schr. Landes. Imm. Bod. Nordrhein-Westfalen Essen, 1971,* pp. 51–57.

Hayatsu, H., and Miura, A. (1970). The mutagenic action of sodium bisulfite. *Biochem. Biophys. Res. Commun.* **39,** 156–160.

Hayatsu, H., Wataya, K., Kai, K., and Iida, S. (1970). Reaction of sodium bisulfite with uracil cytosine, and their derivatives. *Biochemistry* **9,** 2858–2865.

Jacobson, J. S., and Hill, A. C., eds. (1970). "Recognition of Air Pollution Injury to Vegetation: A Pictorial Atlas." Air Pollut. Contr. Ass., Pittsburgh, Pennsylvania.

Jäger, H.-J., and Pahlich, E. (1972). Einfluss von SO_2 auf den Aminosäurestoffwechsel von Erbsenkeimlingen. *Oecologia* **9,** 135–140.

Jäger, H.-J., Pahlich, E., and Steubing, L. (1972). Die Wirkung von Schwefeldioxid auf den Aminosäure und Proteingehalt von Erbsenkeimlingen. *Angew. Bot.* **46,** 199–211.

Jiracek, V., Machackova, I., and Kastir, J. (1972). Nachweis der Bisulfit-Addukte (α-Oxysulfonsäuren) von Carbonylverbindungen in den mit SO_2 behandelten Erbsenkeimlingen. *Experientia* **28,** 1007–1009.

Kamogawa, A., and Fukui, T. (1973). Inhibition of α-Glucan phosphorylase by bisulfite competition at the phosphate binding site. *Biochim. Biophys. Acta* **302,** 158–166.

Lehman, J., and Benson, A. A. (1964). The plant sulfolipid. IX. Sulfosugar synthesis from methyl hexoseenides. *J. Amer. Chem. Soc.* **86,** 4469–4472.

Lüttge, U., Osmond, C. B., Ball, E., Brinckmann, E., and Kinze, G. (1972). Bisulfite compounds as metabolic inhibitors: Nonspecific effects on membranes. *Plant Cell Physiol.* **13,** 505–514.

Ma, T-H., Isbandi, D., Khan, S. H., and Tseng, Y-S. (1973). Low level of SO_2 enhanced chromatid aberrations in *Tradescantia* pollen tubes and seasonal variation of the aberration rates. *Mutat. Res.* **21,** 93–100.

MacDowall, F. D. H., and Cole, A. F. W. (1971). Threshold and synergistic damage to tobacco by ozone and sulfur dioxide. *Atmos. Environ.* **5,** 553–559.

Majernik, O., and Mansfield, T. A. (1970). Direct effect of SO_2 pollution on the degree of opening of stomata. *Nature (London)* **227,** 377–378.

Majernik, O., and Mansfield, T. A. (1971). Effects of SO_2 pollution on stomatal movements in *Vicia Faba. Phytopathol. Z.* **71,** 123–218.

Majernik, O., and Mansfield, T. A. (1972). Stomatal responses to raised atmospheric CO_2 concentrations during exposure of plants to SO_2 pollution. *Environ. Pollut.* **3,** 1–7.

Mansfield, T. A., and Majernik, O. (1970). Can stomata play a part in protecting plants against air pollutants? *Environ. Pollut.* **1,** 149–154.

Menser, H. A., and Heggestad, H. E. (1966). Ozone and sulfur dioxide synergism: Injury to tobacco plants. *Science* **153,** 424–435.

Mudd, S. H. (1968). Determination of sulfite in biological fluids: The usefulness of radioactive *N*-ethyl maleimide. *Anal. Biochem.* **22,** 242–248.

Mukai, F., Hawryluk, I., and Shapiro, R. (1970). The mutagenic specificity of sodium bisulfite. *Biochem. Biophys. Res. Commun.* **39,** 983–988.

Mukerji, S. K., and Yang, S. F. (1974). Phosphoenolpyruvate carboxylase from spinach leaf tissue. *Plant Physiol.* **53,** 829–834.

Murray, D. R., and Bradbeer, J. W. (1971). Inhibition of photosynthetic CO_2 fixation in spinach chloroplasts by α-hydroxy-2-pyridine-methanesulphonate. *Phytochemistry* **10,** 1999–2003.

National Air Pollution Control Administration. (1969). "Control Techniques for Sulfur Oxide Air Pollutants," Publ. No. AP-52. NAPCA, Washington, D.C.

O'Gara, P. J. (1922). Sulfur dioxide and fume problems and their solutions. *Ind. Eng. Chem.* **14,** 744.

Pahlich, E. (1973). Uber den Henn-Mechanismus mitochondrialer Glutamat-Oxalacetat-Transaminase in SO_2-begasten Erbsen. *Planta* **110,** 267–278.

Pahlich, E., Jäger, H.-J., and Steubing, L. (1972). Beeinflussung der Aktivetäten von Glutamatdehydrogenase und Glutaminsynthetase aus Erbsenkeimlingen durch SO_2. *Angew. Bot.* **46,** 183–197.

Rao, D. N., and Le Blanc, F. (1966). Effects of sulfur dioxide on the lichen alga, with special reference to chlorophyll. *Bryologist* **69,** 69–75.

Schiff, J. A., and Hodson, R. C. (1973). The metabolism of sulfate. *Annu. Rev. Plant Physiol.* **24,** 381–414.

Schroeter, L. C. (1966). "Sulfur Dioxide. Applications in Foods, Beverages and Pharmaceuticals." Pergamon, Oxford.

Shapiro, R., Servis, R. E., and Welcher, M. (1970). Reactions of uracil and cytosine derivations with sodium bisulfite. A specific deamination method. *J. Amer. Chem. Soc.* **92,** 422–424.

Syratt, W. J., and Wanstall, P. J. (1969). The effect of sulphur dioxide on epiphytic bryophytes. *Proc. Eur. Congr. Air Pollut., 1st 1968,* pp. 79–85.

Thomas, M. D., Hendricks, R. H., Collier, T. R., and Hill, G. R. (1943). The utilization of sulfate and sulfur dioxide for the nutrition of alfalfa. *Plant Physiol.* **18,** 345–371.

Unsworth, M. H., Biscoe, P. V., and Pinckney, H. R. (1972). Stomatal responses to sulphur dioxide. *Nature* (*London*) **239,** 458–459.

Weigal, J., and Ziegler, H. (1962). Die räumliche Vertilung von ^{35}S und die Art der markierten Verbindungen in Spinatblättern nach der Begasung mit $^{35}SO_2$. *Planta* **58,** 435–447.

Yang, S. F. (1970). Sulfoxide formation from methionine or its sulfide analogs during aerobic oxidation of sulfite. *Biochemistry* **9,** 5008–5014.

Yang, S. F. (1973). Destruction of tryptophan during the aerobic oxidation of sulfite ions. *Environ. Res.* **6,** 395–402.

Yang, S. F., and Saleh, M. A. (1973). Destruction of indole-3-acetic acid during the aerobic oxidation of sulfite. *Phytochemistry* **12,** 1463–1466.

Zelitch, I. (1971). "Photosynthesis, Photorespiration, and Plant Productivity." Academic Press, New York.

Ziegler, I. (1972). The effect of SO_3 on the activity of ribulose-1,5-diphosphate carboxylase in isolated spinach chloroplasts. *Planta* **103,** 155–163.

Ziegler, I. (1973). Effect of sulphite on phosphoenolpyruvate carboxylase and malate formation in extracts of *Zea mays, Phytochemistry* **12,** 1027–1030.

3

OZONE

Robert L. Heath

I. Introduction

It is difficult to write a general chapter on biochemical mechanisms of ozone injury to green plants because there is no consensus on any mechanism at this time. Much data now exist in the literature regarding ozone-induced metabolic changes in plants, but no consistent relationship has been established conclusively.

The first indication that air pollutants do, in fact, injure living tissue arose from observations on green plants. During the classical studies by Haagen-Schmidt *et al.* (1952), it was found that some component(s) in air (around the Los Angeles basin) caused necrosis and flecking on the leafy tissue of plants; however, not all species were injured in the same or uniform manner. Most air pollution work on plant species has been accomplished with only one firm conclusion: some species are much more sensitive to ozone than others. Exposure of very sensitive plants (e.g., bean, tobacco, leafy vegetables, and some ornamentals) under the right conditions to ozone (at a concentration as low as 0.1 ppm for 1 hour) will cause visible injury to appear on the leaf within the next few days. Reports are lacking which might show any form of injury to animals at such low concentrations for such short durations. The fact is that certain plants can now be used as monitoring devices for the presence of air pollutants (Oshima, 1973).

The divergent responses to ozone between plants and animals stem partly from obvious differences between the two. First, plant cells are characterized by the presence of a strong cell wall, which varies in thickness with cell age (Northcote, 1972). Any cellular expansion occurs through wall extension after the cell has accumulated osmotic water to provide a high internal turgor pressure. This turgor pressure can reach many atmospheres and is thought to result from the accumulation of soluble biochemicals or salts within the cell (Cleland, 1971). A second major difference lies in the plant cell's ability to use light photosynthetically; specialized organelles, called chloroplasts, have evolved for this purpose. Normally, light greatly influences plant cell metabolism.

The minimal visible changes attributable to ozone injury are necrosis, chlorosis, and/or flecking of the upper leaf surface (Fig. 1). These visible symptoms are thought to result by way of the following sequence of events: ozone interaction with some component of the cells in leaf tissue; collapse of cell and a general "water-logging" (or localized accumulation of extracellular water) in the vicinity of the interaction; bleaching of the chlorophyll within the injured cell; and breakdown of the leaf structure around the cell. This proposed sequence of events raises several interesting physiological and biochemical questions, which may serve to outline important areas of ozone research:

1. What is the primary site of ozone interaction with the cell—the biochemicals and structures?

2. Does ozone participate in secondary cellular reactions which accentuate the initial damage or lead to further damage? What are the metabolic ramifications of these alterations of the cell's machinery?

3. Following the interactions of ozone with the biochemicals and basic structure of the cell, what metabolic steps transform reversible injury into irreversible, fatal injury?

4. By what mechanism does an "amplification" of this damage occur? (It is noted that small regions of ozone injury occur on the leaf surface with the appearance that all cells within a neighborhood are injured simultaneously).

5. Does the plant possess mechanisms which enable it to repair this injury or prevent further injury?

6. How much leaf damage can a plant sustain without noticeable decline in productivity?

This chapter will present data which will bear on and suggest possible answers to these six questions. In some sections there will be little data and many questions. By breaking the discussion into sections, I hope to stimulate other plant scientists into applying their expertise to the problems of ozone injury. For a more complete picture of the field, other reviews should be consulted (Middleton, 1961; Rich, 1964; Darley and Middleton, 1966; Heggestad, 1968; Heck, 1968b; Dugger and Ting, 1970a,b; Stern *et al.*, 1973; Mudd, 1973; Dugger, 1974).

II. Entry of Ozone into the Injury Region

A. Atmospheres: Chemistry and Physiological Doses of Ozone

The production of ozone in polluted urban atmospheres has been the subject of much controversy and study. Several atmosphere models have recently been proposed, and a specific model for the Los Angeles Basin is now being developed at the Statewide Air Pollution Center at Riverside (Moses, 1969; Larsen, 1969). While a complete discussion is beyond the scope of this chapter, it is accurate to state that the high production of ozone within the Los Angeles basin in particular, and urban air basins in general, is predominantly due to car exhaust components (Odabasi, 1973). Through a reversible reaction among molecular oxygen, ozone, nitrogen dioxide, and nitric oxide, and side reactions between these four with hydrocarbons and metals, the ozone concentration rises in the urban atmosphere. Average ozone levels are the following: clear, unpolluted air, 0.01–0.02

FIG. 1. See facing page for legend.

ppm*; "slightly smoggy" day during the summer in Los Angeles, 0.1–0.2 ppm; and highest levels of ozone measured in urban atmospheres, 0.5–0.8 ppm (Taylor, 1968). The duration of these periods of high ozone concentrations ranges from several minutes to several hours. In addition, it has been found that other air pollutants may also cause an increase of ozone in the atmosphere (Stephens, 1969).

Heck (1968a) has classified plants in the following tabulation into three groups according to sensitivity based on the concentration of ozone producing minimal visible injury after 1 hour exposure.

Sensitive (0.1 ppm O_3)	Intermediate (0.2 ppm O_3)	Resistant (0.35 ppm O_3)
Spinach	Begonia	Zinnia
Radish	Onion	Beet
Muskmelon	Chrysanthemum	Radish
Oat	Dogwood	Poinsetta
Pinto bean	Sweet corn	Black walnut
White pine	Wheat	Strawberry
Potato	Lima bean	Carrot
Tomato		

In more typical experimental plants approximately 0.1 ppm ozone for 2–3 hours will produce acute visible injury in *Nicotiana tabacum, Phaseolus*

* The general measurement of ozone in the literature is in ppm (parts per million) or pphm (parts per hundred million), volume/volume. The conversion to moles/moles uses the perfect gas conversion of 22.4 liters per mole (at 0°C) for ozone.

FIG. 1. (A) Visible ozone injury to cotton plant. Cotton (18-day-old plant, Acala SJ-1) was exposed to 0.9 ppm ozone for 1 hour. Photograph was taken 2 days after exposure. The cotyledons and first primary leaf are severely damaged, (noted by white necrotic areas), but no visible injury has yet appeared in second true leaf. (From Ting and Dugger, 1968. (B) Ozone injury to tobacco cultivars. Two separate cultivars of tobacco (Bel B and W-3) were exposed equally to ozone (0.15 ppm for 1 hour). Note the flecking and extreme injury to the W-3 plant compared to Bel B (for more details, see Ting and Dugger, 1971). (C) Effect of light quality on ozone injury to pinto bean. Pinto bean (14 days old) were exposed to 0.7 ppm ozone for 1 hour. The plants had been grown under varied light conditions. 39C-14A, standard cool white fluorescent light; 39C-14B; V. Ho 0 fluorescent light (containing more red component than A); 39C-14S, sun light. Top: upper surfaces of leaves. Bottom: lower surfaces. (D) Ozone injury to *Impatiens*. *Impatiens* were fumigated for 7 hours at 0.39 ppm. Left is upper surface, right is bottom. (E) Ozone injury to Ponderosa pine needles: Ponderosa pine trees were fumigated for 5 days, (12 hours per day at 0.4 ppm ozone). The ozone induced necrosis is apparent as white areas and patches (Miller *et al.*, 1963). (B)–(E) courtesy of Dr. O. C. Taylor, Statewide Air Pollution Research Center, University of California, Riverside.

vulgaris, Allium cepa, Raphanus sativa, and *Pinus strobus* (Taylor, 1968). It should be kept in mind, however, that the plant's metabolic condition has a pronounced effect upon its possible classification (to be discussed later).

It is unfortunate that many experiments are not performed on the same species grown under the same environmental conditions. The important experimental conditions which must be controlled are (1) species; (2) growth condition with respect to light, humidity, water status (water potential of soil), nutrients present in soil, day–night period, and temperatures; (3) age of leaf on which the experiment is performed; (4) ozone levels and duration; (5) environmental conditions during the exposure and subsequent holding period.

B. Chemistry

The ozone molecule is very reactive with a standard redox potential of approximately +2.1 volts (Thorp, 1954). It is thought that the active species is a slightly ionic form of a bent array of oxygen atoms. One such resonance structure is

$$:\overset{-}{O}-\overset{+}{O}=O$$

Since ozone is so reactive, one would expect it to decompose rapidly in an urban atmosphere. Indeed, it appears that decomposition and back reactions may ultimately limit the final ozone concentration in the atmosphere. However, decomposition is very dependent upon the total atmospheric condition.

Typical methods for generating ozone in the laboratory are to pass pure dry oxygen over either a silent spark discharge or ultraviolet lights (Dugger and Ting, 1970a,b). Under these conditions it is important that pure, dry oxygen be used to prevent the formation of other side products, e.g., NO_2. Much of the chemical nature of ozone has been summarized in two handbooks of ozone technology (Thorp, 1954, Vols. I & II), but its behavior as a gas is not well understood.

C. Stomatal Size and Diffusion of Ozone

The literature concerning the role of stomata in ozone injury contains much that is confusing and contradictory (Mansfield, 1973). Most workers agree, however, that when stomata are closed, no ozone can enter the plant and no injury occurs. Problems arise when the stomatal size varies under changing experimental conditions during ozone exposure of the leaf. For example, the surface application of abscisic acid (which is known to close stomata) prevents or reduces ozone injury (Jones and Mansfield, 1970; Adedipe *et al.,* 1973).

It has been suggested that even if stomata remain open, in some cases the plant may not necessarily be injured. It is this experimental situation (with open stomata) which provides the focal point for further discussion regarding the biochemical site of ozone attack of the cell. Experimentally, it is difficult to maintain open stomata throughout the ozone treatment period. With a high ozone dose, the initial "waterlogging" effect and leaf desiccation normally lead to stomatal closure (Zelitch, 1969; Mansfield, 1973). However, carefully chosen conditions—use of plants which have been grown under conditions of high soil water potential and use of low ozone concentrations and high humidity with a short duration of exposure (Ting and Dugger, 1971)—may help in circumventing this difficulty. In general, it is best to measure stomatal opening if environmental conditions are changed. Seemingly contradictory evidence can be accounted for by the finding that stomatal control seems to be influenced by the plant species as well as its growth conditions (Zelitch, 1969). For example, Ting and Dugger (1971) found no closure of stomata and little stimulation of photosynthesis in pinto bean (*Phaseolus*) which showed visible injury. Hill and Littlefield (1969) found a depression of photosynthesis, reduction of transpiration, no visible injury and, not surprisingly, that stomatal closure had occurred in oats.

Stomatal behavior may be quite complex. Evans and Ting (1974a) studied the water potential, leaf resistance, stomatal spacing, and other leaf characteristics in primary bean leaves in relation to ozone sensitivity and injury. Leaf water potentials decreased during ozone exposure, and the relative water content decreased drastically within 75 minutes after exposure to 0.6 ppm ozone, a change indicative of a large leaf water loss. After ozone treatment, however, abaxial leaf stomatal resistance increased initially, but then decreased. After 1 hour, abaxial resistance returned to its prefumigation level. But at high ozone concentrations, after a short lag period, abaxial leaf resistance steadily decreased. The authors attributed this to general destruction of the palisade and upper epidermal layers. In other words, ozone injury was strictly an intraleaf phenomenon involving the palisade cells in the internal top region of the leaf.

Most of the work on the mechanism of stomatal opening and closure implicates a flow of K^+ ions into the guard cells (Zelitch, 1969). Water flow, following K^+ osmotically, increases guard cell turgidity and leads to stomatal opening. Presumably, any condition which would interfere with K^+ flow or water balance would affect stomatal response. Thus, if the water balance of the leaf is disturbed enough by desiccation, the decreased turgidity of the guard cell leads to closure of stomata. This, in turn, would protect the leaf from further injury.

A related problem which has not yet been satisfactorily answered concerns the diffusion rate of ozone into the leaf. Unfortunately, the rapid breakdown of ozone and lack of suitable tracer isotopes of oxygen preclude any easy measure of diffusion. One has to assume that ozone moves passively down a concentration gradient, much as CO_2 diffusion during active photosynthesis (CO_2 within the cell is at a lower concentration than that in the atmosphere due to photosynthetic CO_2 fixation). In this case, one would expect the net flux of O_3 (in moles/cm^2 second) to be 1000-fold lower than that of CO_2 (because the atmospheric concentration for CO_2 is 300 ppm compared to 0.3 ppm for O_3) or about 0.02–0.05 μmoles/cm^2 leaf area hours [$1/1000$ of best photosynthetic rate, (see Zelitch, 1971)]. These types of calculations are, at best, "ballpark" figures.

D. Architecture of Intercellular Spaces Within Leaf

Once through the stomata, ozone molecules must pass through the leaf into the palisade parenchyma cell, a major area of damage (Dugger and Ting, 1970a,b). Only during severe ozone injury do the mesophyll cells show signs of injury. Most plants used in ozone-damage experiments possess stomata on both sides of the leaf. As easily seen in scanning electron microscope pictures (Fig. 2), the two intraleaf regions—spongy mesophyll and palisade layers—differ in the ratio of exposed cell surface to gaseous intercellular space volume as well as in overall structure. The palisade cells, where ozone injury first appears (Dugger and Ting, 1970a; Evans and Ting, 1974a; Troughton and Donaldson, 1972), possess a higher surface/volume ratio. This surface/volume ratio changes during aging of the plant leaf (Evans and Ting, 1974a). If the surface/volume ratio is a critical factor in ozone damage, this change could affect plant sensitivity.

Many plants do demonstrate leaf age/ozone sensitivity variation, but the reasons for this variation are not yet clear. Age-dependent sensitivity to ozone is shown in Fig. 3 for cotton. Ozone exposure injures only leaves in a certain stage of expansion; thus, the cotyledons are injured in young plants (10 days old), while only the primary leaf is injured in 18- to 20-day-old plants. It is important to note that neither young nor old leaves may show visible injury (Fig. 3), even though it can be shown that their stomata remain open during mild ozone exposure (Ting and Dugger, 1971). The peak of ozone sensitivity appears to be at the physiological age which corresponds to the period (a) following that of maximum leaf expansion rate, (b) just preceding the period of maximum surface to volume ratio, and (c) prior to the final full expansion size where secondary wall synthesis and lignification occur. This period is slightly variable (compare Ting and Mukerji, 1971; Dugger and Ting, 1970b; Tingey *et al.*, 1973a).

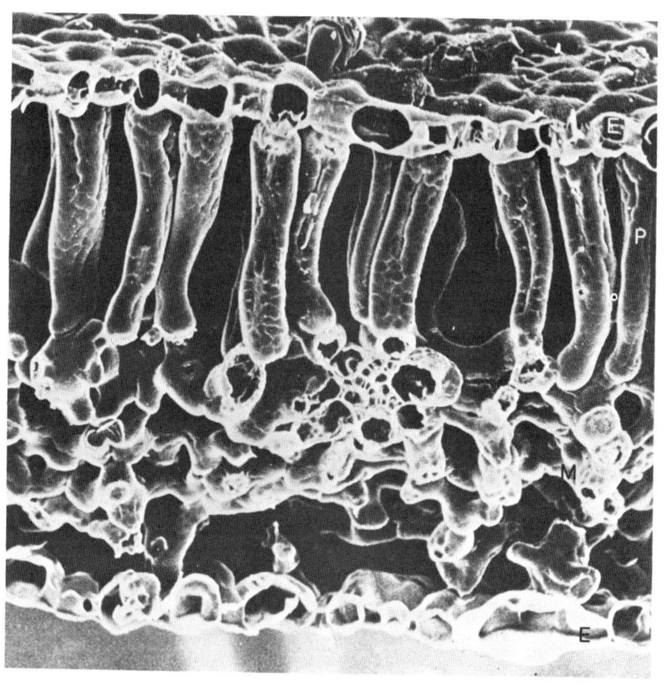

FIG. 2. Transverse view of bean leaf. Scanning electron micrograph (leaf thickness is 0.4 mm) of a mature broad bean leaf. The cell types are epidermis (E), palisade (P) and spongy mesophyll cells (M). From Troughton and Donaldson (1972).

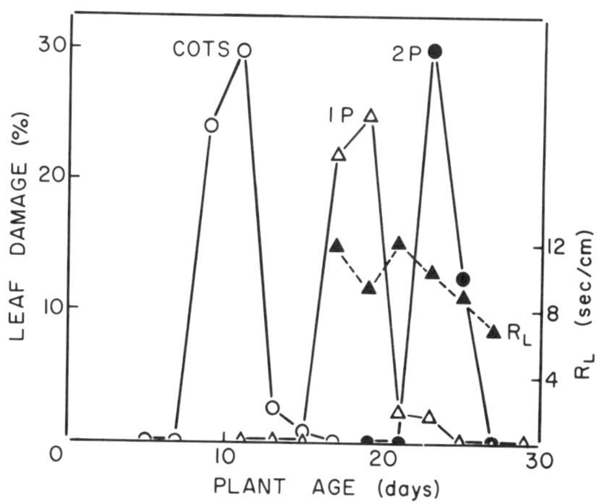

FIG. 3. Visible ozone injury with respect to plant age. Percent of visible leaf injury or damage to cotton leaves of various ages 2 days after a 1-hour exposure to 0.7 ppm ozone. Cots, cotyledons; 1P, first true leaf; 2P, second true leaf. Leaf resistance (R_L), which is an estimation of stomatal opening, is plotted for first true leaf at time of exposure. (From Ting and Dugger, 1968.)

Microscopic examination indicates a maximum amount of the intercellular space in the palisade layer (where most of the visible ozone-induced injury takes place) at the time of maximum ozone sensitivity (Evans and Ting, 1974a); lesser amounts of space before and after this period coincide with less sensitive periods. In contrast, the relative amount of spongy mesophyll intercellular space increases gradually with leaf expansion. Therefore, the period of maximum ozone sensitivity correlates only with maximum palisade intercellular spaces.

E. Hydration and the Role of Water

The area within the leaf is thought to be at nearly 100% relative humidity; in addition, it is believed that a thin layer of water (through which ozone must pass) exists around each of the cells themselves. This water layer could alter the ozone concentration by decomposition or by differential solubility, but data regarding these points are scarce.

The solubility or Bunsen coefficient of ozone in water is known for some conditions, particularly for acidic media where the ozone is relatively stable.* In alkaline solutions ozone rapidly decomposes releasing molecular oxygen (Alder and Hill, 1950; Kilpatrick *et al.*, 1956). Ozone is more soluble than oxygen and at 25°C has a Bunsen coefficient of approximately 0.25 (ml O_3 dissolved/ml H_2O) (Hoather, 1948). The solubility of ozone obeys Henry's law up to at least several percent (Kashtanov and Oleshchuk, 1937).

It is not known, of course, whether the aqueous layer that surrounds the cell wall is acidic or basic; however, the zwitterionic form of ozone might be stabilized by bonding near the cell wall/membrane and oriented by the bound charges within the cell wall/membrane (Somers, 1973). Water in the cell wall is highly involved in the gel structure of matrix material and with the charged ions (notably Ca^{2+}) which neutralize the acidic wall constituents (Northcote, 1972). Also, it is known that water changes its chemical potential near biomembranes (Kuiper, 1972). All the above statements could be related to the ease of ozone penetration into the cell.

"Model" unicellular systems (Mudd, 1973) in liquid media have been used to elucidate the mode of action of ozone. The ozone is usually bubbled into the liquid media for a short time (1–10 minutes). The relative ozone concentration for these experiments is always above 100 ppm. These concentrations are often criticized because they are nearly 100–200 times

* Much of the knowledge concerning ozone in solution has been obtained by researchers (Thorp, 1954) in waste treatment and sanitary engineering, and these technologies commonly use ozone at a concentration of 1–5% to sterilize and decompose chemicals found in waste water.

higher than the highest ambient urban atmospheres; however, nearly all the ozone which enters a liquid medium containing biological organisms will pass through unreacted (Heath *et al.*, 1974; Frederick and Heath, 1975). Only with certain reactive biochemicals (e.g., fatty acids or sulfhydryl reagents) does a relatively high proportion of the ozone react (Mudd *et al.*, 1971b). At room temperatures (using the Bunsen coefficient), 100 ppm ozone in the air above an aqueous medium effectively amounts to about 1 μM ozone concentration in the water solution or 10^{-6} moles O_3 per 55 moles of water (0.02 ppm, in terms of water) (Thorp, 1954). Unfortunately, if the water of the hydrated cellular surface within a leaf has the same chemical potential as the bulk water, then 100 ppm for liquid media may still be too high (see Kuiper, 1972, for arguments against bulk water properties being equal to water properties near surfaces).

III. Primary Site of Injury

At this point it may be helpful to define a few terms. "Visible injury" is a term used to indicate necrosis on the leaf surface. At the cellular level, injury is often used to mean an irreversible alteration of cell metabolism due to destruction of functional biochemicals and/or structures. Researchers speak of the "reversible" stage of injury which is defined as a point at which the cell can halt or repair insults to its structure. "Death" signifies that point in the injury process at which the reversible injury becomes irreversible and the cell cannot recover. Admittedly, these terms are vague. If, for instance, another stress is applied to the injured cell, it may be unable to respond to correct the second stress and therefore may die. In short, the effect of two stresses (e.g., ozone injury plus water loss) may be synergistic (see Heck, 1968b, for a discussion of gas synergism). The notion of "primary site" of injury is also somewhat unclear. For example, a slight ozone-induced alteration in cell membrane function might eventually throw metabolism into complete disarray by a shift away from homeostasis. In this case, the primary site would be the cell membrane, even though this initial reaction alone would not irreversibly alter the cell. We will refer to the primary site as a location in the cell in which the initial ozone interaction can be observed.

A. Dose Response and Depth of Penetration

The production of visible injury is not a simple linear relationship with respect to either amount of ozone, time of exposure, or dose (defined as time of exposure times the amount). As Heck *et al.* (1966) clearly pointed out, ozone injury/dose is best represented by a sigmoidal curve, that is, one for which definitive threshold concentration or exposure time is required before visible injury is initiated. For example, twice the time of exposure at half the concentration does not usually yield the same injury pattern (see Heck *et al.,* 1966, Fig. 1; injury at 10 ppm for 2 hours is not equal to 20 ppm for 1 hour).

Furthermore, Ting and Dugger (1971) have shown that the diurnal time for the dose is an important variable (Fig. 4). For a given set of growth conditions (e.g., day–night period and temperature), bean leaves show injury only after a 3 to 5 hour exposure to light. Injury is maximal for only several hours during the day, decreasing later in the day. As Ting and Dugger (1971) have further shown, plants kept in the dark for long periods (up to 72 hours) prior to ozone treatment show little ozone injury.

Ozone is often used as an analytical tool in organic chemistry for precise, stoichiometric quantitative analysis through bond cleavage. Using NADH as a model system for studying the interaction of ozone with biochemicals, Mudd *et al.* (1974) showed that the amount of NADH attacked and altered by ozone was stoichiometric with the *dose* (over a wide range of ozone concentrations). Likewise, certain amino acids (e.g., tryptophan) react stoichiometrically with respect to ozone *dose.* This is very unlike what is observed with green plants (Heck *et al.,* 1966; Ting and Dugger, 1971). Also in a unicellular model system to study ozone interactions with mem-

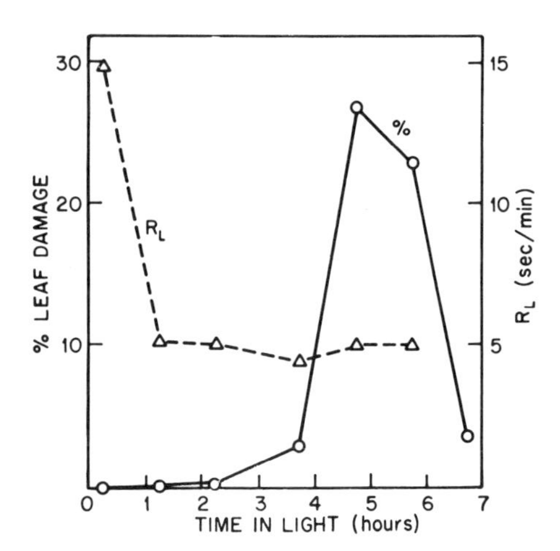

FIG. 4. Diurnal fluctuation in visible ozone injury to leaves. Percent leaf damage to 3-week-old cotton plants (exposed to 0.75 ppm ozone for 1 hour) are expressed as time in light (60 kerg/cm^2 second) after a normal 12-hour period in darkness. Leaf resistance (which is inverse to stomatal opening) was estimated just prior to treatment. The stomates are fully opening after 1 hour of light. (From Ting and Dugger, 1968.)

brane systems, Chimiklis and Heath (1972, 1975) studying the K^+ efflux from the thermophilic alga, *Chlorella sorokiniana,* in a K^+-deficient media of Tris–$CaCl_2$ (pH 8) have shown that the rate of ozone-induced K^+ efflux does not vary linearly with either the ozone concentration or the exposure time. Low ozone concentrations cause little efflux initially; but subsequently, the rate of efflux increases.

Heath *et al.* (1974) have used ultraviolet light absorption by ozone at 254 nm to quantitate the amount of ozone taken up by aqueous solutions of methyl esters of unsaturated fatty acids (in Tris, pH 8). Extending this technique to algae, Heath *et al.* (1974) have found that *Chlorella* reacts with ozone very slowly (e.g., nearly no ozone is removed from the ozone gas stream bubbled into the test solution) for the first 10–15 minutes of treatment. During longer exposures, more ozone is removed from the gas stream (the amount increasing with increasing time of exposure). Only during this period, do the algae show both a decline in viability and an increase in lipid peroxide products (Heath *et al.,* 1974; Frederick and Heath, 1975; and see Section III.C.3.). The decline in viability of the algae is exponential and cannot be related to a linear dose effect.

The extent to which ozone penetrates a cell is another important but inadequately understood problem in ozone research. Since ozone is a very reactive molecule (Kilpatrick *et al.,* 1956), it might be argued that ozone should react completely with the biochemicals with which it first comes in contact, i.e., those of the cell wall or plasmalemma. However, it is known that metabolic processes, including photosynthesis, are grossly affected during ozone treatment (Dugger and Ting, 1970a,b). The question becomes, does ozone penetrate the cell far enough to damage organelles or react with internal biochemical pathways itself? This might have been answered with the data of Coulson and Heath (1974) who have shown that when isolated intact chloroplasts are exposed to ozone at concentrations (100 ppm for 1 minute bubbling, 400 nmoles of ozone) which normally would have inhibited light-induced electron flow (the Hill reaction), no such inhibition occurs. Ozone does not seem to penetrate the outer membranes and stroma phase of the chloroplast into the grana membranes. It might be argued that both the outer membranes, which possess large amounts of high unsaturated fatty acids (Mackender and Leech, 1974), and the stroma, which contains many sulfhydryl groups, merely "mopped up" all the available ozone. Perhaps with larger doses, the ozone would have penetrated. This scheme would account for the observed nonlinear concentration effects: low ozone concentrations react with the cell plasmalemma membrane and do not reach cytoplasmic membranes and biochemicals; high concentrations penetrate the cell more deeply and cause the visible injury damage.

B. Estimation of Number of Reactive Sites

Since ozone has no radioactive isotope and breaks down very rapidly, it is difficult to measure the number of sites at which ozone reacts. Two attempts have been made, however, one using a green plant and the other using algae.

Craker and Starbuck (1973), by subtracting the rate of ozone loss from a chamber without a leaf from the rate with a leaf present, estimated that 10^{-3} μl ozone/liter second were taken up by tobacco leaves at their most sensitive age. Taking the chamber volume into account, 3×10^{-8} moles ozone/minute leaf were taken up. Unfortunately, Craker and Starbuck (1973) give neither the volume nor size of leaf used, but they further suggest that the ozone taken up correlates with both the number of stomata (they were all open) and leaf size. This number should be judged as a "ballpark" estimate.

In Section II,B, we estimated the entry of ozone into the leaf to be about 0.03 μmole/cm^2 hour; for an 80 cm^2 leaf (pinto bean at maximum sensitivity age); this amounts to 4.2×10^{-8} moles ozone/leaf minute. This rate is surprisingly close to the estimation by Craker and Starbuck (1973), given above. Therefore, much of ozone decomposition measured by Craker and Starbuck may be the entrance of ozone into the leaf.

Heath *et al.* (1974) have also determined the amount of ozone taken up by unicellular algae using a technique which measures the ultraviolet spectra of ozone (see above). From the death of algae with ozone exposure, it can be calculated that 3×10^{8} molecules of ozone (5×10^{-17} moles) are required to kill a cell. This number agrees well with the number of molecules of ozone necessary to kill 50% of a culture of *Escherichia coli* (4×10^{7} molecules of ozone per cell), as found by McNair-Scott and Lesher (1963). Using Craker and Starbuck's data (1973), estimating the number of palisade cells per leaf to be not more than 10^{8}, and using the number of cells killed per 60-minute exposure to be not more than 50% (see Fig. 3), the amount of ozone required to cause visible injury is about 4×10^{-14} molecules per injured cell or approximately 10^{3} more ozone reactions per cell than found for the model algae system.

C. Biochemical Components and Cytological Barriers

1. Cell Wall

The cell wall is important in constraining the cell's pressure potential—the force which ultimately induces plant cell expansion (Cleland, 1971).

If the cell wall cannot contain this force, a tremendous alteration of normal metabolism, including water loss, ionic imbalance and destruction of the plasmalemma, would result. Likewise, if the normal flexibility of the wall was altered by any cross-linking reaction, normal expansion and growth could not occur.

There is probably little in the wall that could be altered easily by ozone; mainly, the wall contains carbohydrates as polymerized hexoses in cellulose (Northcote, 1972). There is, however, a large amount of unresearched and unknown material in the wall, including amino acids, galacturonic acid residues, lignic acid, and bound Ca^{2+} which together form a gel-like substance inside the wall (Somers, 1973). The interaction of ozone with this gel could result in profound effects with respect to the wall's ionic exchange properties and water permeability. At present, there is no evidence that either the cell wall or its components react with ozone.

Enzymes which are involved in cell wall synthesis, however, are susceptible. Ordin and co-workers (1969; Ordin and Hall, 1967) have demonstrated that these enzyme systems, including UDP-glucose polysaccharide synthetase, are inhibited by exposure to ozone. Likewise, these systems are inactivated by sulfhydryl reagents. Evans and Miller (1972) have shown wall destruction in mesophyll cells of ponderosa pine after ozone fumigation; however, this was apparent only after appreciable intracellular damage occurred. Increased acid phosphatase activity in the region of the plasmalemma and cell periphery, generally a mark of cellular disruption, was also noted.

2. Plasmalemma and Permeability

Many researchers believe that the primary site of ozone attack on the cell is the plasmalemma. Ozone has been shown to modify amino acids (such as cysteine, methionine, tryptophan, tyrosine, histidine, and phenylalanine), proteins, unsaturated fatty acids, and sulfhydryl residues (Mudd *et al.,* 1969; Heath *et al.,* 1974) all of which are present in the plasmalemma. Furthermore, the plasmalemma is the first major barrier with which ozone comes in contact.

Dugger and Palmer (1969) first showed that, after exposure to ozone, the permeability of lemon leaf discs to radioactive glucose rose to very high levels (nearly a 2-fold increase) during the several days (maximum of 6) following exposure. Furthermore, they noted that in ozone-treated tissue, even nonmetabolizable mannitol was absorbed to a greater extent. Naturally, as the increased permeability of glucose proceeded, the apparent respiration rates (measured by $^{14}CO_2$ release) increased due to increased labeled glucose within the cell.

Evans and Ting (1973) studied ozone-induced water permeability changes using leaf discs from pinto bean plants by: (1) change in net weight measured gravimetrically, and (2) change in the refractive index of the suspending media measured refractometrically. They found that ozone fumigation of plant material (prior to cutting leaf discs) increased water permeability. In addition, the refractive index measurements showed a small leakage of solute materials out of the leaf discs. Also fumigation resulted in a net weight loss, indicating a net movement of water out of the leaf disc. Their studies with radioactive water have shown that the influx of external water into disks at the isotonic point is drastically decreased (60%) by ozone fumigation of the plant and the efflux of internal water is increased about 20–30%. These water permeability changes can be detected within 1 hour (required for handling) after fumigation with a low concentration of ozone (0.3 ppm for 1 hour). Therefore, a net water loss occurs immediately upon exposure to ozone, since influx is decreased and efflux is increased.

The effects of ozone on the permeability of bean plant cell membranes can also be measured by ^{86}Rb tracer fluxes (Evans and Ting, 1974b). The efflux rate for ozone pretreated leaf discs was much higher compared to that for untreated controls, also indicating an increase in the solute permeability of fumigated tissue.

Perchorowicz and Ting (1974) repeated Dugger and Palmer's (1969) work on glucose uptake using *Phaseolus vulgaris* and obtained the same results with an important new finding. Immediately after fumigation no change was observed, but a rise in permeability began after several hours and rose uniformly to a high level (expressed after 24 hours as a 3-fold increase). Expressed as a percentage of total glucose taken up, there was no change in the amount of label within the various fractions of a lipid extract (aqueous, interface, and chloroform layers) or in the amount of CO_2 released. Results with nonmetabolizable 2-deoxyglucose were identical.

Swanson *et al.* (1973) have shown that extreme ozone injury is manifested by a large scale disruption of the palisade cells "with massing of the cytoplasm in the center of extensively damaged cells." This "massing" is most probably due to the extreme loss of water from the cell. Thomson reviews these observations more completely (see Chapter 9).

Unicellular green algae have proved useful in model systems to study the primary site of ozone interaction with plants. Used as investigative systems, these eukaryotes eliminate variables introduced by stomata, cuticle, cellular differentiation, and translocation.

It has been found that the introduction of ozone into an algal culture (*Chlorella sorokiniana*) causes an immediate increased efflux of potassium ions, as measured with the cation electrode (Chimiklis and Heath, 1975;

Heath *et al.,* 1974). Initial amounts of ozone-induced K^+ leakage can be reduced by increased external osmotic potential; e.g., by adding mannitol. If the influx and efflux of K^+ are investigated separately using ^{86}Rb as a tracer, it is found that not only does the efflux increase greatly, but the influx is inhibited by ozone (Heath *et al.,* 1974). The effect on the efflux permeability (as measured with either the cation electrode or ^{86}Rb efflux in preloaded cells) is rapidly reversible in that when the ozone is removed from the solution, the efflux rate returns to the control rate. The effect of ozone on the influx is not reversible (over 30 to 40 minutes). Thus, ozonated cells are more permeable to K^+, though membrane injury at first does not seem to result in failure of osmoregulatory capacities. (Cells will resume growth if replaced in an autotrophic culture medium.) Longer incubations of cells with ozone leads to both failure to grow and bleaching in the light. It can be concluded that, at least initially, ozone injury to membranes may not be a general deterioration, but rather, an impairment of specific permeability sites.

Ozone alterations of membrane permeability, as demonstrated by K^+ leakage in *Chlorella,* indicate that a consideration of other ionic fluxes may further elucidate the injury mechanism. Algae can be manipulated easily by a variety of environmental techniques to produce a statistically highly uniform population exhibiting specific biochemically and physiologically intended diversity. For example, algae grown at low temperatures, in addition to growing slower, increase both in unsaturated lipid and in the storage of starch (Chimiklis *et al.,* 1975). In addition, these algae (grown at 22°C) are less ozone sensitive (compared to those grown at 38°C) with respect to K^+ leakage. This may be due to either an altered permeability in the 22°C cells or a lowered amount of K^+ within the cells.

Coulson and Heath (1974) have used isolated spinach chloroplasts to study the interaction of ozone with plant membranes. Ozone bubbled into a suspension of chloroplasts inhibited electron transport in both photosystems without uncoupling ATP production. Ozone did not appear to act as an energy transfer inhibitor, since the drop in ATP production and high-energy intermediate (measured by amine-induced swelling) as a consequence of ozone was nearly parallel to the decline in electron transport. Coulson and Heath (1974) postulated that ozone disrupts the normal pathway of energy flow from light-excited chlorophyll into the photoacts by "loosening," but not completely disrupting, the membrane. The lack of uncoupling indicates that ozone does not act to increase all permeabilities nonspecifically.

Nobel and Wang (1973) have also observed alterations in permeability of the outer chloroplasts membrane (with intact plastids) upon ozone exposure. The reflection coefficients of erythritol and glycerol were reduced, and these declines were proportional to the dose (with increasing exposure time and ozone concentration over a short range).

3. Cytoplasmic Biochemicals

Ozone-induced visible injury of plants shows a dependence upon the developmental age of leaves primarily affecting leaves of intermediate age (Glater *et al.,* 1962; Dugger *et al.,* 1962; Dugger and Ting, 1970a,b). There is evidence, however, of subtle or hidden ozone injury to younger leaves (Perchorowicz and Ting, 1974). Young leaves show an increased permeability to glucose and an increase in free pool amino acids after ozone exposure (I. P. Ting, unpublished results). Glater *et al.* (1962) concluded that the most recently differentiated cells within the tobacco leaf were the most sensitive to (unspecific) smog, and that the injury was related to the metabolic activity of the cells. They further concluded that older leaves (those not particularly smog-sensitive) were protected by suberized areas which prevented entrance of gaseous material.

Many investigators have tried to relate this specific period of sensitivity to ozone to metabolic events. For example, MacDowell (1965), working specifically with ozone, reported that tobacco leaves were sensitive just after full leaf expansion and that sensitivity was associated with a decline in total protein. On the other hand, the ozone-sensitive developmental stage in cotton leaves can be shown to be correlated with a depletion of soluble sugars (Fig. 5) and a minimum concentration of free amino acids within the leaves (Ting and Mukerji, 1971). But Tingey *et al.* (1973a) reported recently that the low point of soluble sugars and maximal ozone injury were not quite coincident. In the case of bean plants, the period of ozone sensitivity occurred near the developmental stage of the leaf when concentration of sulfhydryl groups was also declining (Dugger and Ting, 1970a,b). Other studies specifically concerned with changes in nucleotides and proteins during leaf development suggest that the susceptible period corresponds to the time when concentration of nucleotides and total proteins are minimal (Leopold, 1967). Thus, it can be reasonably assumed that ozone enters the leaf at all developmental stages and the highest periods of sensitivity must be an integral function of the internal structure, metabolism, and physiology of the specific tissue.

The effect of ozone on metabolism is many-fold. Tomlinson and Rich (1967) reported that in tobacco plants γ-aminobutyrate increased and glutamate decreased as a result of ozone exposure. They suggested that ozonation resulted in the release of glutamate into the cytoplasm where it was decarboxylated. But since certain other amino acids increase, they finally concluded that protein synthesis had probably declined. In contrast,

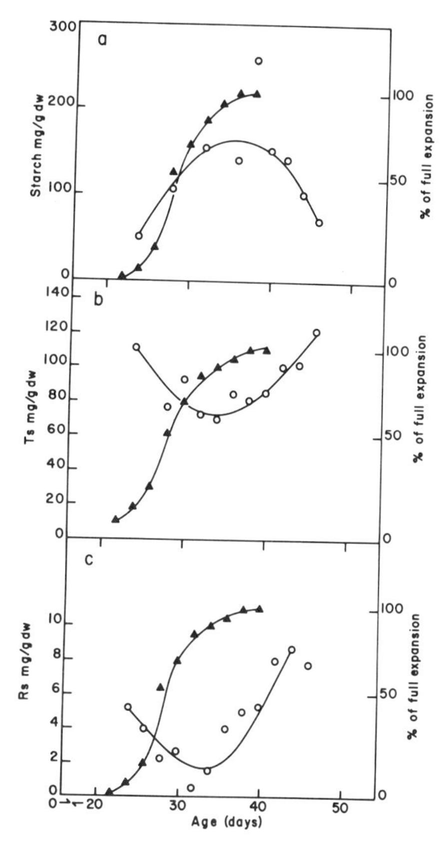

FIG. 5. Sugar and starch content of leaves as a function of age. (a) Starch content. (b) Soluble sugars (T_s). (c) Reducing sugars (R_s). First true leaf from cotton plants grown in greenhouse. Data from 10 pooled leaves. Starch appears to be a maximum and both sugars seem to be a minimum at period of maximal sensitivity to ozone (33–35 days of age). (From Ting and Mukerji, 1971).

Lee (1966) reported that ozone injury to tobacco was more closely correlated with nonprotein than with protein nitrogen. In cotton leaves, Ting and Mukerji (1971) have observed an increase in glycine, aspartate, glutamate, asparagine, β-alanine, threonine, serine, valine, leucine, isoleucine, lysine, histidine, and γ-aminobutyric acid. Nonprotein-associated amino acids (lipid precursors—phosphoserine, phosphoethanolamine, and ethanolamine) decreased immediately after ozone exposure. Therefore, ozone may affect protein metabolism either by enhancing protein hydrolysis, resulting in an increase of free amino acids, or by interfering with protein synthesis without affecting amino acid synthesis. The latter would tend to increase the concentration of free amino acids and decrease soluble protein. In either case results have been ambiguous, possibly because the experiments have been conducted at varying ozone concentrations. The decline in protein synthesis could occur if the endoplasmic reticulum—and thus the protein synthesizing system—were disrupted, or if the general, internal ionic medium was altered (Pestka, 1971).

Alterations in the chloroplasts have been observed by Thomson *et al.* (1966) who found that the first effect of ozone damage which can be detected by the electron microscope is a granulation within the stroma of chloroplasts. They suggested that the initial granulation in chloroplasts is the result of coagulation of fraction I protein since fraction I protein (or carboxydismutase or ribulose-1,5-diphosphate carboxylase) is believed to have approximately 90 sulfhydryls per mole and no disulfides in the native state (Kawashima and Wildman, 1970). Ozone severely inhibits CO_2 fixation, but does not block it completely (Dugger and Ting, 1970a,b), and the sulfhydryls of carboxydismutase are not essential for some carboxylations (Kawashima and Wildman, 1970). The oxidant-induced coagulation of the fraction I proteins could be the result of sulfhydryl oxidation by ozone.

On the other hand, data from Coulson and Heath (1974) suggest that at low ozone concentrations, the high reactivity of ozone with membranes precludes its entrance into the chloroplasts. The granulated material of chloroplasts appearing in electron micrographs after oxidant damage (Thomson *et al.,* 1966) may not occur as a result of a primary interaction of ozone with protein. It is more probable that this granulation results from ionic alterations within the chloroplast or from dehydration of the chloroplast induced by ozone-induced permeability changes elsewhere in the cell.

Water-stressed plants and ozone-injured plants possess several features in common. For example, Chang (1971) found a dissociation of the chloroplast polysomes to single ribosomes after ozone fumigation, a phenomenon also noted in the cytoplasm of root tissue under water stress induced by high osmotic potential in the medium (Hsiao, 1973). Furthermore, soluble amino acids tend to increase in both water-stressed and ozone-injured plants (Taylor, 1968; Ting and Mukerji, 1971). The visible crystalline formations which occur in chloroplast stroma following ozone exposure (Thomson *et al.,* 1966) also resemble desiccation injury (Taylor, 1968). In sum, there is much evidence to indicate a common mechanism for metabolic changes in oxidant-induced and desiccation injury.

Many workers believe that the primary sites of ozone injury are the unsaturated fatty acid residues of the membrane lipids. Lipid peroxides are formed by a cyclic reaction involving (1) abstraction of a hydrogen atom from a methylene bridge carbon between the double bonds, (2) attack of the free radical by molecular oxygen, and (3) a further abstraction of hydrogen from another fatty acid, producing a cycle reaction. Chemical peroxidation has been studied for over 50 years (Lundberg, 1962), and many techniques for measurement have evolved. The most common of these is the thiobarbituric acid (TBA) test (Kwon and Watts, 1963) which measures a breakdown product (malondialdehyde) of polyunsaturated fatty acid peroxides.

However, the term "lipid peroxidation" used in terms of ozone injury is wrong. The first reaction of ozone with unsaturated fatty acids is actually believed to be an ozonide product (ozone addition across the double bond) with one possible breakdown product being malondialdehyde. Teige *et al.* (1974) have recently worked out a series of possible reactions (Fig. 6).

One indication of ozone attack of fatty acid residues would be the loss of their fatty acid materials from the tissue. The losses reported are very low. Swanson *et al.* (1973) showed that while relative concentrations of C16:2 and C16:3 declined slightly (5–10%), the concentrations of C16:0, C16:1, C18:0, and C18:1 increased (ozone-treated plants compared with controls). Tomlinson and Rich (1969) have reported a decline in all fatty acids extracted from ozonated tobacco leaves with the largest decline in

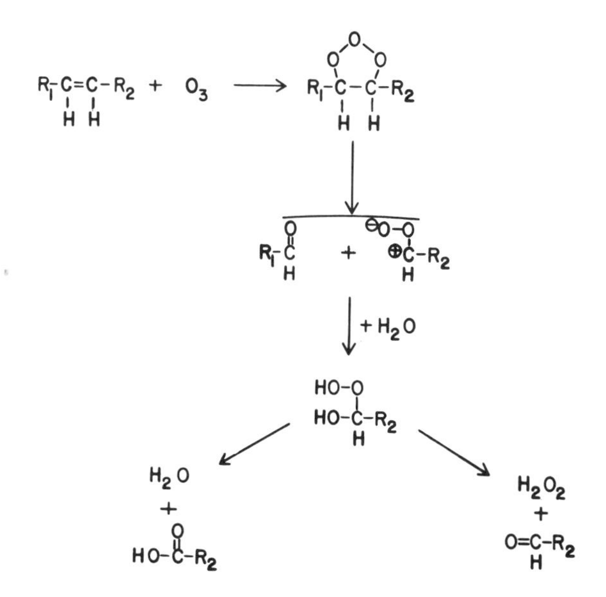

FIG. 6. Ozonation of unsaturated chemical bonds in unsaturated fatty acids. Adopted from Teige *et al.* (1974). R_1 and R_2 are chemical residues. The breakdown of the ozone required water directly or indirectly.

C16:0 and C18:3. It should be remembered that changes in lipid content may be real, but they are difficult to detect since only a small percentage of lipids are involved in the ozone attack (Heath *et al.*, 1974).

Frederick and Heath (1975) and Heath *et al.* (1974), using *Chlorella sorokiniana* var. *pacificensis,* have found that there was an increase in the thiobarbituric acid test reactant (TBAR) concurrent with decrease in algae viability. The viability curve was found to have two major kinetic phases: an initial lag from 5–10 minutes after initiation of ozone exposure, followed by an exponential decline in cell viability. The production of TBAR was a mirror image of this phenomenon; no TBAR was detected for the first 5–10 minutes of ozone exposure, but it appeared and rose rapidly during further exposure. In addition, a small decline in C18:3 fatty acid was detected, which corresponded nearly stoichiometrically with TBAR production. These authors also noted that while lipid oxidation products did occur, they arose only after cellular death occurred and the cell may have "opened up" to the environment.

Since sulfhydryl groups are necessary for lipid synthesis and degradation (Hitchcock and Nichols, 1971), it is possible that all the lipid changes noted above might arise via sulfhydryl oxidations. Tappel and his co-workers have postulated a role for sulfhydryls in decreasing the amount of lipid oxidation in both aged animals (Tappel, 1965) and lung tissue subjected to ozone stress (Chow and Tappel, 1972). Mudd *et al.* (1971a) have shown that galactolipid synthesis can be as easily inhibited by ozone as by sulfhydryl reagents, and in much the same manner.

D. Water Loss and Ionic Alterations

Loss of water from a plant cell due to an efflux of ions and metabolites into the intercellular space of the leaf could profoundly affect the normal functioning of the plant. Many enzymatic processes are controlled by the ionic environment. For example, respiration (Rains, 1972) increases with increasing salt content of the cell. Electron transport and photophosphorylation, measured *in vitro,* can be depressed by high ionic strength (Gross *et al.,* 1969). The amount and type of photosynthesis intermediaries are dramatically altered by H^+ and Mg^{2+} concentrations (Baldry and Coombs, 1973). Protein synthesis also is affected by Mg^{2+} and K^+ concentrations (Pestka, 1971).

Both Tomlinson and Rich (1968), and Pell and Brenner (1973) have observed a fall in ATP levels in bean plants exposed to ozone. This decline is observed immediately and is thought, by the authors, to be a primary response. However, ionic imbalances within the plant could easily have caused this decline, since a reduction of the ATP–ADP levels can be ob-

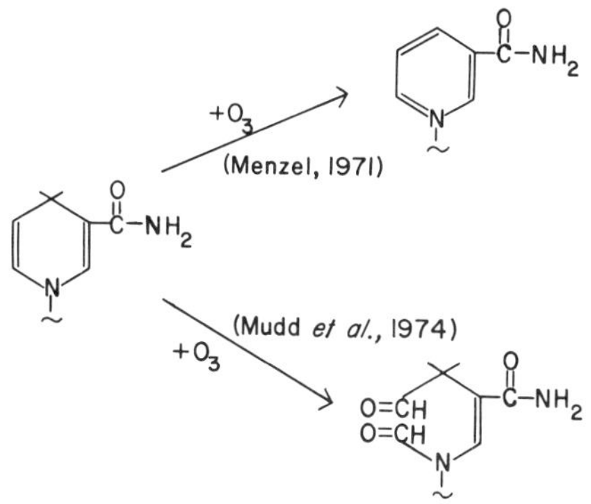

FIG. 7. Ozonation of the nicotinimide ring in NAD(P)H. As adopted from Mudd *et al.* (1974) and Menzel (1971).

served as salinity within the plant is increased (Hasson-Porath and Poljakoff-Mayben, 1971).

But, on the other hand, Mudd *et al.* (1974) have shown (contrary to Menzel, 1971) that ozone bubbled through an aqueous chemical system with NADH cleaves the nicotinamide ring (Fig. 7). Since the ratios of NADH/NAD, NADPH/NADP, and ATP/adenylates are carefully regulated by the cell, one wonders what the effect on metabolism would be if these nucleotides were lost to the cell. Certainly loss of the reduced nucleotide can be compensated by faster operation of the tricarboxylic acid cycle, but the cell can only make up for a net loss of all nucleotides by an increase in synthesis.

Tingey *et al.* (1973b) observed that ozone depressed the levels of nitrate reductase measured *in vivo* in soybean. *In vitro* levels were not depressed as opposed to results for water-stressed plants. A likely explanation is that the levels of NAD(P)H are altered by ozone so that only *in vivo* nitrate reduction is slowed.

IV. Spreading of Injury or Secondary Reactions

A standard review of ozone injury might include a long discussion of papers which generally address the question: how does ozone injury change *X* metabolic pathway (where the *X* is a function being studied in the laboratory)? Most of these reports are concerned with what could be secondary reactions. If the primary ozone reaction is a change in permeability and subsequent collapse in turgor pressure, many subsequent effects are possible. Such studies are informative and interesting but they do not consider the important question of what should constitute the first line of defense to prevent primary ozone injury to plants.

A. Secondary Reactions

1. Sulfhydryls

Sulfhydryls have long been postulated to be a primary site of ozone attack due to their high reaction rates in chemical systems (Mudd, 1973). Not only are sulfhydryls (—SH) oxidized reversibly to disulfides (—S—S) and sulfenic acid groups (—SO_2H), but the high redox potential of ozone facilitates the total, irreversible oxidation of sulfhydryls to sulfonic acid groups (—SO_3H). Furthermore, Mudd *et al.* (1971b) suggested that sulfhydryls will protect lipids from oxidation in chemical systems. In the presence of both glutathione + fatty acids, they found that sulfhydryls were lost before lipids. Heath (unpublished data), on the other hand, has noted that while thiobarbituric acid reactant was reduced when fatty acids (free acids) were bubbled with ozone in the presence of dithiothreitol (disufhydryl), the amount of ozone taken up by both components was additive.

Chow and Tappel (1972) have postulated a role for reduced sulfhydryls in ozone stress of lung tissue. Furthermore, Tomlinson and Rich (1968) believe that sulfhydryl compounds are critically involved in injury, based upon evidence that an ozone-resistant variety of tobacco had less sulfhydryls than that of an ozone-susceptible variety. They found a slight drop in total sulfhydryls after ozone exposure of bean and spinach plants (about 5–10%), but only in bean did a sizable decline in sulfhydryls occur after a 30 minute exposure (0.3 out of 1.2 μEq—SH per gram fresh weight). Spinach and both varieties of tobacco showed no such change. Both Chang (1971) and Tingey *et al.* (1973b) found no change in cytoplasmic sulfhydryl after ozone exposure in bean plants, although Chang (1971) found a decline in sulfhydryls of chloroplast ribosomes (which he did not explain).

So while it appears that sulfhydryls are essential to biological processes and are certainly easily oxidized, there is insufficient evidence that they play a primary role in ozone injury. But as in the case of fatty acids, a few essential sulfhydryls oxidized may be leading to injury, but being such a small fraction of the total they can not be detected.

2. Amino Acids and Proteins

Ting and Mukerji (1971) suggested that free amino acids play a role in ozone sensitivity, based on an observed decline in their concentration at about the same leaf age as that for maximum ozone sensitivity. But while a considerable rise in free amino acids occurred 24 hours after ozone exposure, only a small rise was observed immediately after fumigation (in cotton exposed to 0.8 ppm for 1 hour). On the other hand, Tingey *et al.* (1973b) found a rise in amino acid concentrations immediately after exposure in soybean, which continued for at least 24 hours (Fig. 8). With higher ozone concentrations (0.5 ppm for 2 hours for soybean) the initial rise remained high for several days. These results with high ozone concentrations are similar to those described by Ting and Mukerji (1971). Craker and Starbuck (1972) also noted a rise in the free amino acids but only after 24 hours with low ozone concentrations (0.25 ppm for 1 hour, for bean).

Tingey *et al.* (1973b) noted that the level of soluble protein only rose 24 hours following exposure (at high ozone concentration); there was no change with lower ozone concentrations. Craker and Starbuck (1972) claimed that protein content declines in bean following exposure to ozone, but this is difficult to see from their data (see their Fig. 5). In another publication (1972), however, Craker confirmed this. Again these results are confusing. But it does appear that the change in total protein, if

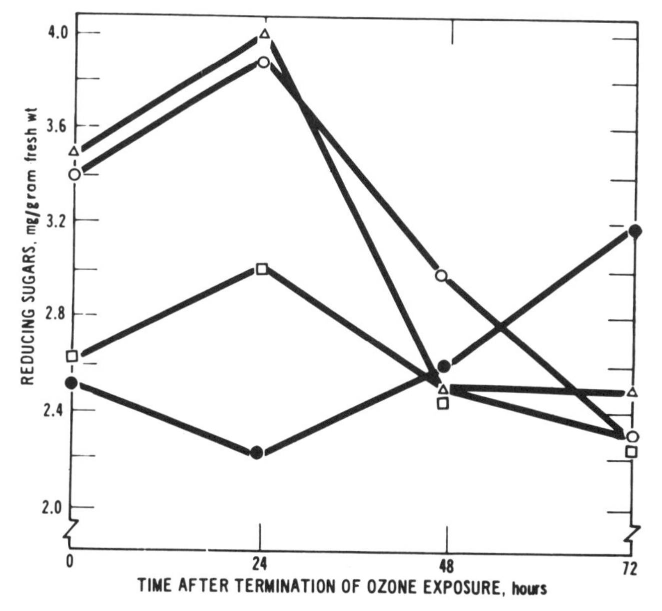

FIG. 8. Alteration in the level of free amino acids in soybean leaves after ozone exposure. Young soybean plants [*Glycine max* (L.), 14 days old] were exposed for 2 hours to ozone. (○, 0 ppm; △, 0.125 ppm; □, 0.25 ppm; ●, 0.50 ppm). Each point is the average of 4 observations with a LSD of 0.4. (From Tingey *et al.*, 1973a.)

present at all, is small and occurs only after many hours. Larger changes might be observed for specific classes of proteins, especially several hours after ozone fumigation.

3. Reducing Sugars

Dugger *et al.* (1962) first noted that the concentrations of reducing and soluble sugars in bean were the lowest at the leaf age of greatest ozone sensitivity. External application of a simple hexose solution reduced the ozone sensitivity of leaves, although it was not clear how much entered the leaf. Tingey *et al.* (1973b) showed that the levels of reducing sugars were elevated greatly (50%) immediately after exposure of soybean to ozone and remained high for 24 hours. Oddly, exposure to higher concentrations of ozone (0.5 ppm for 2 hours) slightly depressed the reducing sugar levels.

It has long been known that K^+ and reducing sugars are inversely related with respect to translocation (Epstein, 1972). Since the maximum expansion rate of the plant leaf cell is reached midway in the age sequence of development, it is not surprising that the amount of K^+ reaches a maximum (corresponding to maximum turgidity) as the reducing sugars reach a minimum (Heath *et al.,* 1974). Recently Neales and Incoll (1968) reviewed the subject of sugar/K^+ translocation and, unfortunately, could not determine any causal relationship between the transport of sugar and K^+. Certainly the interesting relationship between maximum K^+, minimum sugar, and free amino acid concentrations and maximum ozone sensitivity is not accidental.

B. Plant Nutrition

There are scattered reports in the literature regarding the effect of plant nutrient status on ozone injury. Unfortunately, there is little that is consistent among the papers—usually only one element is studied at a time—and different plant species, physiological age, and ozone levels are used. No methodical investigations of nutrient levels within the plant and ozone injury have been done.

Lower levels of nitrogen (I. P. Ting, unpublished data), sulfur (Adedipe *et al.,* 1972), and potassium (Dunning *et al.,* 1975) in the nutrient solution have been shown to alter the levels of ozone sensitivity. Further, these lowered levels have been shown to significantly affect metabolic events within the cell, such as water potential and stomatal responses. Adedipe *et al.* (1972) have shown that lowered sulfur levels within the nutrient solution accentuate ozone injury. However, lowered sulfur levels affect normal metabolism: the dry weight and carbohydrate content of the leaves decrease whereas the chlorophyll levels are unaffected. There was no measurement of stomatal response, age/ozone sensitivity dependence, or water potential in this study. But the lowered carbohydrate level and greater ozone sensitivity are consistent with other data.

Dunning *et al.* (1975) have found that with lowered K^+ levels, foliar injury is decreased but, as they noted, the lowered K^+ may have induced more rapid stomatal response.

I. P. Ting (unpublished data) has also shown that nitrogen deficiency led to lessened ozone injury but also lowered the plant's net productivity. Interestingly, nitrogen deficiency shifted the age of the leaf at which maximum injury occurred, once again demonstrating the importance of physiological age. In this instance, nitrogen deficiency accentuated ozone injury at certain ages and lowered it at others.

In addition, there is always the possibility that many of these effects are due to a double stress phenomenon. For example, a nutrient deficiency may critically hamper the plant's ability to deal with further stress, such as oxidant injury. The synergism of two stressful conditions may be enough to kill the plant.

C. "Spotting," "Flecking," or Cell–Cell Interactions

Visible ozone injury is manifested as regions of necrosis and chlorosis (diameter, 0.1–1 mm). This observation strongly suggests that whole areas of cells within the plants are disabled or killed outright and leads to a concept of rapid cell–cell interaction within the leaf. The death of one cell must quickly lead to the death of many surrounding cells, resulting in a sort of "amplification" effect.

One external cell interaction might be the leakage of ions (e.g., K^+) from one cell causing an osmotic stress in the surrounding cells such that they too would lose water. Since the further water loss would decrease the net osmotic effect of external K^+, this interaction would not be expected to provide much of an "amplification" effect.

A possible external–internal interaction could be the release of peroxides or aldehydes from ozonides of lipid fatty acids (see Section III,C,3). Peroxides themselves are known to be highly reactive (Cortesi and Privett, 1972). In addition, the altered lipids might be expected to act as detergents and disrupt other cellular membranes.

The third and possibly most likely mode of interaction is the influence of an ion–water inbalance acting through the plasmadesmodia connections between cells (Roland, 1973). It is not totally known what sort of materials these plasmadesmodia will pass nor how much. A "hole" induced by ozone in one cell, with a loss of water and turgor pressure, might "suck" the water out of surrounding cells in this manner. The plant possesses mechanisms (e.g., callose deposition) which can stop water movement, but this really depends upon the "sensing" or "feedback" mechanism of the callose deposition. This is an interesting, unexplored area of ozone research.

Little work has been done on the role of translocation in ozone injury. If ozone injury involves water or ion loss, does translocation increase to replace these needed substances? Again there seems to be no answer in the literature.

V. Conclusions

Although past research has contributed to a better understanding of the injury phenomena initiated by ozone, the precise biochemical mechanisms of photochemical oxidant damage to organisms have not yet been satisfactorily characterized. The observation that leaves pass through a particular developmental stage during which they are extremely susceptible to ozone damage suggests that particular metabolic and/or physiological conditions exist which contribute to sensitivity. As these conditions are more precisely defined, it should be possible to prevent ozone injury by applying specific materials to retard injury, by developing particular strains or varieties of plants with minimal sensitivity, or by genetic manipulations. Furthermore, many of these studies furnish valuable information on the developmental processes of plant leaves necessary in understanding the total concept of plant productivity.

A basic concern of future research involves an effort to answer the following questions: (1) what reactions take place at the surface and within the membrane; (2) which cellular components are attacked and in what time sequence; (3) how does this injury affect the functional capacity of the cell to carry out ion transport and maintain structure; (4) what mechanisms exist for the repair and preservation of the cell; (5) how deep within the cell do ozone and its products travel; (6) how do cells deal with internal toxic products; and (7) how can the knowledge obtained in these investigations be applied to counteract the oxidizing environment in which organisms live?

Air pollution has become a critical threat in urban areas throughout much of the United States, and unless effective controls are developed, will increase greatly in the not too distant future. The effects of air pollution already represent a serious health menace in many of the nation's cities, and damage to agricultural crops alone has been estimated at more than $500 million annually. Curiously, despite the seriousness of the air pollution problem, we remain largely ignorant of the basic biochemical and biophysical extent of cellular and organismic damage to living systems resulting from these oxidants.

Acknowledgments

The author would like to express his appreciation to Drs. W. M. Dugger and I. P. Ting for many hours of interesting discussions and to Ms. P. Frederick for reviewing the manuscript. Furthermore, financial aid (Grant No. R-301811) from U.S. Environmental Protection Agency for support of much of the research described here is gratefully acknowledged.

References

Adedipe, N. O., Hofstra, G., and Ormrod, D. P. (1972). Effects of sulfur nitrition on phytotoxicity and growth responses of bean plants to ozone. *Can. J. Bot.* **50,** 1789–1793.

Adedipe, N. O., Khatamian, H., and Ormrod, D. P. (1973). Stomatal regulation of ozone phytotoxicity in tomato. *Z. Pflanzenphysiol.* **68,** 323–328.

Alder, M. G., and Hill, G. R. (1950). The kinetics and mechanism of hydroxide ion catalyzed ozone decomposition in aqueous solution. *J. Amer. Chem. Soc.* **72,** 1884–1886.

Baldry, C. W., and Coombs, J. (1973). Regulation of photosynthetic carbon metabolism by pH and Mg^{++}. *Z. Pflanzenphysiol.* **69,** 213–216.

Chang, C. W. (1971). Effect of ozone or ribosomes in pinto bean leaves. *Phytochemistry* **10,** 2863–2868.

Chimiklis, P., and Heath, R. L. (1972). Effluxes of K^+ and H^+ from *Chlorella sorokiniana* as affected by O_3. *Plant Physiol.* **49,** p. 3.

Chimiklis, P., and Heath, R. L. (1975). Ozone-induced K^+ leakage from *Chlorella sorokiniana. Plant Physiol.* (in press).

Chimiklis, P., and Heath, R. L., and Platt, K. (1974). Internal osmolarity and K^+ concentrations of *Chlorella sorokiniana* stressed by growth at a low temperature. *J. Exp. Bot.* (submitted for publication).

Chow, C. K., and Tappel, A. L. (1972). An enzymatic protective mechanism against lipid peroxidation damage to lungs of ozone-exposed rats. *Lipids* **7,** 518–524.

Cleland, R. (1971). Cell wall extension. *Annu. Rev. Plant Physiol.* **22,** 197–222.

Cortesi, R., and Privett, O. S. (1972). Toxicity of fatty ozonides and peroxides. *Lipids* **7,** 715–721.

Coulson, C., and Heath, R. L. (1974). Inhibition of the photosynthetic capacity of isolated chloroplasts by ozone. *Plant Physiol.* **53,** 32–38.

Craker, L. E. (1972). Influence of ozone on RNA and protein content of *Lemna Minor* L. *Environ. Pollut.* **3,** 319–323.

Craker, L. E., and Starbuck, J. S. (1972). Metabolic changes associated with ozone injury of bean leaves. *Can. J. Plant Sci.* **52,** 589–597.

Craker, L. E., and Starbuck, J. S. (1973). Leaf age and air pollutant susceptibility: Uptake of ozone and sulfur dioxide. *Environ. Res.* **6,** 91–94.

Darley, E. F., and Middleton, J. T. (1966). Problems of air pollution in plant pathology. *Annu. Rev. Phytopathol.* **4,** 103–118.

Dugger, W. M., Jr., ed. (1974). "Symposium on Air Pollution Related to Plant Growth," A.C.S. Ser. Amer. Chem. Soc., Washington, D.C. Number 3.

Dugger, W. M., Jr., and Ting, I. P. (1970a). Air pollution oxidants—Their effects on metabolic processes in plants. *Annu. Rev. Plant Physiol.* **21,** 215–234.

Dugger, W. M., Jr., and Ting, I. P. (1970b). Physiological and biochemical effect of air pollution oxidants on plants. *Recent Advan. Phytochem.* **3,** 31–58.

Dugger, W. M., Jr., Taylor, O. C., Cardiff, E., and Thompson, C. R. (1962). Relationship between carbohydrate content and susceptibility of pinto plants to ozone damage. *Proc. Amer. Soc. Hort. Sci.* **81,** 304–315.

Dugger, W. M., Jr., and Palmer, R. L. (1969). Carbohydrate metabolism in leaves of rough lemon as influenced by ozone. *Int. Proc. Citrus Symp.* **2,** 711–714.

Dunning, J. A., Tingey, D. T., and Heck, W. W. (1975). Foliar sensitivity of legumes to ozone as affected by temperature, potassium nutrition and doses. *Amer. J. Bot.* (in press).

Epstein, E. (1972). "Mineral Nutrition of Plants: Principles and Perspectives," p. 308. Wiley, New York.

Evans, L. S., and Miller, P. R. (1972). Ozone damage to ponderosa pine: A histological and histochemical appraisal. *Amer. J. Bot.* **59,** 297–304.

Evans, L. S., and Ting, I. P. (1973). Ozone-induced membrane permeability changes. *Amer. J. Bot.* **60,** 155–162.

Evans, L. S., and Ting, I. P. (1974a). Ozone sensitivity of leaves: Relationship to leaf water content, gas transfer resistance, and anatomical characteristics. *Amer. J. Bot.* **61,** 592–597.

Evans, L. S., and Ting, I. P. (1973b). Effect of ozone on ^{86}Rb-labeled potassium transport in leaves of *Phaseolus vulgaris* L. *Atmos. Environ.* **8,** 855–861.

Frederick, P. E., and Heath, R. L. (1975). Ozone-induced fatty acid and viability changes in *Chlorella*. *Plant Physiol.* **55,** 15–19.

Glater, R. B., Solberg, R. A., and Scott, R. M. (1962). A developmental study of the leaves of *Nicotiana glutinosa* are related to their smog-sensitivity. *Amer. J. Bot.* **49,** 954–970.

Gross, E., Dilley, R. A., and San Pietro, A. (1969). Control of electron flow in chloroplasts by cations. *Arch. Biochem. Biophys.* **134,** 450–462.

Haagen-Smit, A. J., Darley, E. F., Zaitlin, M., Hull, H., and Nobel, W. (1952). Investigation on injury to plants from air pollution in the Los Angeles area. *Plant Physiol.* **27,** 18–34.

Hasson-Porath, E., and Poljakoff-Mayben, A. (1971). Contents of adenosine phosphate compounds in pea roots grown in saline media. *Plant Physiol.* **47,** 109–113.

Heath, R. L., Chimiklis, P., and Frederick, P. (1974). Role of potassium and lipids in ozone injury to plant membranes. *In* "Symposium on Air Pollution Related to Plant Growth" (W. M. Dugger, ed.), A.C.S. Ser. Amer. Chem. Soc., Washington, D.C. Number 3: 58–75.

Heck, W. W. (1968a). Comments made after Taylor's paper (1968). *J. Occup. Med.* **10,** 496–499.

Heck, W. W. (1968b). Factors influencing expression of oxidant damage to plants. *Annu. Rev. Phytopathol.* **6,** 165–188.

Heck, W. W., Dunning, J. A., and Hindawi, I. J. (1966). Ozone: Nonlinear relation of dose and injury in plants. *Science* **151,** 577–578.

Heggestad, H. E. (1968). Diseases of crops and ornamental plants incited by air pollutants. *Phytopathology* **58,** 1089–1097.

Hill, A. C., and Littlefield, N. (1969). Ozone: Effect on apparent photosynthesis, date of transpiration and stomatal closure in plants. *Environ. Sci. Technol.* **3,** 52–56.

Hitchcock, C., and Nichols, B. W. (1971). "Plant Lipid Biochemistry," Chapters 5 and 8. Academic Press, New York.

Hoather, R. C. (1948). Some experimental observations on the solubility of ozone and the ozonization of a variety of water. *J. Inst. Water Eng.* **2,** 358–368.

Hsiao, T. C. (1973). Plant response to water stress. *Annu. Rev. Plant Physiol.* **24,** 544 (reference to Ramagopal and Hsiao, 1975).

Jones, R. J., and Mansfield, T. A. (1970). Suppression of stomatal opening in leaves treated with abscisic acid. *J. Exp. Bot.* **21,** 714–719.

Kashtanov, L. I., and Oleschuk, O. N. (1937). The question of the solubility of ozone in water and sulphuric acid at different concentrations. *J. Gen. Chem. USSR* **7,** 839–42.

Kawashima, N., and Wildman, S. G. (1970). Fraction I protein. *Annu. Rev. Plant Physiol.* **21,** 325–358.

Kilpatrick, M. L., Herrick, C. C., and Kilpatrick, M. (1956). The decomposition of ozone in aqueous solution. *J. Amer. Chem. Soc.* **78,** 1784–1789.

Kuiper, P. J. C. (1972). Water transport across membranes. *Annu. Rev. Plant Physiol.* **23,** 158–172.

Kwon, T., and Watts, B. M. (1963). Determination of malonaldehyde by ultraviolet spectrometry. *J. Food Sci.* **28,** 627–630.

Larsen, R. I. (1969). A new mathematical model of air pollution concentration averaging time and frequency. *J. Amer. Pollut. Centr. Ass.* **19,** 24–30.

Lee, T. T. (1966). Research Report 1965–66. Research Station, Delhi, Ontario.

Leopold, A. C. (1967). "Plant Growth and Development," p. 466. McGraw-Hill, New York.

Lundberg, W. O., ed. (1962). "Autoxidation and Antioxidants," Vol. II. Wiley, New York.

MacDowell, F. D. H. (1965). Predisposition of tobacco to ozone damage. *Can. J. Plant Sci.* **45,** 1–12.

Mackender, R. O., and Leech, R. M. (1974). The galactolipid, phospholipid, and fatty acid composition of the chloroplast envelope membranes of *Vicia faba* L. *Plant Physiol.* **53,** 496–502.

McNair-Scott, D. B., and Lesher, E. C. (1963). Effect of ozone on survival and permeability of *E. coli. J. Bacteriol.* **85,** 567–576.

Mansfield, T. A. (1973). The role of stomata in determining the response of plants to air pollutants. *Cur. Advan. Plant Sci.,* **2,** 11–20.

Meznel, D. B. (1971). Oxidation of biologically active reducing substances by ozone. *Arch. Environ. Health* **23,** 149–153.

Middleton, J. T. (1961). Photochemical air pollution damage to plants. *Annu. Rev. Plant Physiol.* **12,** 431–440.

Miller, P. R., Parmeter, J. R., Taylor, O. C., and Cerdiff, E. A. (1963). Ozone injury to foliage of *Pinus ponderosa. Phytopathology* **53,** 1072–1076.

Moses, H. (1969). Mathematical urban air pollution models. *62nd Annu. Meet. Air Pollut. Contr. Ass.* Paper No. 69-31.

Mudd, J. B. (1973). Biochemical effects of some air pollutants on plants. *Advan. Chem. Ser.* **122,** 31–47.

Mudd, J. B., Leavitt, R., Ongun, A., and McManus, T. T. (1969). Reaction of ozone with amino acids and proteins. *Atmos. Environ.* **3,** 669–682.

Mudd, J. B., McManus, T. T., Ongun, A., and McCullogh, T. E. (1971a). Inhibition of glycolipid biosynthesis in chloroplasts by ozone and sulfhydryl reagents. *Plant Physiol.* **48,** 335–339.

Mudd, J. B., McManus, T. T., and Ongun, A. (1971b). Inhibition of lipid metabolism in chloroplasts by ozone. *Proc., Int. Clean Air Congr., 2nd, 1970. Academic Press,* New York. Pp. 256–260.

Mudd, J. B., Leh, F., and McManus, T. T. (1974). Reaction of peroxyacetyl nitrate with cysteine, cystine, methionine, lipoic acid, papain and lysozyme. *Arch. Biochem. Biophys.* **161,** 408–419.

Neales, T. E. and Incoll, L. D. (1968). The control of leaf photosynthesis rate by the level of assimilate concentration in the leaf: A review of the hypothesis. *Bot. Rev.* **34,** 107–125.

Nobel, P. S., and Wang, C. T. (1973). Ozone increases the permeability of isolated pea chloroplasts. *Arch. Biochem. Biophys.* **157,** 388–394.

Northcote, D. H. (1972). Chemistry of the plant cell wall. *Annu. Rev. Plant Physiol.* **23,** 113–132.

Odabasi, H. (1973). Atmospheric problems, air pollution—A comprehensive review. *In* "Environmental Problems and their International Implications" (H. Odabasi and S. E. Ulug, eds.), pp. 57–110. Colorado Associated University Press, Boulder.

Ordin, L., and Hall, M. A. (1967). Studies on cellulose synthesis by a cell-free oat coleoptile enzyme system: Inactivation by airborne oxidants. *Plant Physiol.* **42,** 205–212.

Ordin, L., Hall, M. A., and Kindinger, J. I. (1969). Oxidant-induced inhibition of enzymes involved in cell wall polysaccharide synthesis. *Arch. Environ. Health* **18,** 623–626.

Oshmia, R. J. (1973). "Final Report to California Air Resources Board Under Agreement ARB-287," Dept. Food & Agr., Sacramento, California; also partially described in: A viable system of biological indicators for monitoring air pollutants. [*J. Amer. Pollut. Centr. Ass.* **24,** 575–578 (1974).]

Pell, E. J., and Brennan, E., (1973). Changes in respiration, photosynthesis, adenosine 5′-triphosphate, and total adenylate content of ozonated pinto bean foliage as they relate to symptom expression. *Plant Physiol.* **5,** 378–381.

Perchorowicz, J., and Ting, I. P. (1974). Ozone effects in plant cell permeability. *Amer. J. Bot.* **61,** 787–793.

Pestka, S. (1971). Protein biosynthesis: Mechanism, requirements and potassium dependency. *In* "Membranes and Ion Transport" (E. E. Bittar, ed.), Vol. 3, p. 279. Wiley (Interscience), New York.

Rains, D. W. (1972). Salt transport by plants in relation to salinity. *Annu. Rev. Plant Physiol.* **23,** 367–388.

Rich, S. (1964). Ozone damage to plants. *Annu. Rev. Phytopathol.* **2,** 253–266.

Roland, J. (1973). The relationship between the plasmalemma and cell wall. *Int. Rev. Cytol.* **36,** 45–91.

Somers, G. F. (1973), The affinity of onion cell walls for calcium ions. *Amer. J. Bot.* **60,** 987–990.

Stephens, E. R. (1969). Chemistry of atmospheric oxidants. *J. Amer. Pollut. Contr. Ass.* **19,** 181–185.

Stern, A. C., Wohlers, H. C., Boubel, R. W., and Lowry, W. P. (1973). "Fundamentals of Air Pollution." Academic Press, New York.

Swanson, E. W., Thomson, W. W., and Mudd, J. B. (1973). The effect of ozone on leaf cell membranes. *Can. J. Bot.* **51,** 1213–1219.

Tappel, A. L. (1965). Free radical lipid peroxidation damage and its inhibition by Vitamin E and selenium. *Fed. Proc., Fed. Amer. Soc. Exp. Biol.* **24,** 73–78.
Taylor, O. C. (1968). Effects of oxidant air pollutants. *J. Occup. Med.* **10,** 485–496.
Teige, B., McManus, T. T., and Mudd, J. B. (1974). Reaction of ozone with phosphatidylcholine liposomes and the lytic effect of products on RBC. *Chem. Phys. Lipids* **12,** 153–171.
Thomson, W. W., Dugger, W. M., Jr., and Palmer, R. L. (1966). Effect of ozone on the fine structure of the palisade parenchyma cells of bean leaves. *Can. J. Bot.* **44,** 1677–1682.
Thorp, C. E. (1954). "Bibliography of Ozone Technology." Armour Res. Found., J. S. Swift, Co., Chicago, Illinois. Vols. I and II.
Ting, I. P., and Dugger, W. M., Jr., (1968). Factors affecting ozone sensitivity and susceptibility of cotton plants. *J. Amer. Pollut. Contr. Ass.* **18,** 810–813.
Ting, I. P., and Dugger, W. M., Jr., (1971). Ozone resistance in tobacco plants: possible relationship to water balance. *Atmos. Environ.* **5,** 147–150.
Ting, I. P., and Mukerji, S. K. (1971). Leaf ontogeny as a factor in susceptibility to ozone: Amino acid and carbohydrate changes during expansion. *Amer. J. Bot.* **58,** 497–504.
Tingey, D. T., Fites, R. C., and Wicklift, C. (1973a). Foliar sensitivity of soybeans to ozone as related to several leaf parameters. *Environ. Pollut.* **4,** 183–192.
Tingey, D. T., Fites, R. C., and Wickliff, C. (1973b). Ozone alteration of nitrate reduction in soybean. *Physiol. Plant.* **29,** 33–38.
Tomlinson, H., and Rich, S. (1967). Metabolic changes in free amino acids of bean leaves exposed to ozone. *Phytopathology* **57,** 972–974.
Tomlinson, H., and Rich, S. (1968). The ozone resistance of leaves as related to their sulfhydryl and ATP content. *Phytopathology* **58,** 808–810.
Tomlinson, H., and Rich, S. (1969). Relating lipid content and fatty acid synthesis to ozone injury of tobacco leaves. *Phytopathology* **59,** 1284–1286.
Troughton, J., and Donaldson, L. A. (1972). "Probing Plant Structure," Plates 38 and 40. McGraw-Hill, New York.
Zelitch, I. (1969). Stomatal control. *Annu. Rev. Plant Physiol.* **20,** 329–350.
Zelitch, I. (1971). "Photosynthesis, Photorespiration, and Plant Productivity." Academic Press, New York.

4

FLUORIDES

Chong W. Chang

I. Introduction

In preparing this chapter biochemical analyses and physiological observations were combined to explain basic mechanisms associated with effects of fluoride as an air pollutant. Both *in vivo* and *in vitro* studies are reviewed. Experimental materials include not only intact leaves of higher plants as natural entries for gaseous fluoride, but also other plant parts, segments of tissue, isolated subcellular organelles, and some lower organisms. A literature survey on effects of atmospheric fluoride on vegetation and soil under field conditions is beyond the scope of this chapter.

II. Absorption, Accumulation, Translocation, and Injury-Effects of Fluoride

Although all plants normally contain fluorine, the concentration varies greatly (Webster, 1967). For example, Zimmerman and Hitchcock (1956) found that leaves of dogwood (*Cornus florida*), apple (*Malus* sp.), and peach (*Prunus persica*) accumulated different amounts of fluoride, 40.3, 7.9, and 7.5 ppm of fluoride, respectively, although the plants grew in the same location. In addition, certain varieties of one species showed dissimilar responses to fluoride. Schneider and MacLean (1970) reported that fumigation of seven grain sorghum hybrids (*Sorghum* sp.) with gaseous hydrogen fluoride (3.9–7.0 μg F/m^3 for 8 days) caused necrosis of leaf tips and chlorotic mottle of subtending tissue. However, the severity of injury was variable. DeKalb $C_{44}C$ and RS 608 were most resistant to hydrogen-induced necrosis, while Northrup King 222A, RS 671, and RS 625 were more susceptible. There appeared to be no relationship between foliar injury and fluoride accumulation.

Fluoride injury often is characterized by tip burn of leaves (Fig. 1). Hitchcock *et al.* (1962) reported 10 years' work (1951–1960) on various effects of fluorides on gladiolus (*Gladiolus* sp.). Tip burn and accumulation of fluoride in control and HF-fumigated plants varied considerably with va-

FIG. 1. Acute fluoride injury on apricot, which is characterized by tip burn. (Photo courtesy by E. F. Darley.)

riety of gladiolus, cultural site, age of plant, and age of leaf. Control plants in the area where experimental fumigation with fluorides was carried out had more tip burn and accumulated more fluoride than plants 0.5–1.3 miles away. This reduced damage correlated with less atmospheric fluoride at the distant locations. Gladiolus in the 5- to 7-leaf stage had more tip burn and accumulated more fluoride than plants in the 3-leaf stage. Middle-aged leaves generally had the most tip burn, but old leaves accumulated more fluoride than young leaves. Zimmerman and Hitchcock (1956) reported that *Camellia japonica* growing in a greenhouse, contained 1500–2000 ppm fluoride in old leaves but only about 100 ppm fluoride in young leaves. In addition, Hitchcock *et al.* (1962) showed that in general the more resistant varieties accumulated most fluoride and susceptible varieties accumulated least. This applied to nonfumigated plants and plants fumigated with HF. Tip burn and accumulation of fluoride were greater at higher concentrations of HF. Increase in tip burn correlated with dosage up to 2 to 4 times the threshold level. In Hitchcock's experiments, the dosage–response curve for the moderately resistant variety, Elizabeth the Queen, was significantly flatter than the curve for the susceptible Snow Princess variety. There was a highly significant regression of tip burn on fluoride content of leaves in both varieties. However, there was no correlation between susceptibility and number and location of stomata. This indicated that the routes of fluoride absorption by the leaf had not been fully elucidated.

Hitchcock also found that at levels of HF up to about two parts per billion (ppb) of air, which caused 5 to 8 cm of tip burn, there was no significant effect on dry weight of tops or corn yield in the varieties Snow Princess and Elizabeth the Queen, but production of flower spikes was significantly less in treated plants. In addition, within 1 week after exposure to HF, Snow Princess lost up to 40 to 50% of the fluoride from aboveground parts of the plant, while control plants gained fluoride. Elizabeth the Queen lost little (average of 10%) or no fluoride. The loss generally was greater from small than from large plants. Other investigators also reported apparent loss of fluoride from plants (Thomas, 1951; Zimmerman and Hitchcock, 1946). Loss of fluoride that was absorbed into plant leaves seems to be a general phenomenon.

The mechanism of fluoride absorption into and desorption from plant tissues was first studied by Venkateswarlu *et al.* (1965) for barley (*Hordeum vulgare*) roots. Plant tissues are associated with outer and inner spaces in absorbing ionic elements. Outer space is the volume of tissue to which ions have free and ready access by diffusion, while inner space is the region to which ions are transported by active mechanisms and from which they are not removed by diffusion or exchange in an ambient solution. Venkateswarlu *et al.* (1965) reported that fluoride was present in the outer space of barley roots and that the absorption occurred by diffusion and did not require an active metabolic mechanism. Therefore, absorbed fluoride is essentially almost completely desorbed from barley roots by water. Such a mechanism of fluoride transport contrasted with that for chlorine, which required oxidation energy for absorption.

Jacobson *et al.* (1966) studied accumulation and distribution of fluoride in gladiolus (*Gladiolus* sp. var. Snow Princess and Elizabeth the Queen) leaves. They indicated that fluoride from the air can be adsorbed to the surface of leaves as well as accumulated internally, and that fluoride in leaves can be translocated outward to the surface as well as upward to the tips. Fluoride injury and accumulation can be induced in any desired location on a gladiolus blade by restricting gas exchange of the blade, indicating that all parts of a blade are sensitive to fluoride-induced injury. Fluoride also remains in a soluble form in leaves and maintains the chemical properties of free, inorganic fluoride. The solubility and mobility of fluoride and the ease of its removal from plant tissues indicated that irreversible binding to cellular components did not occur. However, the fact that loss of fluoride generally is greater from small than from large plants (Hitchcock *et al.,* 1962) indicated that some part of the fluoride probably becomes associated with subcellular organelles as the plant develops. This assumption seems to be compatible with the finding of Ledbetter *et al.,* (1960) that distribution of ^{19}F applied as $H^{19}F$ to tomato (*Lycopersicon esculentum*) leaves after long-term accumulation was, in order of decreasing concentration, cell wall, chloroplasts, soluble protein, mitochondria, and microsomes. Similar findings were obtained for navel orange trees (*Citrus sinensis*) grown under the influence of fluoride-polluted air in the vicinity of Kaiser Steel, Riverside (Chang and Thompson, 1966a). Woltz (1964a) demonstrated that fluoride was translocated freely with the transpiration stream in gladiolus (*Gladiolus* sp. var. Friendship and Orange Gold) leaves and did not necessarily move to the upper leaf margins in the manner commonly noted in fluoride leaf scorch of gladiolus. If the transpiration stream was interrupted e.g. by cutting a notch in one side of the leaf, necrosis resulting from fluoride fumigation was greater on that side of the leaf. Solberg and Adams (1956), on the other hand, presented histological data indicating that if fluoride entered a leaf in subphytotoxic concentrations, it was transported acropetally.

Several investigators discussed the effects of root-acquired fluoride on leaves. Brewer *et al.* (1959) reported that leaves of mature navel orange trees absorbed substantial amounts of fluoride from nutrient solutions containing up to 25 ppm fluoride. Growth and yield also were markedly depressed. Bovay *et al.* (1969) also indicated that fluorine entered the roots of test plants grown hydroponically and accumulated in leaves as a function of the nature of the fluoride compounds added to the solution. These compounds also determined the extent of necrosis, which depended on fluorine content. Boron enhanced fluorine uptake, but greatly decreased the causticity, if fluorine was in the complex form BF_4. The distribution of fluoride in plant systems differs, depending on the entry of fluoride. Fluoride absorbed from the air was concentrated in the leaves, while fluoride acquired by root absorption was distributed in both leaves and roots (Daines *et al.*, 1952). After fumigation of plants with fluoride, fluoride content was high in leaves and low in stems and roots. McCune (1969) indicated that fluoride originating in the soil tended to accumulate in roots. However, some fluoride in the foliage had its origin in the soil and this was not removed by light washings in water. Root-acquired fluoride caused necrosis of internal leaf tissues, whereas leaf-acquired fluoride caused marginal and tip a necrotic pattern involving more of the internal tissues of the gladiolus necrosis. Woltz (1964a) also reported that soil-acquired fluoride produced leaf in contrast to atmosphere-acquired fluoride, which affected marginal areas almost entirely. These findings were in accord with those of Woltz (1964b) and McCune (1969).

Many studies deal with the influence of mineral nutrition on sensitivity of plants to hydrogen fluoride. However, the results vary considerably depending on experimental conditions and materials. Brennan *et al.* (1950) investigated effects of N, Ca, and P on susceptibility of tomato plants to injury from NaF applications to roots and from HF fumigation of plant tops. They found that medium levels of these minerals favored absorption and translocation of fluorine in sufficient quantities to cause visible leaf injury by root treatments and also by absorption of toxic quantities of fluorine in plants fumigated with higher amounts of fluorine. However, a low or deficient supply of these three minerals aided in preventing absorption of toxic amounts of fluorine through the roots or after fumigation. Root injury occurred in the same plants which showed foliage injury in the root-treated series. However, fumigation did not injure the roots. When NaF was added to the substrate, fluorine was translocated to the leaves but more fluorine was accumulated by roots than by leaves. In contrast, fumigated plants accumulated large amounts of fluorine in leaves and normal amounts in roots, suggesting, a lack of downward translocation of fluorine from the leaves. McCune *et al.* (1966) studied the effects of mineral solutions on sensitivity of gladiolus (*Gladiolus* sp. var. Snow Princess and Elizabeth the Queen) to HF. Tip burn was associated with K and P deficiency in Snow Princess variety and with Mg deficiency in Elizabeth the Queen. Decreased tip burn was correlated with Ca and N deficiencies in Snow Princess. Adams and Sulzbach (1961) also treated bean (*Phaseolus vulgaris*) seedlings grown in nutrient solution with fluoride (4 to 7 μg of F/m³). They observed symptoms of foliar fluorosis in nitrogen-deficient plants fumigated for 20 days. MacLean *et al.* (1969) found that in Mg deficient tomato plants, HF exposure caused reduction in dry weight increment of both tops and roots, and increase in severity of Mg deficiency symptoms. Foliar accumulation of fluoride was suppressed by Mg deficiency. However, in Ca-deficient plants, HF treatment caused a reduction in top growth and increased leaf symptoms. Fluoride accumulation was not affected by Ca nutrition. The response of K-deficient plants to HF was limited to increase in F accumulation and in necrosis. Leone *et al.* (1948) investigated the influence of fluoride (10 to 400 ppm) in the nutrient solution on leaves of peach, tomato, and buckwheat (*Fagopyrum* sp.). Peach and buckwheat were moderately injured within 13 days by 10, 25, or 50 ppm fluorine, whereas tomato required up to 48 days for symptoms to appear. Fluorine concentrations below 10 ppm did not cause injury in any species. In all three species, fluorine accumulation in aerial tissues varied with the amount of fluorine in the nutrient substrate. Applegate and Adams (1960c) indicated that more fluoride was absorbed by leaves of phosphorus-deficient plants than by plants deficient in N, K, Fe, or Ca. Hansen *et al.* (1958) reported that fluoride uptake by plants from soils varying in pH, lime, clay, and organic matter increased with addition of sodium fluoride and sodium fluorosilicate. Uptake, however, was less when soils contained higher levels of these additives. Yields generally were reduced when fluoride absorbed from soil exceeded 60 ppm in the shoots. Under field conditions, fluoride contamination of plants apparently occurred directly from the air rather than indirectly via the soil.

III. Effects of Fluoride on Various Metabolites, Injury (Subcellular), and Recovery

Weinstein (1961) conducted time-course experiments on effects of atmospheric fluoride on metabolic constituents of tomato and bean leaves. Plants were fumigated with fluoride (12.4 ppb and 1.3 ppb) for 24 hours. During fumigation, the plants were harvested at 3-day intervals for 9 days. The treatment was stopped after 9 days. Plants were again harvested at the same intervals during the recovery period. Weinstein noted that total fluoride content and fluoride concentration of both tomato and bean leaf tissues decreased during the recovery period. The free sugar content of leaves was reduced in both species during the fumigation and recovery periods. In general, nonvolatile organic acids increased after 3 days of fumigation. By 6 days, the level approached that of the controls and then dropped to a minimum at 9 days (or after 3 days of recovery in beans).

During the postexposure period, the organic acid level recovered and, in tomato, exceeded that of the controls. RNA phosphorus in fumigated tomato leaves was lower than in controls. These changes in metabolic constituents occurred in leaves which showed a minimum amount of fluoride injury. Weinstein (1961) found that the level of nonvolatile organic acids increased, remained constant, or decreased, depending on the duration of fumigation.

McCune *et al.* (1964) fumigated bean and Milo maize plants (*Sorghum vulgare*) with fluoride (1.7 to 7.6 $\mu g/m^3$) for 10 days and determined activities of enzymes associated with glycolysis, pool sizes of various keto acids, and catalase activity during and after treatment. Leaves of bean plants exposed to fluoride showed increased levels of pyruvate and α-ketoglutarate. Phosphoenolpyruvate carboxylase activity and oxalacetate were unaffected. Leaves of Milo maize plants showed increased levels of enolase and pyruvate kinase activity, and a decreased level of pyruvate. Oxalacetate and α-ketoglutarate levels were unaffected. Catalase activity, however, first increased, then decreased after HF fumigation. These HF-induced changes were greatest 6 to 10 days after initiation of fumigation and disappeared or decreased during the postfumigation period. The treated plants did not show visible injury or growth reduction. McCune *et al.* (1964) claimed that the changes in activities of enolase and pyruvate kinase and in concentration of the α-keto acids indicated an altered glycolytic pathway by HF. Also, they proposed that the increased activity of catalase indicated that HF induced general increase in metabolic activity.

Adams and Emerson (1961) fumigated plants with fluoride (0.5 to 10 $\mu g/m^3$) for various periods and analyzed effects on starch and nonpolysaccharide content of *Pinus ponderosa* needles. The study suggested that one possible effect of fluoride may be to alter the relationship between nonpolysaccharides and starch. The data further indicated the possibility that plants can adapt to atmospheric fluoride in the range of 5 to 10 μg fluoride/m^3, if provided a recovery period between each exposure. Conversely, plants might not adapt so readily to lower total amounts of more nearly continuously supplied fluoride. Similar observations of the relative effects of daily fumigations at the lowest fluoride concentration (1.5 μg fluoride/m^3) as compared with the highest concentrations (5 and 10 μg fluoride/m^3) on a twice weekly fumigation basis were previously observed, when comparing onset of initial leaf symptoms at three levels of fluoride concentration (Adams *et al.*, 1957). Adams and Emerson (1961) concluded that the apparent ability of plants to adapt or to show greater resistance to intermittent, higher fluoride fumigation levels was important because plants in the field were exposed to such intermittent fluoride fumigation (Adams and Koppe, 1959). This contrasts with the long-term, low atmospheric concentrations (less than 1 μg fluoride/m^3), which previously had been assumed to exist in the field (Hill *et al.*, 1959b). They reemphasized that the sequence of fluoride exposure might outweigh the influence of the actual fluoride level within a fumigation concentration range of 0.5 to 10 μg fluoride/m^3.

McCune *et al.* (1970) found that hydrogen fluoride in the atmosphere at concentrations of 4.8 to 10.7 fluoride/m^3 for 4 to 12 days had no consistent effect on levels or composition of acid-soluble nucleotide pools in leaves of Pinto bean, tomato, or corn (*Zea mays*). Distribution of ^{32}P among the various nucleotide fractions was not affected by hydrogen fluoride. However, incorporation of ^{32}P by the whole acid-soluble nucleotide pool was reduced in corn leaves during both the fumigation and post fumigation periods. Pack and Wilson (1967) also investigated the influence of hydrogen fluoride (14 μg/m^3) on acid-soluble phosphorus compounds in bean seedlings. They reported that the seedlings accumulated 275 ppm of fluoride during a period of 4 days, but showed no symptoms of injury. There also was no change in acid-soluble phosphorus compounds.

Yang and Miller (1963a) studied metabolic changes during development of leaf necrosis in soybean plants (*Glycine max*). They reported inhibition of sucrose synthesis by fluoride by demonstrating an increase in reducing sugars and a decrease in sucrose. Also, they noted that necrotic leaves contained increased concentrations of organic acids, mostly malic, malonic, and citric. The greater increase in malic acid, relative to that of citric acid, was the reverse of what happened in chlorotic tissue. Necrotic leaves also had a high free amino acid content. The greatest increase in concentration of asparagine was correlated with increased respiration rate of necrotic leaves. Whereas pipecolic acid accumulated in large quantities in necrotic tissue, it was not detectable in normal leaves. Accumulation of organic acids and amino acids in leaves during fluoride fumigation was related to a lowered respiratory quotient.

Yang and Miller (1963c) conducted experiments to determine if the accumulated organic acids and amino acids in fluoride-necrotic leaves observed in their previous studies (1963a) were due to increased dark CO_2 fixation, and to ascertain how fluoride might affect the carboxylating enzyme system. They found that necrotic leaves of soybean had a higher rate of dark CO_2 fixation than control leaves both *in vivo* and *in vitro*. Phosphoenolpyruvate carboxylase activity also was higher in necrotic leaves than in control leaves. By summarizing the studies of Yang and Miller (1963a, 1963c), they concluded that accumulation of organic acids and amino acids in necrotic leaves resulted from an increased rate of dark CO_2 fixation, because of increase in total organic acids plus amino acids, enhanced rate of dark CO_2 fixation, equal or greater amounts of protein, and a lower respiratory quotient in necrotic leaves. They proposed that some abnormal protein or enzyme was actively synthesized by utilizing adversely accumulated free amino acids at an early stage of fluoride fumigation. This protein could have induced the subsequent altered metabolism which finally led to leaf necrosis.

Lovelace and Miller (1967) made histochemical investigations of the *in vivo* effects of fluoride on tricarboxylic acid cycle dehydrogenases in *Pelargonium zonale*. Plants were exposed to 17 ppb HF, and enzyme activities in treated plants were compared to those in controls. Leaves also were incubated in 5×10^{-3} *M* NaF. Injuries to fumigated leaves and those incubated in NaF solution were similar. Leaf tissue subjected to HF or NaF had less succinic *p*-nitro blue tetrazolium reductase activity than control tissue. Other TCA cycle dehydrogenase enzymes were not noticeably affected by the fluoride concentrations used. Excised leaves cultured in 5×10^{-3} *M* NaF exhibited less succinic *p*-nitro blue tetrazolium reductase activity after 24 hours than the leaves cultured in 5×10^{-3} *M* NaCl.

Miller and Wei (1971) studied subcellular injury in leaf cells of soybean plants treated with fluoride. Symptoms of general fluoride leaf injury appeared as vesiculation of the tonoplast and presence of black, lipidlike granules. The appearance of these granules followed membrane disruption. At this stage of injury, the middle lamella of the cell wall had started to deteriorate. The nuclear envelope was intact but some dilation occurred. At later stages of injury, electron-dense particles were deposited on the mitochondria, endoplasmic reticulum, plasmalemma, and dictyosomes. The endoplasmic reticulum showed a tendency to vesiculate. In advanced stages of injury, enlarged organelles occurred and clumping of various cell organelles followed.

The effects of fluoride range from alteration at the subcellular level to tissue destruction. Damage of the latter may be an outcome of serial and multiple metabolic events in the former. Our control program requires further investigation of basic mechanisms linked with the action of fluoride in plant systems.

IV. Effects of Fluoride on Respiration and Related Biochemical Reactions

Both stimulatory and inhibitory effects of fluoride on respiration have been extensively studied. Although many factors influence respiration, the concentration of fluoride and plant species are of prime importance (Yu and Miller, 1967).

McNulty and Newman (1957) demonstrated high levels of respiration after fumigation with fluoride in apparently healthy gladiolus (*Gladiolus* sp. var. Snow Princess) leaf tissue adjacent to areas injured by fluoride. Hill *et al.* (1959a) reported that such high respiration rates could be caused by leaf tissue damaged by fluoride, since the increase in respiration was greatest near the injured tissue and apparently proportional to the amount of damage produced. Hill's findings (1959a) were in accord with earlier studies of Audus (1935) and Whiteman and Schomer (1945). Audus (1935) found that the rate of respiration in laurel (*Prunus* sp.) leaf blades was increased by merely rubbing or bending them. A wound stimulation of respiration in sweet potato (*Ipomoea batatas*) roots also was reported by Whiteman and Schomer (1945. Yang and Miller (1963a) found that accumulation of organic acids resulted in a lowered respiratory quotient in damaged tissues of necrotic leaves of soybean plants that had been treated with fluoride.

McNulty (1964) reported that concentrations of 1×10^{-3} to 5×10^{-3} *M* sodium fluoride increased the respiratory rate of bean leaf discs from healthy plants by about 25% without causing visible damage. Higher concentrations of fluoride reduced oxygen uptake long before visible symptoms appeared. These findings agreed with those of Luštinec *et al.* (1960), for wheat (*Triticum vulgare*) seedlings treated with fluoride. Stimulation of respiration by fluoride was reported for pea (*Pisum sativum* var. Alaska), epicotyl tissue (Christiansen and Thimann, 1950), tomato and bean (Weinstein, 1961), leaf discs of grape (*Vitis vinifera*), and root tips of *Lens culinaris* (Pilet, 1963, 1964).

In 1945, Borei demonstrated inhibition of respiration in yeast treated with 2.5×10^{-3} *M* sodium fluoride. McNulty and Newman (1956) showed that sublethal concentration of fluoride significantly inhibited respiration of Italian prune trees (*Prunus* sp.). Applegate *et al.* (1960) noted that high concentrations of fluoride (1×10^{-1} *M*) inhibited oxygen uptake, whereas low concentrations (1×10^{-4} *M*) accelerated respiration rate in bush bean (*Phaseolus vulgaris* var. *humilis*) seedlings. Inhibitory effects of fluoride on respiration also were reported for intact bush bean plants (Applegate and Adams, 1960a,b). Yu and Miller (1967) made detailed studies of the effects of various concentrations of fluoride (5×10^{-3} to 1×10^{-2} *M* KF and 54 ppb HF) on respiration of leaves of *Chenopodium murale* and soybean. Fluoride treatment included both excised leaves cultured in nutrient solutions and leaves from plants fumigated with HF. Tissues treated with low fluoride concentrations initially showed increased oxygen uptake, but eventually oxygen consumption decreased below normal rates. Tissues treated with high concentrations of fluoride showed increased oxygen uptake if analyzed soon after initiation of treatment. Increase in respiration generally preceded visible damage. Also, if the tissue was damaged, decrease in respiration could be correlated with the degree of injury. However, Hill *et al.* (1958) did not observe any effect of fluoride on respiration of tomato leaves that had been fumigated with fluoride at concentrations less than about 70 $\mu g/m^3$.

Both fluoride and DNP (2,4-dinitrophenol) stimulated respiration of *Chlorella pyrenoidosa* by about 60% (Lords, 1960). However, if both reagents were added together, the stimulatory effect was weak. Yu and Miller (1967) reported that both fluoride and DNP similarily influenced respiration of soybean leaves. Also, DNP stimulated respiration of fluoride-treated leaves less than in control leaves.

Dinitrophenol uncouples phosphorylation from oxidation and promotes the flow of electrons through the electron transport system. The effect of DNP on respiration would be greater in older tissues than in younger ones, because the aged tissue has a lower ATP turnover and, therefore, is released from the retarding effect of coupled phosphorylation (Yu and Miller, 1967). If fluoride-treated tissue was associated with more ATP turnover because of stimulated respiration, it behaved as metabolically active younger tissue in response to DNP (Yu and Miller, 1967).

Although previous studies indicate some similarities in the ways fluoride and DNP stimulate respiration, these agents appear to differ with respect to production of phosphorylated nucleotides. If fluoride stimulated respiration by uncoupling phosphorylation in a way similar to the action of DNP, fluoride would decrease the content of ATP. However, McNulty and Lords (1960) demonstrated that concentrations of fluoride of 1.05×10^{-4}, 1.05×10^{-3}, and 1.05×10^{-2} *M* increased oxygen consumption of *Chlorella pyrenoidosa* and at the same time increased total phosphorylated nucleotides. Further, experiments with corn roots showed that fluoride induced accumulation of total triphosphate nucleotides, with the level of ATP enhanced most (Chang, 1968).

Fluoride affects a number of enzyme systems. The best known effect of fluoride on enzymes, inhibition of enolase, was first demonstrated by Warburg and Christian (1942). The enzyme may be inhibited by formation of a magnesium fluorophosphate complex. However, it is likely that fluorophosphate itself is not inhibitory (Peters *et al.*, 1964). Miller (1957, 1958) reported that enolase isolated from peas was sensitive to fluoride and required phosphate and magnesium for inhibition to occur.

In an effort to clarify the metabolic pathways associated with the function of respiration as oxygen consumption, Bonner and Wildman (1946) reported inhibition of respiration by fluoride in spinach (*Spinacia oleracea*) leaves and reversibility of such inhibiton by sodium pyruvate, but not by glucose or sucrose. Similar results were repoted for *Avena sativa,* var. Siegeshafer coleoptiles (Bonner, 1948) and *Hordeum* roots (Laties, 1949). These reports indicate that the site responsible for fluoride-influenced respiration is associated with the glycolytic pathway. On the other hand, McNulty (1964) treated leaf discs with pyruvate, fluoride, and pyruvate plus fluoride. Fluoride (8×10^{-3} M) increased the rate of respiration by about 70%, and pyruvate increased it by approximately 40%. When both reagents were present the effects were additive, with a 107% increase in respiration. The data indicated that stimulation of respiration by pyruvate and fluoride was distinct. Possibly, therefore, a secondary catabolic pathway was operative through the pentose phosphate cycle. If enolase is inhibited by fluoride, the substrate preceding the enolase-mediated reaction might accumulate and enhance the activity of the pentose phosphate cycle.

To compare the rate of relative respiratory activity through the pentose phosphate cycle with that by the glycolytic pathway, the ratio of $^{14}CO_2$ from [6-^{14}C] glucose metabolized through the glycolytic pathway, to $^{14}CO_2$ from [1-^{14}C] glucose metabolized by the pentose phosphate cycle was determined. Therefore, the ratio of C_6/C_1 should indicate relative respiratory activities of two metabolic schemes. Lee *et al.* (1966) demonstrated marked stimulation of glucose-6-phosphate dehydrogenase activity in fluoride-fumigated leaves. This confirmed earlier findings that the pentose phosphate pathway was important in fluoride-treated tissue (Lovelace and Miller, 1967). Overall oxygen consumption, therefore, may be greater in treated tissue because of increased concentration of reduced pyrimidine nucleotides. Ross *et al.* (1962) studied the influence of fluoride on the pathway of glucose breakdown as estimated by use of C_6/C_1 ratios in normal and injured leaves of *Polygonum orientale* and *Chenopodium murale.* The ratios were decreased in fluoride-treated leaves, indicating a relative increase in importance of the pentose phosphate pathway. Such decreases were verified in excised leaves exposed to KF and in leaves from plants that were exposed to atmospheric HF. The decreased ratios are believed to be due to inhibition of glycolysis, probably by enolase inactivation. Since the leaves were damaged, however, it was not clear that the decreased C_6/C_1 ratios were due to secondary effects. In contrast with the report of a study similar to that of Ross *et al.* (1962), Luštinec and Pokorná (1962) determined the rate of respiration and the C_6/C_1 ratio after exposing young wheat leaves to sodium fluoride and to ^{14}C-labeled glucose. They reported a 27% increase in respiration and an increase in the C_6/C_1 ratio rather than a decrease. They concluded that fluoride stimulated the glycolytic pathway. To determine if sensitivity of plants to hydrogen fluoride could be related to these metabolic schemes, Ross *et al.* (1960) conducted experiments with [6-^{14}C] glucose and [1-^{14}C] glucose. They showed that leaves of fluoride-sensitive varieties of gladioli (*Gladiolus* sp.) usually had higher C_6/C_1 ratios than the leaves of fluoride-resistant varieties. They concluded that sensitive varieties apparently respire mainly by the glycolysis pathway whereas resistant varieties depend more on the pentose phosphate pathway.

Kravitz and Guarine (1958) showed that 0.02 *M* phosphate inhibited operation of the pentose phosphate cycle, but not that of the glycolytic pathway. However, a limited level of phosphate favored oxidation through the former cycle at the expense of the latter pathway. They concluded that the effect of fluoride on respiration would be modified in the presence of phosphate. Such an intimate interrelationship between fluoride and phosphorus was also supported by Brennan *et al.* (1950). They investigated effects of nitrogen, calcium, and phosphorus nutrition on sensitivity of tomato plants to injury from sodium fluoride applied to the roots and from hydrogen fluoride fumigation of shoots. Low phosphorus had the least inhibitory effect. On the other hand, Rapp and Sliwinski (1956) reported that sodium monofluorophosphate inhibited both phosphorylase and phosphatase. The activity of phosphorylase was inhibited because monofluorophosphate competed with orthophosphate in the reaction. Tietz and Ochoa (1958) found that fluorophosphate could be formed by "fluorokinase," which, in rabbit muscle, is identical to pyruvic kinase. Weinstein (1961) theorized that the reaction

$$\text{ATP} + \text{fluoride} \xrightarrow{CO_2 + Mg^{2+}} \text{ADP} + \text{fluorophosphate}$$

would result in formation of an enzyme inhibitor and scission of a high energy bond. If fluoride exerts its effect by reduced phosphorylative activity, this could result in a reduced level of sugar and Krebs cycle acids, which were found in tomatoes and beans (Weinstein, 1961). However, respiration was stimulated rather than inhibited. Weinstein stated that the observed effects of fluoride on metabolism were not evidence for inhibition of enolase and suggested that fluoride may exert its effect by interfering with phosphorus metabolism. To clarify the effect of the interaction of fluoride and phosphate on respiration, I. B. McNulty (personal communication, 1960) conducted experiments with bean plants grown in phosphate-deficient solutions and observed the influence of both fluoride and phosphate on respiration. Leaf discs were sampled before they showed visible signs of phosphate deficiency and were treated with fluoride, phosphate, or a combination of the two. Fluoride had no measurable effect on respiration; phosphate alone caused a 13% increase in oxygen consumption, and phosphate plus fluoride increased respiration by 38%. Applegate and Adams (1960c) studied the role of nutrition and water balance on fluoride uptake and respiration of bean seedlings. Plants were grown in an atmosphere containing about 2.0 $\mu g/m^3$ (1.6 ppb) fluoride. The effect of N, P, K, Fe, and Ca deficiencies and the effects of osmotic pressure of 0,

1.5, 3.0, 4.5, 6.0, and 7.5 lb on fluoride uptake and fluoride-mediated respiration were studied. Phosphorous deficient plants absorbed more fluoride than plants deficient in any of the other elements studied. Fluoride-mediated respiration was phosphorus-dependent, however. Plants low in Fe also showed inhibition of oxygen uptake accentuated by fluoride. The data also indicated that neither fluoride uptake nor fluoride-mediated respiration were directly linked to plant water balance. In 1960, Miller reported that the activity of cytochrome oxidase, one of the components of the respiratory chain, depends on phosphorus concentration. Further, Penot and Buvat (1967a,b) showed that fluoride simultaneously increased respiration and foliar uptake of orthophosphate. Earlier Ducet and Rosenberg (1953) emphasized the relationship between the rate of respiration and a series of oxidative reactions associated with energy conservation.

The most commonly accepted control mechanism of respiration is that advanced by Hackett (1959). He proposed that the overall rate of respiration in plants depends on concentration of the acceptor, ADP, rather than the donor, ATP. In addition, Krebs (1957) indicated that control of respiration depended on interrelations between inorganic phosphate, ADP, and ATP; ADP or the phosphate acceptor was the most significant factor in the control mechanism. Beevers (1953) and Lehninger (1965) also concluded that the ratio of ADP to ATP was critical in control of respiration, with a low ratio limiting and a high ratio accelerating respiration. The action of fluoride on nucleotide metabolism caused changes in amounts (McNulty and Lords, 1960) and ratios (Chang, 1968) of nucleotides and inhibited activity of ATPase (Akazawa and Beevers, 1957).

The change in respiratory response noted in the experiments with additional orthophosphate (Yu and Miller, 1967) could be associated with change in the ratios in nucleotides and orthophosphate. The stimulation of respiration in fluoride-injured tissue (Hill *et al.,* 1959a) also may be consistent with the theory of nucleotides-orthophosphate-respiration. Stimulation of respiration in such tissues may be in response to an energy requirement for alleviation of cellular injury. Increased metabolism would be associated with more oxidative phosphorylation. Yang and Miller (1963a) proposed that increased respiration in necrotic leaves was related to the high concentration of asparagine, since Webster and Varner (1955) reported that synthesis of asparagine from aspartate and ammonia required ATP and generated ADP. Therefore, the increased rate of respiration in fluoride-necrotic leaves could be partly attributable to this energy-consuming process in which asparagine is formed and ADP generated.

Newman (1962) studied effects of fluoride on leaf catalase in an effort to explain the mechanism associated with fluoride-induced stimulation and inhibition of respiration. He used bush beans of the same variety used by

McNulty (1964) to demonstrate fluoride-stimulated leaf oxygen consumption. The catalases generate oxygen by attacking hydrogen peroxide (H_2O_2). According to Agner and Theorell (1946), the mechanism of anion-inhibiting catalase may be explained by iron atoms of catalase having hydroxyl groups that may be replaced by one of low molecular weight anions, such as fluoride. When such a substitution occurred, catalase was inhibited. The net result of decomposition of hydrogen peroxide by catalase or by other types of iron porphyrin compounds was reduction of the rate of apparent respiration. Newman therefore assumed that catalase activity would be greatly inhibited during the early period of fluoride-stimulated respiration and greatly increased during the later period, when fluoride-induced respiration is decreased. However, his data did not support this assumption. McNulty (1959) found stimulation of respiration up to 48 hours in bush bean leaf discs treated with 1×10^{-3} *M* NaF. During this first period, inhibition of catalase by fluoride was insufficient. Four days were required for about 50% inhibition (see Section VII).

V. Effects of Fluoride on Apparent Photosynthesis, Hill Reaction, and Pigments

Very little is known about the influence of fluoride on photosynthesis in comparison to what is known about effects on respiration. Most studies dealt with gas exchange measurements of apparent photosynthesis until Ballantyne (1972) demonstrated fluoride inhibition of the Hill reaction in bean (*Phaseolus vulgaris* var. Tendergreen Improved) chloroplasts.

Thomas and Hendricks (1956) fumigated several varieties of gladiolus intermittently with 1–10 ppb HF and found that the resulting decrease in carbon dioxide assimilation was directly correlated with the amount of injury produced by the fumigation. They also reported temporary reduction of carbon dioxide assimilation followed by recovery when the plants were fumigated at a much higher fluoride concentration (20–50 ppb), but for a shorter period of time (a few hours). Similar responses occurred in fruit trees (20–50 ppb), and some other crops (40–250 ppb). Smith (1961) treated bean leaves and *Chlorella* with fluorides by tissue infiltration, root uptake, and fumigation at sublethal fluoride concentrations. He found no effect of fluorides on photosynthesis as determined by oxygen evolution. Smith (1961) also fumigated bush beans with 10–15 ppb HF for 5 days at which time leaf injury began. When fumigation was stopped, the leaves lost their chlorotic appearance but the rate of photosynthesis was not measurably affected. On the other hand, Woltz and Leonard (1964) reported that removal of fluoride from the atmosphere around orange trees caused an increase in photosynthesis and in chlorophyll content. However, brief immersion (15 seconds) of leaves in dilute hydrofluoric acid solution induced stomatal closure and depression of photosynthesis. Woltz (1964a) also found that fluoride imbibed from solution by detached leaves reduced photosynthetic efficiency in Orange Gold variety of gladiolus, but not in Friendship. The fluoride content of leaves in which photosynthesis was inhibited ranged from 26–48 ppm on a dry weight basis. Fluoride also caused a marked increase in stomatal opening, with the effect more pronounced in Orange Gold than Friendship.

Warburg and Krippahl (1956) reported rapid evolution of carbon dioxide by *Chlorella* exposed to fluoride. However, Bishop and Gaffron (1958), Warburg *et al.* (1957), and Warburg and Krippahl (1956) found this was due to decarboxylation of glutamic acid and not associated with photosynthesis. Clendenning and Ehrmantraut (1950) found 65% inhibition of the Hill reaction in *Chlorella* subjected to 0.02 *M* NaF. Vennesland and Turkington (1966) measured rates of the Hill reaction and photosynthesis in blue-green algae poisoned by flouride. An increasing concentration of KF caused progressive decrease of photosynthetic activity, followed by loss of their capacity to perform the quinone Hill reaction. At intermediate concentrations of KF (0.003 *M*), the Hill reaction depended strongly on the presence of added CO_2 and chlorine ions. Similar fluoride inhibition of the Hill reaction was found in *Euglena gracilis* chloroplasts (Satoh *et al.*, 1970).

Chang and Thompson (1966a), in agreement with Ledbetter *et al.* (1960), found that fluoride could accumulate in chloroplast organelles. Treshow (1970), in addition, indicated a possible combination of fluoride with magnesium in chlorophyll. These reports supplied an impetus for direct tests of effects of fluoride on photochemical activity in isolated light-dependent organelles, the chloroplasts.

Ballantyne (1972) showed that at pH 4.8–5.7 fluoride (added as KF) inhibited the Hill reaction, with the effect detectable when bean chloroplasts were exposed to KF before or after exposure to light. He stated that this relationship could be expected if undissociated HF was the effective inhibitor. The threshold concentration of KF was approximately 2 m*M* and inhibition was related to fluoride concentration to 35 m*M* KF, at which maximum inhibition was detected. KF was employed as $KF \cdot 2\ H_2O$ rather than NaF because $KF \cdot 2\ H_2O$ was highly soluble and suitable for preparing the high concentrations required.

Ballantyne (1972) also found that magnesium ions offset inhibition of the Hill reaction by KF. Both $MgCl_2$ and $MgSO_4$ were effective. Magnesium appeared to affect the fluoride concentration in the reaction mixture. When 35 m*M* KF was added to a 22 m*M* $MgSO_4$ solution, the concentration of fluoride was 20 m*M* and rates of oxygen evolution were equivalent to those of a 20 m*M* KF solution. However, when $MgSO_4$ was injected into a reaction mixture that was treated with KF for 150 seconds, KF inhibition of the Hill reaction was not affected. When KF was added to chloroplasts, its inhibiting effect appeared to be irreversible. Because he believed phosphate might have an appreciable effect on the Hill reaction (Good, 1962), Ballantyne tested its effect in the absence of KH_2PO_4. Despite a decrease in the rate of oxygen evolution, the KF inhibition and the relationship between KF and $MgSO_4$ differed very little from the situation when KH_2PO_4 was present. Adding KCl rather than KF to the reaction mixture had little effect on oxygen evolution. However, KCl was effective in overcoming inhibition caused by KF. Altering the ferricyanide concentration had little effect on the results. Adding chloride or magnesium did not result in additional activity except when fluoride was present.

The effect of sodium monofluoroacetate was also investigated. Fluoroacetate is virtually ineffective in inhibiting the Hill reaction. Neither adenosine diphosphate nor ammonium chloride (an uncoupler of photophosphorylation) affected the fluoride response. There was little difference in fluoride inhibition when chlorophyll concentrations were 100 or 250 μg. Ballantyne's findings (1972) that fluoride inhibited the Hill reaction of bean chloroplasts agreed with data of Spikes *et al.* (1955) for chloroplasts of Swiss chard (*Beta vulgaris*).

Some reports show effects of fluoride on chlorophylls which participate in photosynthesis but are not a rate limiting factor. Wander and McBride (1955, 1956) reported that citrus leaves in the vicinity of a plant manufacturing triple-superphosphate in Florida contained high levels of fluorine and were chlorotic. Weinstein (1961) treated bean and tomato plants with fluoride (1.3–12.4 ppb) and determined chlorophyll content during and after a 24-hour fumigation. He reported that fluoride caused slight inhibition of formation of chlorophylls a and b but no inhibition was evident by the end of the recovery period.

McNulty and Newman (1956, 1961; Newman and McNulty, 1959) investigated the mechanism of fluoride-induced chlorosis. McNulty and Newman (1956) sprayed 14% calcium hydroxide solution on peach and Italian Prune leaves in the Utah Valley area, where sublethal concentrations of fluoride were present. The spray was applied to provide a protective layer of calcium hydroxide on the leaf surfaces. Calcium combined readily with fluorides to form insoluble calcium fluoride. Thus, the sprayed trees acted as controls for unsprayed trees. In the presence of atmospheric fluoride, chlorophyll content was higher in the sprayed than in unsprayed leaves. Apparently sublethal concentrations of fluoride appeared to affect the chlorophyll concentration of citrus leaves. When Newman and McNulty (1959) floated leaf discs on a nutrient-fluoride solution they found reduction in chlorophyll a, chlorophyll b, and protochlorophyll content over leaf discs floated on nutrient solution only. Reduction in leaf protochlorophyll induced by fluoride was evident before reduction in chlorophyll. When the stems of bean plants were immersed in a sodium fluoride solution, leaf chlorophylls were also reduced. Newman and McNulty (1959) concluded that aqueous sodium fluoride caused a reduction of chlorophyll synthesis. McNulty and Newman (1961) also conducted experiments with leaf tissue cultured in a calcium-free mineral nutrient solution, with and without added fluoride, under constant light and temperature. Analyses were made of chlorophylls, protochlorophyll, carotenes, and other soluble magnesium compounds. The rate of change in the chlorophylls, with respect to the rate of change in suspected chlorophyll precursors and related compounds, was measured to investigate the mechanism of fluoride-induced chlorosis. Sodium fluoride prevented accumulation of chlorophyll a, chlorophyll b, and protochlorophyll in bean leaves etiolated at initiation of fluoride treatment. However, transformation of protochlorophyll to chlorophyll apparently was not affected. Chlorophyll a, chlorophyll b, carotenes, and total ether-soluble magnesium-containing compounds were affected proportionally. The similarity in amount of reduction of both chlorophylls and total ether-soluble magnesium-containing compounds indicated that there was no inhibition of pigment synthesis following attachment of the magnesium atom to the ring structure. They reasoned that if fluorides caused inhibition of pigment synthesis, the inhibition must have occurred very early in the synthesis of pigment components or in some phase of basic metabolism necessary for their synthesis. The apparent dissolution of chloroplast structure that occurred concurrently with appearance of chlorosis was cited to explain the similarity in decrease of all pigments. Thus, fluoride may have affected early stages of pigment synthesis or induced degradation of chlorophyll structure.

Two distinct primary photochemical reactions, photosystem I and photosystem II, are involved in photosynthesis. Light absorbed by photosystem I causes reduction of NADP and oxidation of a cytochrome, while light absorbed by photosystem II causes reduction of cytochrome and oxidation of water to oxygen, coupled with photophosphorylation. Since fluoride inhibits both apparent photosynthesis and the Hill reaction, these findings should be extended by investigating the responses of these two photosysman and Bonner, 1947).

VI. Effects of Fluoride on Growth, Aging, and Associated Biochemical Metabolites

Growth and vigor of young navel orange trees were appreciably reduced in an atmosphere containing as little as 2–3 ppb by volume of hydrogen fluoride, even though easily distinguishable visible symptoms were absent. The most pronounced effect of hydrogen fluoride on growth was a 25 to 35% reduction in average leaf size (Brewer *et al.*, 1960). Fluoride-induced growth reduction was also associated with reduction of quantity and quality of crops (Brewer *et al.*, 1959). Weinstein (1961) found that fumigation of tomato and bean plants with 12.4 ppb fluoride also decreased growth and yield of tomato leaves.

Growth inhibition of yeast by fluoride was due to blocking of phosphoglucomutase (Chung and Nickerson, 1954). This enzyme also was inhibited in oat (*Avena sativa*) coleoptile segments treated with 0.01 *M* fluoride solution (Ordin and Altman, 1965). However, Ordin and Altman did not show a relation between growth rate and inhibition of phosphoglucomutase. Bonner and Thimann (1950) noted that fluoride inhibited growth of isolated sections of *Avena* coleoptiles and *Pisum* stems. Since the inhibition was not affected by organic acids, magnesium ions, pyruvate, or ATP, they concluded that fluoride did not inhibit growth by interfering with the enolase system. Moreover, the growth-inhibiting concentration of fluoride (2.5×10^{-3} *M* NaF) did not affect respiration of coleoptile sections, although the rate of respiration was inhibited by higher concentrations of fluoride (4×10^{-3} *M* and above). That growth was not controlled by the glycolytic pathway in higher plants also was demonstrated by Ordin and Skoe (1963). Their experiments with *Avena* coleoptile segments treated with 0.01 *M* NaF and labeled glucose showed that growth inhibition was not associated with a change in glycolytic activity. Miller (1957) showed further that hydrogen fluoride-fumigated plants sprayed with pyruvate exhibited restored enolase activity. Normal growth rates were only partially restored to pyruvate-sprayed plants, suggesting that growth inhibition was not due to reduced enolase activity alone. A literature survey indicated that the amount of respiration necessary for growth was a very small fraction of the whole, as suggested early by Commoner and Thimann (1941).

As shown previously (Section IV), a number of phosphatase enzymes are inhibited by fluoride (Wildman and Bonner, 1947; Bonner, 1948; Plaut and Lardy, 1949; Giri, 1937; Heppel and Hilmore, 1951; Reiner *et al.*, 1955; Akazawa and Beevers, 1957). For example, phosphatase of spinach leaves was inhibited by the same fluoride concentration (2.5×10^{-3} *M*) that inhibited growth of *Avena* coleoptile segments and *Pisum* stems (Wildman and Bonner, 1947).

McNulty and Lords (1960) reported that fluoride caused accumulation of phosphorylated nucleotides in *Chlorella pyrenoidosa.* Chang (1968) showed that fluoride modified the ratios of acid-soluble nucleotide species in corn roots, resulting in accumulation of RNA (Brown, 1965). Alteration in RNA metabolism influences growth, since RNA content plays a vital role in protein synthesis (Ingle and Hageman, 1964) and this synthesis controls growth rate (Neidhardt and Magasanik, 1960). Such data led to investigation of effects of fluoride on metabolic constituents associated with growth.

Chang and Thompson (1966b) demonstrated that fluoride inhibited growth of corn seedling roots by a third to a half. In another experiment growth was reduced by as much as two-thirds (Chang, 1968). Treatment of corn seeds with a relatively high fluoride concentration was required to suppress germination and growth, because of the diffusion barrier of cutinized seed coats.

Chang and Thompson (1966b) found that rates of both cell elongation and cell multiplication of seedling roots were reduced by fluoride, but the rate of cell multiplication was inhibited more after a short period of fluoride treatment. As the time of fluoride treatment was increased, rates of cell elongation and cell multiplication were reduced to about the same degree. In addition, fluoride reduced the rate of cell division in roots. Chang and Thompson (1966b) further reported that the rate of cell elongation was highest at 3 mm from the root tip and decreased in a basipetal direction in both control and fluoride-treated roots. A reduction of the rate of cell elongation was initiated by fluoride about 3 mm from the root tip, and this reduction rate paralleled the elongation rate of control plants during later developmental stages of seedling roots. The inhibition of cell elongation was not reversed during the subsequent developmental period.

Chang and Thompson (1966b) investigated the relation of fluoride-induced growth inhibition of corn roots to RNA. Growth rates of seedling roots after corn seeds were treated with fluoride for various times were correlated with total RNA content of the 3-mm root tips. To determine if this relationship was related to cell number in root tips, the content of RNA was analyzed on a cell basis. Cell volumes were determined by macerating the root tip tissue in 5% chromic acid and by counting the cell number using a hemacytometer. Total RNA content per root tip cell was correlated with growth rate per cell.

Chang (1970a) further determined the amount of ribosomal RNA in corn roots treated with fluoride (5×10^{-4} to 4×10^{-3} *M*). Fluoride decreased ribosomal RNA in proportion to fluoride concentration. Therefore he assumed the data showing the decrease in total RNA content after fluoride treatment (Chang and Thompson, 1966b) reflected a decrease in ribosomal RNA content, since the ribosomal RNA represented the bulk of, and a constant proportion of, the total RNA content of the cell (Ingle and Hageman, 1964).

According to Bonner (1965), ribosomes are free in some types of meristematic cells and become associated with membranes as the cells elongate and mature. Chang (1970a) demonstrated operation of such a system in corn seedling roots by showing the ratios of free to bound ribosomes to be nearly 5.0 and about 0.90 in tissue segments at distances of 0–25 mm and 25–100 mm, respectively, from the root tip. Following the report of Key *et al.* (1961) that bound ribosomes of corn roots were metabolized during the elongation process, Chang (1970a) studied the influence of fluoride on free and bound ribosomes in corn seedling roots. Fluoride (5×10^{-4} to 4×10^{-3} M) reduced the amounts of both free and bound ribosomes. However, fluoride did not modify the ratio of free to bound ribosomes. Since there was no accumulation of free ribosomes, Chang proposed that the decreased amount of bound ribosomes was not caused by blockage of conversion of free to bound ribosomes. However, the amount of bound ribosomes possibly might have been restricted by the free ribosomes simultaneously suppressed by fluoride, since ATP accumulated in corn seedling roots that had been treated with fluoride (Chang, 1968). Fluoride also decreased RNA–protein ratios in free and bound ribosomes, with the latter influenced more than the former. This observation was in accord with the previously observed inhibition of growth elongation in corn seedling roots treated with fluoride (Chang and Thompson, 1966b).

To better understand the decreases in RNA–protein ratios in free and bound ribosomes, Chang (1970a) investigated the effect of fluoride on structural composition of ribosomes. Fluoride did not modify base composition of ribosomal RNA. Rather it influenced the structure of ribosomal proteins. This was demonstrated by electrophoretic separation of acid-soluble ribosomal proteins from fluoride-treated corn seedling roots. The decreases in RNA–protein ratios in free and bound ribosomes therefore could be due to reduction of ribosomal RNA and possible accumulation of structurally altered ribosomal protein components.

During seed germination, corn plants depend on endosperm and scutellar tissues for reserve foods for germination and growth. Three major types of hydrolytic enzymes, including amylases, proteinases, and esterases (lipase and phosphatase), are responsible for hydrolysis of carbohydrates, proteins, and fats and phosphates, respectively (Poljakoff-Mayber, 1953). Seeds contain 50–88% of the total organic phosphate as phytin (Sobolev, 1962). Ergle and Guinn (1959) showed that increasing orthophosphate content coupled with diminishing phytin levels in germinating seeds paralleled increasing RNA and DNA contents during growth of embryonic plants.

Chang (1967) studied the effect of fluoride on phytase during seed germination and subsequent growth of corn plants. The highest total phytase activity was in the 1700 *g* fraction of the endosperm-scutellar tissue homogenate of seedlings. The Michaelis constant (K_m) of this enzyme was 0.91×10^{-4} moles/liter. In addition, the action of various salts on phytase activity was tested. Enzyme activity was inhibited most by sodium fluoride. To determine the minimum range of fluoride concentrations which inhibit phytase activity, Chang assayed enzyme activity in the presence of various concentrations of fluoride. One-tenth to 10 m*M* sodium fluoride reduced enzyme activity *in vitro* to 46–87% of the controls. He suggested that the action of such a small amount of fluoride indicated possible direct inhibition of phytase activity by fluoride *in vivo*. Fluoride also prevented release of phytin phosphorus in endosperm-scutellar tissues during seed germination and seedling growth. This also indicated inhibitory action of fluoride on phytase activity *in vivo*. The breakdown of organic phosphate decreased as fluoride concentration in the germination medium increased. Release of orthophosphate from phytin was blocked by fluoride most during the 24- to 48-hour germination period. Thus the kinetics of phytin phosphorus during seed germination were related to the general pattern of growth inhibition caused by fluoride.

In 1966, Chang and Thompson (1966b) showed that fluoride accelerated aging in corn roots and inhibited root growth as well. According to Srivastava (1967), a decrease in RNA, protein content, and the capacity to synthsize these metabolic constituents occurred during aging. Srivastava (1965) also suggested that protein synthesis was directly related to polysome content. Therefore, Chang (1970b) investigated the amount and particle distribution of ribosomes and the alterations associated with ribosomal components in fluoride-treated corn roots. Increasing fluoride concentrations caused reduction of the sum of the RNA and protein content of ribosomes to about two-thirds the control level. Figure 2 and Table I show that corn roots demonstrate a progressive reduction in the level of ribosomal particles heavier than the 80 S monosome with increasing fluoride concentrations (column I in Table I). However, the relative number and concentration of monosomes, subunit A, and subunit B combined increase correspondingly (column VI in Table I). These data imply that the sum of monosomes and breakdown products changed at the expense of polysomes. Reduction of polysome level in each experiment, however, reflected absolute decrease in polysome material, because fluoride decreased total RNA and protein of ribosomes per unit fresh weight of roots as indicated above.

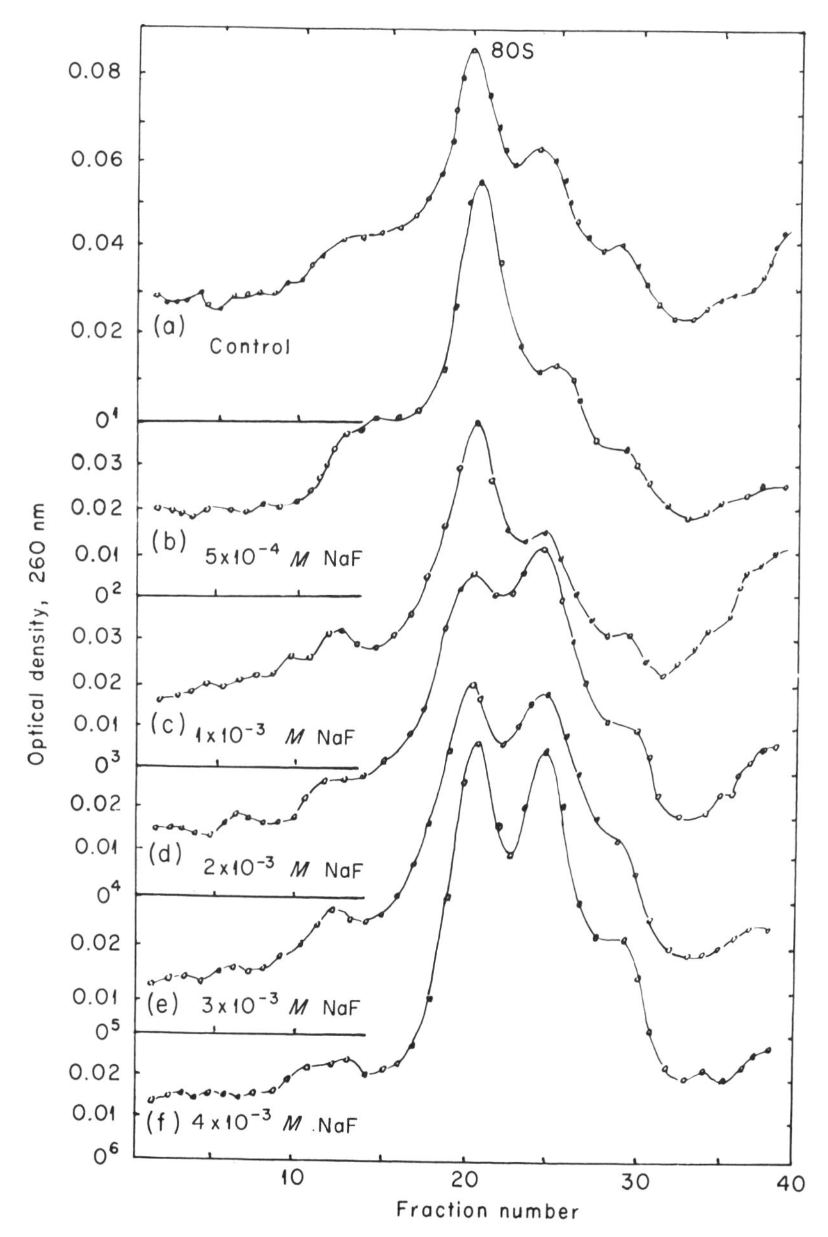

FIG. 2. Sucrose gradient sedimentation profiles of ribosomes isolated from (a) control corn roots, and from those treated with (b) 5×10^{-4} *M* NaF, (c) 1×10^{-3} *M* NaF, (d) 2×10^{-3} *M* NaF, (e) 3×10^{-3} *M* NaF, (f) 4×10^{-3} *M* NaF. 0^1, 0^2, 0^3, 0^4, 0^5, and 0^6 are positions of the abscissa for sedimentation profiles of samples a, b, c, d, e, and f, respectively. Each experiment is the result from 1 ml of ribosome sample having an absorbance of 3 optical density units at 260 nm. Direction of sedimentation is from right to left.

TABLE I
RELATIVE AMOUNT OF DIFFERENT RIBOSOMAL COMPONENTS ISOLATED FROM CONTROL AND FLUORIDE-TREATED CORN ROOTS

Treatment with NaF (10^{-3} *M*)	Percentage absorbance of total components (V) at 260 nm					
	I[a]	II	III	IV	V	VI
0.0	44.4	26.5	15.7	13.4	100	55.6
0.5	39.1	34.0	15.4	11.5	100	60.9
1.0	37.9	30.3	19.1	12.7	100	62.1
2.0	34.1	21.9	31.7	12.3	100	65.9
3.0	30.6	26.4	27.2	15.8	100	69.4
4.0	26.3	31.3	25.3	17.1	100	73.7

[a] (I) Particles heavier than monomer (fractions 1–17), (II) monosomes (fractions 18022), (III) subunit A (fractions 23–26), (IV) subunit B (fractions 27–32), (V) sum of four components, total components (I + II + III + IV), (VI) sum of three components (II + III + IV).

Chang (1970b) also reported that fluoride modified specific activities of subcellular ribonuclease. Assays of subcellular distributions of ribonuclease activity in fluoride-treated materials showed that all subcellular components, except the plastid fraction, showed progressive increases in activity with increasing fluoride concentrations. However, ribonuclease specific activity of the microsomal component was increased most by fluoride. Activity of plastid components was not compatible with fluoride concentrations. The level of this enzyme activity also was not related to that of the soluble fraction. The data indicated a negligible amount of enzyme contamination between the plastid component (presumably also other particulates) and the soluble fraction.

Ribonuclease activity *in vivo* is known to be directly controlled by the ratio of potassium, calcium, and magnesium ions in corn roots (Hanson, 1960). It is also known that fluoride forms a complex with magnesium ions in inhibition of enolase activity (Miller, 1958). Therefore, fluoride possibly influences ribonuclease activity by altering the ratio of these free ions required for the normal level of this enzyme activity *in vivo*. The enhanced ribonuclease specific activity would be due to enzyme activation by fluoride.

Using sucrose density gradients, Chang (1970b) analyzed the size distribution of ribosomal RNA as influenced by fluoride. The control ribosomal RNA had two major peaks, the first peak of heavier ribosomal RNA particles, 23 S, and the second peak of lighter ribosomal particles, 16 S. Fluoride caused a fall in the first peak and a rise in a second one, which was spread over a wide range beyond the 16 S position. Such results imply that the increase in ribonuclease activity caused by fluoride degraded the first peak and induced the second peak of heterogeneous ribosomal RNA particles. This finding indicated a possible destruction of messenger RNA, which could be one of the factors linked with disintegration of polysomes in fluoride-treated materials.

Dissociation of polysomes also could be caused by lack of an energy supply, since energy is involved in maintenance in the polysome complex (Lin and Key, 1967). Fluoride could curtail an energy supply and cause decrease in the level of polysomes, because fluoride alters carbohydrate metabolism by inhibiting sucrose synthesis (Yang and Miller, 1963a) and enolase (Miller, 1958), and also because fluoride induces accumulation of ATP (Chang, 1968). However, these factors would be secondary to the reduction in polysome level. The dissociation of polysomes in aged tissue, as in fluoride-treated corn roots, is consistent with findings in higher plants aged by other factors. For example, Srivastava and Arglebe (1967) found that polysomes and ribosomes were lost from senescing barley leaves.

The data cited above provided evidence that fluoride-caused growth retardation and aging are associated with changes linked with the site of protein synthesis. Fluoride treatment decreased the content of total and ribosomal RNA, altered ribosomal components, and shifted ribosomal distribution from polysomes to smaller ribosomal particles. Accumulation of ATP and activation of ribonuclease were responsible for these changes.

VII. Fluoride and Fluoroorganic Compounds

A number of naturally growing plants that are toxic to animals are known to contain fluoroorganic compounds. Most such plants grow in Australia and are in the genera *Gastrolobium* and *Oxylobium* (Aplin, 1968a,b, 1969a,b, 1971a,b; McEwan, 1964a,b). Other plants which contain fluoroorganic compounds include *Acacia georginae* (Oelrichs and McEwan, 1961), *Dichapetalum cymosum* (Marais, 1944), and *D. toxicarium* (Ward *et al.*, 1964) in Africa and *Palicourea marcgravii* (De Oliveira, 1963) in South America.

The presence of an organic compound of fluorine in these plants has been known since Marais (1944) located fluoroacetate in leaves of *Dichapetalum cymosum*. Bartlett and Barrón (1947) reported that fluoroacetate led to accumulation of acetate in tissue slices. Saunders (1947) found that toxicity was related to increasing length of the carbon chain in the ω-fluoroester; compounds with an even number of carbon atoms were toxic. Liebecq and Peters (1948, 1949) reported that fluoroacetate caused depression of oxygen uptake without increase of acetate, but with accumulation of citrate at the same time, oxidation of added citrate was inhibited by the poison. From these data they proposed a hypothesis of fluoroacetate poisoning in animals. According to their theory, sodium fluoroacetate can be activated and condensed with oxaloacetate to form a fluorotricarboxylic acid. This acid or a derivative inhibits the tricarboxylic acid cycle and so causes accumulation of citrate. The toxic substance, fluorocitrate, was isolated and further studied by Buffa *et al.* (1951). Since then many additional investigations have been carried out on the toxic mechanism of fluoroacetate in animals. These were reviewed by Chenoweth (1949), Peters (1957), Peters and co-workers (1960), Pattison (1959), and others. The toxic mechanism in animals involved conversion of a nontoxic substance, fluoroacetate, by enzymatic synthesis to fluorocitrate which interferes with citric acid metabolism (termed "lethal synthesis" by Peters, 1957). In connection with toxicity, Morrison and Peters (1954) found that fluorocitrate was a potent inhibitor of the enzyme aconitase. However, a different view of this mechanism was proposed by Williamson *et al.* (1964) based on experiments with perfused rat heart tissue. They reported that although the initial effect of fluoroacetate was to give rise to fluorocitrate, secondary inhibition of phosphofructokinase by the accumulated citrate was lethal, since it deprived cells of pyruvate which would eventually overcome the inhibition of aconitase. Lowry and Passonneau (1964) indicated that citric acid accumulation caused by fluorocitrate inhibited phosphofructokinase in a cell-free system.

Oelrichs and McEwan (1962) isolated the toxic principle in a pure form from seeds of *Acacia georginae* and identified the compound as fluoroacetic acid. Peters *et al.* (1965a) studied the conditions under which *A. georginae* synthesized fluoroacetate and contributed to the understanding of the biochemical paths of C—F bond synthesis. Peters *et al.* (1965b) also found the conditions in which the homogenate of *A. georginae* synthesized fluoroacetate *in vitro*. Preuss *et al.* (1970a) and Preuss *et al.* (1970b) made a biosynthesis of fluoroacetate in culturing tissues of *A. georginae.* Although these investigators succeeded in synthesizing fluoroacetate, they used only plants known to synthesize fluoroorganic compounds naturally. Therefore they did not establish that the capacity to synthesize fluoroacetate was a general property of plants.

The first discovery that plants other than those known to contain fluoroacetate naturally can convert inorganic fluoride to an organically combined form was made by Miller and colleagues (Cheng *et al.,* 1968; Lovelace *et al.,* 1968; Yu and Miller, 1970). Cheng *et al.* (1968) conducted experiments with soybean plants fumigated by HF and grown in nutrient solution containing sodium fluoride or fluoroacetate. Fluoroorganic compounds were demonstrated in these plants after analyzing infrared spectra and paper chromatographic R_f values of organic acid extracts. Further confirmation of the presence of fluorocitrate in fluoride-treated leaves was obtained by demonstrating inhibition of aconitase with the organic acid fraction isolated from the treated plants. An increase in the concentration of these organic acid fractions in the reaction mixture (expressed as an increase in micrograms of fluoride) resulted in decreased activity of both pig heart and soybean leaf aconitase. An aliquot of a comparable extract from control plants equal in volume to the extract from HF-treated plants, that caused maximum aconitase inhibition, did not cause significant inhibition. Inhibition by the organic acid fractions containing fluoride equalled or exceeded inhibition by commercial fluorocitrate at equivalent concentrations based on fluoride content. Inorganic fluoride did not affect aconitase activity until 11 mg were added to the mixture. The presence of only 20 μg of fluoride in an inhibitory organic extract from fluoride-fumigated plants strongly indicated the presence of the inhibitory fluorocitrate. Since soybean aconitase is sensitive to fluorocitrate *in vitro,* this enzyme appeared to be inhibited *in vivo.*

The content of fluoroacetate was about 6 μg per gram of fresh weight (approximately 40 μg per gram on a dry weight basis). The fluorocitrate content was much greater (140 μg per gram of dry weight of tissue). In contrast, *Dichapetalum cymosum* may contain 15 mg of fluoroacetate per gram of dry leaves (Marais, 1944). Fluorocitrate was not found in these plants, indicating that fluoroacetate is not further metabolized. Cheng *et al.* (1968) suggested that the nontoxicity of this high concentration of fluoroacetate in *Dichapetalum cymosum* might be a result of cellular separation of fluorocitrate. Also, they proposed that the possible inability of the acetyl kinase or the citrate-condensing enzyme to utilize fluoroacetate might account for the lack of toxicity in *Dichapetalum cymosum.*

Lovelace *et al.* (1968) analyzed forage crops from a pasture mix which included *Medicago sativa* and *Agropyron cristatum* grown within 2 miles of a phosphate plant. Horses grazing in this area on plants which contained up to 1000 ppm in total fluoride showed symptoms of severe fluoride poisoning. However, the plants showed no injury. Analyses of these plants indicated considerable accumulation of fluoroacetate and fluorocitrate. The presence of these compounds was also established by chromatographic techniques, inhibition of aconitase, and infrared spectral analysis (Cheng *et al.,* 1968).

Yu and Miller (1970) also analyzed forage crops near a phosphate plant for fluoroacetate and fluorocitrate and confirmed the findings of Cheng *et al.* (1968) and Lovelace *et al.* (1968). In agreement with these studies, using [^{14}C] fluoroacetate, Ward and Huskisson (1969) reported conversion of [^{14}C] fluoroacetate to fluorocitrate in lettuce. Peters and Shorthouse

(1972) also demonstrated synthesis of fluoride into fluorocitrate in culturing a single cell from soybean plants, which were used previously by Miller and colleagues (Cheng *et al.,* 1968; Lovelace *et al.,* 1968; Yu and Miller, 1970).

These reports, which show synthesis of fluoroorganic compounds from fluoride in plants other than those known naturally to produce this toxic compound, indicate that incorporation of atmospheric fluoride into an organic form is characteristic of most plants. The mechanism of such metabolism is not fully understood. Cheng *et al.* (1968) suggested possible blocking of the TCA cycle resulting from fluorocitrate inhibition of aconitase, since Yang and Miller (1963a) showed accumulation of organic acids, especially citrate and malate, in fluoride-injured plant tissues.

To understand the relative nontoxicity of toxic fluorocitrate in plant systems, Preuss *et al.* (1968) and Preuss and Weinstein (1969) conducted experiments on metabolism of fluoroacetate in *A. georginae,* castor bean, and Pinto bean. They reported that CO_2 was evolved and ^{14}C was incorporated into water-soluble fractions and lipids when ^{14}C-labeled sodium fluoroacetate was supplied to sterile seedlings. They noted that these plants contained an enzyme capable of cleaving the carbon–fluorine bond. However, in a study of metabolism of [2-^{14}C] fluoroacetate in rats Gal *et al.* (1961) reported little evolution of $^{14}CO_2$. In contrast, bacteria are known to release CO_2 when incubated with fluoroacetate (Goldman, 1965). Preuss *et al.* (1968) suggested that if the products of defluorination reaction in higher plants are similar to those in bacteria, the relative nontoxicity of fluoroacetate in plants would be explained. Plants could utilize fluoroacetate by removing fluorine and producing a metabolite. Relative toxicity associated with fluorocitrate may also be linked to the relative sensitivity of aconitase to fluoroorganic compounds. Treble *et al.* (1962) found that 200 times more fluorocitrate was needed to inhibit the aconitase extracted from a cambial tissue suspension of *Acer pseudoplatanus* than was required to inhibit aconitase isolated from pig heart. Further studies are needed to clarify the relationship between formation of toxic fluoroorganic compounds and the mechanisms of detoxification in plants.

VIII. Effects of Fluoride on Enzymes *in Vitro*

Borei (1945) summarized the action of fluoride ions in enzyme systems in four different catagories.

a. A general ionic effect on the enzyme protein, according to its position in Hofmeister's lyotropic series ($F^- < SO_4^{2-} < HPO_4^{2-} < CH_3COO^- < Cl^- < Br^- < NO_3^- < I^- < SCN^- < OH^-$). (However, in glycolysis, it is known that the fluoride ion has a decidedly stronger action than its position in the series indicated. Its effect is usually greater than that of iodine. Also, fluoride greatly reduces the free phosphate content even at low concentration.)

b. Precipitation of calcium, resulting in formation of CaF_2.

c. Binding to calcium or magnesium incorporated into enzyme molecules or substrates and formation of a complex with other ions.

d. Formation of a metal complex with active groups other than calcium or magnesium in one or several members of the enzymic chain (i.e., the prosthetic groups of the hemin proteins active in cellular respiration).

The above properties of fluoride indicate that it may influence plant systems by modifying the balance of general mineral status and by acting specifically on a number of enzymes. A survey of enzymes which are affected by fluoride *in vitro* will be helpful as a future research guide.

Fluoride inhibition of enolase has been explained in Section IV. Magnesium is required for the enzymatic reaction, and in the presence of phosphate or arsenate, an inactive dissociable magnesium fluorophosphate complex is formed. The product of the function $(Mg)(PO_4)(F^{2+}) \times$ (fractional activity)/inhibition is a constant equal to about 3.2×10^{-12}. As a consequence, the fluoride concentration producing 50% inhibition of enolase decreases as magnesium concentration increases.

Phosphoglucomutase was first studied by Najjar (1948). This enzyme catalyzed the reversible reaction between glucose 1-phosphate and glucose 6-phosphate with glucose 1,6-diphosphate as a coenzyme. Here, a magnesium fluoride complex with glucose 1-phosphate was formed by the enzyme. Fluoride inhibition of this enzyme also depended on magnesium concentration. Ordin and Altman (1965) extracted a crude enzyme system from oat coleoptiles and found that it converted glucose 1-phosphate to glucose 6-phosphate and was reversibly inhibited by fluoride *in vitro*.

The magnesium-dependent 5′-nucleotidases are inhibited by fluoride. Inhibition of alkaline phosphatase by fluoride may be a nonspecific anion effect, since phosphate, arsenate, and borate also inhibited this enzyme (Hoppel and Hilmore, 1951).

Both acid and alkaline pyrophosphatases of potato described by Naganna *et al.* (1955) were inhibited (86 and 91%, respectively) by 10^{-3} *M* fluoride, although the acid pyrophosphatase was inhibited by magnesium, while the same metal was specifically required by the alkaline pyrophosphatase.

The mechanism of acid phosphatase inhibition by fluoride may be quite complex (Reiner *et al.*, 1955). Maximum fluoride inhibition occurs at intermediate concentrations of about 0.01 *M*; higher and lower concentrations produce markedly less inhibition. The magnitude of the inhibition and the effect of fluoride concentration also vary with the nature and concentration of the phosphate ester substrate. Inhibition of phosphatase by fluoride also was reported by Wildman and Bonner (1947), Plaut and Lardy (1949), and Giri (1937).

Flouride inhibition of succinic dehydrogenase (Slater and Bonner, 1952) shows another aspect of the complexity of fluoride inhibition. Unlike enolase or acid phosphatase, where two or more fluorine atoms are involved, inhibition of succinic dehydrogenase involved only one fluorine ion (F^-) per enzyme molecule. Inhibition by fluoride alone was weak as was that by phosphate alone. Inhibition by fluoride and phosphate together in a 1:1 ratio was very strong and competitive with succinate. It may be that the ferric iron atom in succinic dehydrogenase studied by Massey (1958) was the combining center and formed a ferric fluorophosphate complex analogous to that between enolase and magnesium, but involving only a single dissociable fluoride ion.

Fluoride also combined with cytochrome oxidase in the ferric iron state (Borei, 1945). Apparently fluoride prevents hemoprotein from taking part in the preceding and succeeding links in the reaction chain. The blocking action leaves the prosthetic group of the hemoprotein completely unchanged. The data indicated that formation of a fluorophosphoprotein complex was analogous to that found in the case of enolase. Magnesium may possibly play a part in this process (Borei, 1945).

Not all enzymes with activity dependent on magensium are sensitive to fluoride. The polynucleotide phosphorylase enzymes, which depend particularly on magnesium, are entirely insensitive to fluoride (Grunberg-Manago *et al.*, 1956). Alkaline phosphatases often were unaffected by fluoride unless only magnesium was present (Roche and Thoai, 1950).

Yang and Miller (1963b) made a detailed *in vitro* investigation of responses to fluoride of three enzymes involved in sucrose synthesis. Phosphoglucomutase, uridine diphosphate glucose pyrophosphorylase, and uridine diphosphate glucose-fructose transglycosylase were extracted from higher plants, purified, and characterized with respect to activation by metal ions and inhibition by fluoride. These enzymes required Mg^{2+} ions for maximum activation but differed in their sensitivity to fluoride. Phosphoglucomutase was inhibited most by fluoride. Uridine diphosphate glucose-fructose transglucosylase was inhibited slightly at higher fluoride concentration, whereas uridine diphosphate glucose pyrophosphorylase was completely insensitive to fluoride. Mn^{2+} ions provided twice as great an activation of phosphoglucomutase as did Mg^{2+} ions. Inhibition of phosphoglucomutase was proportional to fluoride concentration, was enhanced by increasing the substrate concentration, and was almost independent of the concentration of activating metal ions. Phosphoglucomutase was inhibited seven times more by fluoride when Mg^{2+} ion served as the activator than when Mn^{2+} ion activated the reaction.

The review of these enzymes sensitive to fluoride *in vitro* indicated that there were many enzymes to be tested by *in vivo* experiments. Many of the enzymes were associated with phosphate metabolism, which is implicated in fluoride-affected respiration, as explained previously. Particularly, investigators should study the effect of fluoride on cytochrome oxidase, a terminal respiratory chain proper, after fumigating plants with fluoride, or by treating mitochondria (the site of this enzyme) with fluoride. Such experiments would provide direct information on the effect of fluoride on respiration.

References

Adams, D. F., and Emerson, M. T. (1961). Variations in starch and total polysaccharide content of *Pinus ponderosa* needles with fluoride fumigation. *Plant Physiol.* **36,** 261–265.

Adams, D. F., and Koppe, R. K. (1959). Automatic atmospheric fluoride pollutant analyzer. *Anal. Chem.* **31,** 1249–1254.

Adams, D. F., and Sulzbach, C. W. (1961). Nitrogen deficiency and fluoride susceptibility of bean seedlings. *Science* **133,** 1425–1426.

Adams, D. F., Hendrix, J. W., and Applegate, H. G. (1957). Relationship among exposure periods, foliar burn, and fluorine content of plants exposed to hydrogen fluoride. *Agr. Food Chem.* **5,** 108–116.

Agner, K., and Theorell, H. (1946). On the mechanism of the catalase inhibitor by anions. *Arch. Biochem.* **10,** 321–338.

Akazawa, T., and Beevers, H. (1957). Some chemical components of mitochondria and microsomes from germinating castor beans. *Biochem. J.* **67,** 110–114.

Aplin, T. E. H. (1968a). Poison plants of Western Australia. The toxic species of the genera *Gastrolobium* and *Oxylobium*. Heart-leaf poison (*Gastrolobium bilobum* R. Br.), river poison (*Gastrolobium forrestii* A. J. Ewart.), Stirling Range poison (*Gastrolobium velutinum* Lindl.). *J. Agr., West Aust.* [4] **9,** 69–74.

Aplin, T. E. H. (1968b). Poison plants of Western Australia. The toxic species of the genera *Gastrolobium* and *Oxylobium*. Crinkle-leaf poison (*Gastrolobium vilosum* Benth.), runner poison (*Gastrolobium ovalifolium* Henfr.), horned poison and Hill River poison (*Gastrolobium polystachyum* Meissn.), woolly poison (*Gastrolobium tomentosum* C. A. Gardn.). *J. Agr., West. Aust.* [4] **9,** 356–362.

Aplin, T. E. H. (1969a). Poison plants of Western Australia. The toxic species of the genera *Gastrolobium* and *Oxylobium*. Champion Bay poison (*Gastrolobium oxylobiodes* Benth.), sandplain poison (*Gastrolobium microcarpum* Meissn.), cluster poison (*Gastrolobium bennettsianum* C. A. Gardn.), Gilbernine poison (*Gastrolobium rotundifolium* Meissn.). *J. Agr., West. Aust.* [4] **10,** 248–257.

Aplin, T. E. H. (1969b). Poison plants of Western Australia. The toxic species of the genera *Gastrolobium* and *Oxylobium*. Berry poison (*Gastrolobium parvi-*

folium Benth.), spike poison (*Gastrolobium glaucum* C. A. Gardn.), hook-point poison (*Gastrolobium hamulosum* Meissn.), scale-leaf poison (*Gastrolobium appressum* C. A. Gardn.). *J. Agr., West. Aust.* [4] **10,** 517–522.

Aplin, T. E. H. (1971a). Poison plants of Western Australia. The toxic species of the genera *Gastrolobium* and *Oxylobium.* Thickleaf poison (*Gastrolobium crassifolium* Benth.), narrow-leaf poison (*Gastrolobium stenophyllum* Turez.), mallet poison (*Gastrolobium densifolium* C. A. Gardn.), wall-flower poison (*Gastrolobium grandiflorum* F. Muell). *J. Agr., West. Aust.* [4] **12,** 12–18.

Aplin, T. E. H. (1971b). Poison plants of Western Australia. The toxic species of the genera *Gastrolobium* and *Oxylobium.* Wodjil poison (*Gastrolobium floribundum* S. Moore), breelya or kite-leaf poison (*Gastrolobium laytonii* J. White), Roe's poison (*Oxylobium spectabile* Endl.), granite poison (*Oxylobium graniticum* S. Moore). *J. Agr., West. Aust.* [4] **12,** 154–159.

Applegate, H. G., and Adams, D. F. (1960a). Effect of atmospheric fluoride in bush beans. *Bot. Gaz.* (*Chicago*) **121,** 223–227.

Applegate, H. G., and Adams, D. F. (1960b). Invisible injury of bush beans by atmospheric and aqueous fluorides. *Int. J. Air Pollut.* **3,** 231–248.

Applegate, H. G., and Adams, D. F. (1960c). Nutritional and water effect on fluoride uptake and respiration of bean seedlings. *Phyton* **14,** 111–120.

Applegate, H. G., Adams, D. F., and Carriker, R. C. (1960). Effect of aqueous fluoride solutions on respiration of intact bush bean seedlings. *Amer. J. Bot.* **47,** 339–345.

Audus, L. J. (1935). Mechanical stimulation and respiration rate in the Cherry laurel. *New Phytol.* **34,** 386–402.

Ballantyne, D. J. (1972). Fluoride inhibition of the Hill reaction in bean chloroplasts. *Atmos. Environ.* **6,** 267–273.

Bartlett, G. R., and Barrón, E. S. G. (1947). The effect of fluoroacetate on enzymes and on tissue metabolism. Its use for the study of the oxidative pathway of pyruvate metabolism. *J. Biol. Chem.* **170,** 67–82.

Beevers, H. (1953). 2,4-Dinitrophenol and plant respiration. *Amer. J. Bot.* **40,** 9–96.

Bishop, N. I., and Gaffron, H. (1958). The inhibition of photosynthesis by sodium fluoride. I. The sodium fluoride-induced carbon dioxide burst from *Chlorella. Biochem. Biophys. Acta* **28,** 35–44.

Bonner, J. (1948). Biochemical mechanisms in the respiration of *Avena* coleoptile. *Arch. Biochem.* **17,** 311–326.

Bonner, J. (1965). Ribosomes, *In* "Plant Biochemistry" (J. Bonner and J. E. Varner, eds.), 2nd ed., pp. 34–35. Academic Press, New York.

Bonner, J., and Wildman, S. G. (1946). Enzymatic mechanisms in the respiration of spinach leaves. *Arch. Biochem.* **10,** 497–518.

Bonner, W. D., Jr., and Thimann, K. V. (1950). Studies on the growth and inhibition of isolated plant parts. III. The action of some inhibitors concerned with pyruvate metabolism. *Amer. J. Bot.* **37,** 66–75.

Borei, H. (1945). Inhibition of cellular oxidation by fluoride. *Arkiv För Kemi, Mineralogi O. Geologi.* Bd 20 A. No 8, 1–215.

Bovay, E., Bolay, A., Zuber, R., Desbaumes, P., Collet, G., Quinche, J. P., Neury, G., and Jacot, B. (1969). Accumulation de fluor dans les vegetaux sous influence de certains engrais combines boriqués. *Recherche Agronomique en Suisse* 8(1): 49–64.

Brennan, E. G., Leone, I. A., and Daines, R. H. (1950). Fluorine toxicity in tomato as modified by alterations in the nitrogen, calcium and phosphorus nutrition of the plant. *Plant Physiol.* **25,** 736–747.

Brewer, R. F., Chapman, H. D., Sutherland, F. H., and McCulloch, R. B. (1959). Effect of fluorine additions to substrate on navel orange trees grown in solution cultures. *Soil Sci.* **87,** 183–188.

Brewer, R. F., Sutherland, F. H., Guillemet, F. B., and Creveling, R. K. (1960). Some effects of hydrogen fluoride gas on bearing navel orange trees. *Proc. Amer. Soc. Hort. Sci.* **76,** 208–214.

Brown, G. N. (1965). Temperature controlled growth rates and ribonucleic acid characteristics in mimosa epicotyl tissue. *Plant Physiol.* **40,** 557–561.

Buffa, P., Peters, R. A., and Wakelin R. W. (1951). Biochemistry of fluoroacetate poisoning isolation of an active tricarboxylic acid fraction from poisoned kidney homogenates. *Biochem. J.* **48,** 467–477.

Chang, C. W. (1967). Study of phytase and fluoride effects in germinating corn seeds. *Cereal Chem.* **44,** 129–142.

Chang, C. W. (1968). Effect of fluoride on nucleotides and ribonucleic acid in germinating corn seedling roots. *Plant Physiol* **43,** 669–674.

Chang, C. W. (1970a). Effect of fluoride on ribosomes from corn roots. Changes with growth retardation. *Physiol. Plant.* **23,** 536–543.

Chang, C. W. (1970b). Effect of fluoride on ribosomes and ribonuclease from corn roots. *Can. J. Biochem.* **48,** 450–454.

Chang, C. W., and Thompson, C. R. (1966a). Site of fluoride accumulation in navel orange leaves. *Plant Physiol.* **41,** 211–213.

Chang, C. W., and Thompson, C. R. (1966b). Effect of fluoride on nucleic acids and growth in germinating corn seedling roots. *Physiol. Plant.* **19,** 911–918.

Cheng, J. Y. O., Yu, M. H., Miller, G. W., and Welkie, G. W. (1968). Fluoroorganic acids in soybean leaves exposed to fluoride. *Environ. Sci. Technol.* **2,** 367–370.

Chenoweth, M. B. (1949). Monofluoroacetic acid and related compounds. *J. Pharmacol. Exp. Ther.* II, **97,** 383–424.

Christiansen, G. S., and Thiman, K. V. (1950). The metabolism of stem tissue during growth and its inhibition. II. Respiration and ether-soluble material. *Arch. Biochem.* **26,** 248–259.

Chung, C. W., and Nickerson, W. J. (1954). Polysaccharide syntheses in growing yeast. *J. Biol. Chem.* **208,** 395–407.

Clendenning, K. A., and Ehrmantraut, H. C. (1950). Photosynthesis and Hill reactions by whole *Chlorella* cells in continuous and flashing light. *Arch. Biochem.* **29,** 387–403.

Commoner, B., and Thimann, K. V. (1941). On the relation between growth and respiration in the *Avena* coleoptile. *J. Gen. Physiol.* **24,** 279–296.

Daines, R. H., Leone, I., and Brennan, E. (1952). The effect of fluorine on plants as determined by soil nutrition and fumigation studies. *In* "Air Pollution" (L. C. McCabe, ed.), pp. 97–105. McGraw-Hill, New York.

De Oliveira, M. M. (1963). Chromatographic isolation of monofluoroacetic acid from *Palicourea marcgravii. Experientia* **19,** 586–587.

Ducet, G., and Rosenberg, A. J. (1953). Activités Réspiratoires chez les vegetaux supérieurs. VII. Glycolyse aeuobie provoguée par l'oxyde de carbone. *Bull. Soc. Chim. Biol.* **35,** 467–476.

Ergle, D. R., and Guinn, G. (1959). Phosphorus compounds of cotton embryos and their changes during germination. *Plant Physiol.* **34,** 476–480.

Gal, E. M., Drewes, P. A., and Taylor, N. F. (1961). Metabolism of fluoroacetic acid-2-C^{14} in the intact rat. *Arch. Biochem. Biophys.* **93,** 1–14.

Giri, K. V. (1937). Vegetable phosphatases. I. The phosphatase of sprouted soybean (*Glycine hispida*). *Hoppe-Seyler's Z. Physiol. Chem.* **245,** 185–196.

Goldman, P. (1965). The enzymatic cleavage of the carbon-fluorine bond in fluoroacetate. *J. Biol. Chem.* **240,** 3434–3438.

Good, N. E. (1962). Uncoupling of the Hill reaction from photophosphorylation by anions. *Arch. Biochem. Biophys.* **96,** 653–661.

Grunberg-Manago, M., Oritz, P. J., and Ochoa, S. (1956). Enzymatic synthesis of polynucleotides. *Biochim. Biophys. Acta* **20,** 269–285.

Hackett, D. P. (1959). Respiratory mechanisms in higher plants. *Annu. Rev. Plant Physiol.* **10,** 113–146.

Hansen, E. D., Wiebe, H. H., and Thorne, W. (1958). Air pollution with relation to agronomic crops. VII. Fluoride uptake from soils. *Agron. J.* **50,** 565–568.

Hanson, J. B. (1960). Impairment of respiration, ion accumulation, and ion retention in root tissue treated with ribonuclease and ethylenediaminetetraacetic acid. *Plant Physiol.* **35,** 372.

Heppel, L. A., and Hilmore, R. J. (1951). Purification and properties of 5′-nucleotidase. *J. Biol. Chem.* **188,** 665–676.

Hill, A. C., Transtrum, L. G., Pack, M. R., and Winters, W. S. (1958). Air pollution with relation to agronomic crops. VI. An investigation of the "hidden injury" theory of fluoride damage to plants. *Agron. J.* **50,** 562–565.

Hill, A. C., Pack, M. R., Transtrum, L. G., and Winters, W. S. (1959a). Effects of atmospheric fluorides and various types of injury on the respiration of leaf tissue. *Plant Physiol.* **34,** 11–16.

Hill, A. C., Transtrum, L. G., Pack, M. R., and Holloman, A., Jr. (1959b). Facilities and techniques for maintaining a controlled fluoride environment in vegetation studies. *J. Air Pollut. Contr. Ass.* **9,** 22–27.

Hitchcock, A. E., Zimmerman, P. W., and Cooe, R. R. (1962). Results of ten years' work (1951–1960) on the effect of fluorides on gladiolus. *Contrib. Boyce Thompson Inst.* **21,** 303–344.

Ingle, J., and Hageman, R. H. (1964). Studies on the relationship between ribonucleic acid content and rate of growth of corn roots. *Plant Physiol.* **39,** 730–734.

Jacobson, J. S., Weinstein, L. H., McCune, D. C., and Hitchcock, A. E. (1966). The accumulation of fluoride by plants. *J. Air Pollut. Contr. Ass.* **16,** 412–417.

Key, J. L., Hanson, J. B., Lund, H. A., and Vatter, A. E. (1961). Changes in cytoplasmic particulates accompanying growth in the mesocotyl of *Zea mays*. *Crop Sci.* **1,** 5–8.

Kravitz, E. A., and Guarino, A. J. (1958). On the effect of inorganic phosphate on hexose phosphate metabolism. *Science* **128,** 1139–1140.

Krebs, H. A. (1957). Control of metabolic processes. *Endeavour* **16,** 125.

Laties, G. G. (1949). The role of pyruvate in the aerobic respiration of barley roots. *Arch. Biochem.* **20,** 284–299.

Ledbetter, M. C., Mavrodineanu, R., and Weiss, A. J. (1960). Distribution studies of radioactive fluorine-18 and stable fluorine-19 in tomato plants. *Contrib. Boyce Thompson Inst.* **20,** 331–348.

Lee, C. J., Millor, G. W., and Welkie, G. W. (1966). The effects of hydrogen fluroide and wounding on respiratory enzymes in soybean leaves. *Int. J. Air Water Pollut.* **10,** 169–181.

Lehninger, A. L., ed. (1965). "Bioenergetics," p. 109. Benjamin, New York.

Leone, I. A., Brennan, E. G., Daines, R. G., and Robbins, W. R. (1948). Some effects of fluorine on peach, tomato and buckwheat when absorbed through the roots. *Soil Sci.* **66,** 259–266.

Liebecq, C., and Peters, R. A. (1948). The inhibitory effect of fluoroacetate and the tricarboxylic cycle. *J. Physiol.* (*London*) **108,** 11–12.

Liebecq, C., and Peters, R. A. (1949). The toxicity of fluoroacetate and the tricarboxylic acid cycle. *Biochim. Biophys. Acta* **3,** 215–230.

Lin, C. Y., and Key, J. L. (1967). Dissociation and reassembly of polysomes in relation to protein synthesis in the soybean root. *J. Mol. Biol.* **26,** 237–247.

Lords, J. L. (1960). Some effects of fluoride on the respiratory metabolism of *Chlorella pyrenoidosa.* Master's Thesis, University of Utah, Salt Lake City.

Lovelace, J., and Miller, G. W. (1967). *In vitro* effects of fluoride on tricarboxylic acid cycle dehydrogenase and oxidative phosphorylation. Part I. *J. Histochem. Cytochem.* **15,** 195–201.

Lovelace, J., Miller, G. W., and Welkie, G. W. (1968). The accumulation of fluoroacetate and fluorocitrate in forage crops collected near a phosphate plant. *Atmos. Environ.* **2,** 187–190.

Lowry, O. H., and Passonneau, J. V. (1964). Fructokinase and the control of glycolysis. *Proc. Int. Congr. Biochem., 6th., 1964,* pp. 705–706.

Luštinec, J., and Pokorná, V. (1962). Alternation of respiratory pathways during the development of wheat leaf. *Biol. Plant. Acad. Sci. Bohemoslov.* **4,** 101–109.

Luštinec, J., Krekule, J., and Pokorná, V. (1960). Respiratory pathways in gibberellin-treated wheat. The effect of fluoride on the respiration rate. *Biol. Plant. Acad. Sci. Bohemoslov.* **2,** 223–226.

McCune, D. C. (1969). A symposium—The technical significance of air quality standard. *Environ. Sci. Technol.* **3,** 720–735.

McCune, D. C., Weinstein, L. H., Jacobson, J. S., and Hitchcock, A. E. (1964). Some effects of atmospheric fluoride on plant metabolism. *J. Air Pollut. Contr. Ass.* **14,** 465–468.

McCune, D. C., Hitchcock, A. E., and Weinstein, L. H. (1966). Effect of mineral nutrition on the growth and sensitivity of gladiolus to hydrogen fluoride. *Contrib. Boyce Thompson Inst.* **23,** 295–300.

McCune, D. C., Weinstein, L. H., and Mancini, J. F. (1970). Effects of hydrogen fluoride on the acid-soluble nucleotide metabolism of plants. *Contrib. Boyce Thompson Inst.* **24,** 213–226.

McEwan, T. (1964a). Isolation and identification of the toxic principle of *Gastrolobium grandiflorum. Nature (London)* **201,** 827.

McEwan, T. (1964b). Isolation and identification of the toxic principle of *Gastrolobium grandiflorum. Queensl. J. Agr. Sci.* **21,** 1–14.

MacLean, D. C., Roark, O. F., Folkerts, G., and Schneider, R. E. (1969). Influence of mineral nutrition on the sensitivity of tomato plants to hydrogen fluoride. *Environ. Sci. Technol.* **3,** 1201–1204.

McNulty, I. B. (1959). The influence of fluoride on leaf respiration. *Proc. Int. Bt. Congr., 9th, 1959,* Vol. 2, p. 245.

McNulty, I. B., and Lords, J. L. (1960). Possible explanation of fluoride-induced respiration in *Chlorella pyrenoidosa. Science* **132,** 1553–1554.

McNulty, I. B., and Newman, D. W. (1956). Effects of lime spray on the respiration rate and chlorophyll content of leaves exposed to atmospheric fluorides. *Utah Academy Proceedings* **33,** 73–79.

McNulty, I. B., and Newman, D. W. (1957). Effects of atmospheric fluoride on the respiration rate of bush bean and gladiolus leaves. *Plant Physiol.* **32,** 121–124.

McNulty, I. B., and Newman, D. W. (1961). Mechanisms of fluoride induced chlorosis. *Plant Physiol.* **36,** 385–388.

Marais, J. S. C. (1944). Monofluoroacetatic acid, the toxic principle of "gifflaar" *Dichapetalum cymosum* (Hook) Engl. *Onderstepoort J. Vet. Sci. Anim. Ind.* **20,** 67–73.

Massey, V. (1958). The role of iron in beef-heart succinic dehydrogenase. *Biochim. Biophys. Acta* **30,** 500–509.

Miller, G. W. (1957). The effect of atmospheric hydrogen fluoride (HF) on plant growth and enolase activity. Annual Report on Utah's Contributing Project to Regional Project W-39. *Utah, Agr. Exp. Sta., Annu. Rep* **457,** 11.

Miller, G. W. (1958). Properties of enolase in extracts from pea seed. *Plant Physiol.* **33,** 199–206.

Miller, G. W. (1960). Carbon dioxide-bicarbonate absorption, accumulation, effects on various plant metabolic reactions, and possible relations to lime-induced chlorosis. *Soil Sci.* **89,** 241–245.

Miller, G. W., and Wei, L. L. (1971). The fine structure of leaves from *Glycine max* Merr in relation to fluoride injury. *4th Annu. Conf. Int. Soc. Fluoride Res., 1971,* p. 3.

Morrison, J. F., and Peters, R. A. (1954). Biochemistry of fluoroacetate poisoning: The effect of fluorocitrate on purified aconitase. *Biochem. J.* **58,** 473–479.

Naganna, B., Raman, A., Venugopal, B., and Sripathi, C. E. (1955). Potato pyrophosphatases. *Biochem. J.* **60,** 215–223.

Najjar, V. A. (1948). The isolation and properties of phosphoglucomutase. *J. Biol. Chem.* **175,** 281.

Neidhardt, N. G., and Magasanik, B. (1960). Studies on the role of ribonucleic acid in the growth of bacteria. *Biochim. Biophys. Acta* **42,** 99.

Newman, D. W. (1962). Effects of sodium fluoride on leaf catalase activity. *Ohio J. Sci.* **62,** 281–288.

Newman, D. W., and McNulty, I. B. (1959). Fluoride effects on chloroplast pigments. *Proc. Int. Bot. Congr., 9th, 1959,* Vol. 2, p. 281.

Oelrichs, P. B., and McEwan, T. (1961). Isolation of the toxic principle in *Acacia georginae*. *Nature* (*London*) **190,** 808–809.

Oelrichs, P. B., and McEwan, T. (1962). The toxic principle of *Acacia georginae. Queensl. J. Agr. Sci.* **19,** 1–16.

Ordin, L., and Altman, A. (1965). Inhibition of phosphoglucomutase activity in oat coleoptiles by air pollutants. *Physiol. Plant.* **18,** 790–797.

Ordin, L., and Skoe, B. P. (1963). Inhibition of metabolism in *Avena* coleoptile tissue by fluoride. *Plant Physiol.* **38,** 416–421.

Pack, M. R., and Wilson, A. M. (1967). Influence of hydrogen fluoride fumigation on acid-soluble phosphorus compounds in bean seedlings. *Environ. Sci. Technol.* **1,** 1011–1013.

Pattison, F. L. M., ed. (1959). "Toxic Compounds." Amer. Elsevier, New York.

Penot, M., and Buvat, R. (1967a). Action du fluorure de sodium sur l'absorption des phosphates par des disques de feuille de *Nicotiana tobacum:* Inhibition, stimulation. *C. R. Acad. Sci.* **264,** 926–928.

Penot, M., and Buvat, R. (1967b). Action due pretraitement de disques de feuille de *Nicotiana tobacum* par le fluorure de sodium: Absorption de p^{32} taux respiratoire, activité phosphatasique. *C. R. Acad. Sci.* **264,** 1169–1171.

Peters, R. A. (1957). Mechanism of the toxicity of the active constituent of *Dichapetalum cymosum* and related compounds. *Advan. Enzymol.* **18,** 113–159.

Peters, R. A., and Shorthouse, M. (1972). Formation of monofluorocarbon compounds by single cell cultures of *Glycine max* growing on inorganic fluoride. *Phytochemistry* **11,** 1339.

Peters, R. A., Hall, R. J., Ward, P. F. V., and Sheppard, N. (1960). The chemical nature of the toxic compounds containing fluorine in the seeds of *Dichapetalum toxicarium*. *Biochem. J.* **77,** 17–23.

Peters, R. A., Shorthouse, M., and Murray, L. R. (1964). Enolase and fluorophosphate. *Nature (London)* **202,** 1331–1332.

Peters, R. A., Murray, L. R., and Shorthouse, M. (1965a). Fluoride metabolism in *Acacia georginae* Gidyea. *Biochem. J.* **95,** 724–730.

Peters, R. A., Shorthouse, M., and Ward, P. T. V. (1965b). The synthesis of the carbon-fluorine bond by *Acacia georginae in vitro. Life Sci.* **4,** 749–752.

Pilet, P. E. (1963). Action du fluor et de !'acide B-indolylacetique sur la respiration de disques de feuilles. *Bull. Soc. Vaudoise Sci. Natur.* **68,** 359–360.

Pilet, P. E. (1964). Action du fluor et de l'acide β-indolylacetique sur la respiration des tissues radicularires. *Rev. Gen. Bot.* **71,** 12–21.

Plaut, G. W., and Lardy, H. A. (1949). The oxalacetate decarboxylase of *Azotobacter vinelandii. J. Biol. Chem.* **180,** 13–27.

Poljakoff-Mayber, A. (1953). Germination inhibitors and plant enzyme systems. III. Hydrolytic enzymes. *Palestine J. Bot.* **6,** 101–106.

Preuss, P. W., and Weinstein, L. H. (1969). Studies on fluoro organic compounds in plants. II. Defluorination of fluoroacetate. *Contrib. Boyce Thompson Inst.* **24,** 151–155.

Preuss, P. W., Lemmens, A. G., and Weinstein, L. H. (1968). Studies on fluoroorganic compounds in plants. I. Metabolism of 2-C^{14}-fluoroacetate. *Contrib. Boyce Thompson Inst.* **24,** 25–31.

Preuss, P. W., Birkhahn, R., and Bergmann, E. D. (1970a). The effect of sodium fluoride on the growth and metabolism of tissue culture of *Acacia georgimae* and tomato. *Isr. J. Bot.* **19,** 609–619.

Preuss, P. W., Colavito, L., and Weinstein, L. H. (1970b). The synthesis of monofluoroacetic acid by a tissue culture of *Acacia georginae. Experientia* **26,** 1059–1060.

Ramagopal, S., Welkie, G. W., and Miller, G. W. (1969). Flouride injury of roots and calcium nutrition. *Plant Cell Physiol.* **10,** 675–685.

Rapp, G. W., and Sliwinski, R. A. (1956). The effect of sodium monofluorophosphate on the enzyme phosphorylase and phosphatase. *Arch. Biochem. Biophys.* **60,** 379–383.

Reiner, J. M., Tsuboi, K. K., and Hudson, P. B. (1955). Acid phosphatase. IV. Fluoride inhibition of prostatic acid phosphatase. *Arch. Biochem. Biophys.* **56,** 165–183.

Roche, J., and Thoai, N. V. (1950). Phosphatase alkaline. *Advan. Enzymol.* **10,** 83–122.

Ross, C. W., Wiebe, H. H., and Miller, G. W. (1960). Respiratory pathways in various plants as related to susceptibility to fluoride injury. *Plant Physiol.* **35,** Suppl., XXIX.

Ross, C. W., Wiebe, H. H., and Miller, G. W. (1962). Effect of fluoride on glucose catabolism in plant leaves. *Plant Physiol.* **37,** 305–309.

Satoh, K., Katoh, S., and Takamiya, A. (1970). Effects of chloride ion on Hill reaction in *Euglena* chloroplasts. *Plant Cell Physiol.* **11,** 453–466.

Saunders, B. C. (1947). Toxic properties of ω-fluorocarboxylic acid and derivatives. *Nature (London)* **160,** 179–180.

Schneider, R. E., and MacLean, D. C. (1970). Relative susceptibility of seven grain *Sorghum* hybrids to hydrogen fluoride. *Contrib. Boyce Thompson Inst.* **24,** 241–244.

Slater, E. C., and Bonner, W. D. (1952). Effect of fluoride on succinic oxidase system. *Biochem. J.* **52,** 185–196.

Smith, B. N. (1961). The effects of fluorides on basic plant processes. Master's Thesis, University of Utah, Salt Lake City.
Sobolev, A. M. (1962). The distribution, production and utilization of phytin in higher plants. *Usp. Biol. Khim.* **4,** 248–261.
Solberg, R. A., and Adams, D. F. (1956). Histological responses of some plant leaves to hydrogen fluoride and sulfur dioxide. *Amer. J. Bot.* **43,** 755–760.
Spikes, J. D., Lumry, R., and Rieske, J. S. (1955). Inhibition of the photochemical activity of isolated chloroplasts. *Arch. Biochem. Biophys.* **55,** 25–37.
Srivastave, B. I. S. (1965). Effect of kinetin on the *Ecteola* cellulose elution profile and other properties of RNA from the excised first seedling leaves of barley. *Arch. Biochem. Biophys.* **110,** 97.
Srivastava, B. I. S. (1967). Cytokinins in plants. *Int. Rev. Cytol.* **22,** 349.
Srivastava, B. I. S., and Arglebe, C. (1967). Studies on ribosomes from barley leaves. Changes during senescence: *Plant Physiol.* **42,** 1497.
Thomas, M. D. (1951). Gas damage to plants. *Annu. Rev. Plant Physiol.* **2,** 293–322.
Thomas, M. D., and Hendricks, R. H. (1956). Effect of air pollution on plants. *In* "Air Pollution Handbook" (P. L. Magill, F. R. Holden, and C. Ackley, eds.), Sect. 9, p. 44. McGraw-Hill, New York.
Tietz, A., and Ochoa, S. (1958). Fluorokinase and pyruvic kinase. *Arch. Biochem. Biophys.* **78,** 477–493.
Treble, D. H., Lamport, D. T. A., and Peters, R. A. (1962). The inhibition of plant aconitate hydratase (aconitase) by fluorocitrate. *Biochem. J.* **85,** 113–115.
Treshow, M., ed. (1970). "Environment and Plant Response," pp. 267–301. McGraw-Hill, New York.
Venkateswarlu, P., Armstrong, W. D., and Singer, L. (1965). Absorption of fluoride and chloride by barley roots. *Plant Physiol.* **40,** 255–261.
Vennesland, B., and Turkington, E. (1966). The relationship of the Hill reaction to photosynthesis: Studies with fluoride-poisoned blue-green algae. *Arch. Biochem. Biophys.* **116,** 153–161.
Wander, I. W., and McBride, J. J., Jr. (1955). A chlorosis produced by fluorine on citrus in Florida. *Proc. Fla. State Hort. Soc.* **68,** 23–24.
Wander, I. W., and McBride, J. J., Jr. (1956). Chlorosis produced by fluorine on citrus in Florida. *Science* **123,** 933–934.
Warburg, O., and Christian, W. (1942). Isolierung und Kristallisation des Gärungsferment Enolase. *Biochem. Z.* **310,** 385–421.
Warburg, O., and Kirppahl, G. (1956). Functional carbon dioxide of *Chlorella. Z. Naturforsch. B.* **11,** 718–726.
Warburg, O., Klotzach, H., and Krippahl, G. (1957). Glutaminsäure in *Chlorella. Z. Naturforsch. B* **12,** 622–628.
Ward, P. F. V., and Huskisson, N. S. (1969). The metabolism of fluoroacetate by plants. *Biochem. J.* **113,** 9p.
Ward, P. F. V., Hall, R. J., and Peters, R. A. (1964). Fluorofatty acids in the seeds of *Dichapetalum toxicarium. Nature (London)* **201,** 611–612.
Webster, C. C. (1967). "The Effects of Air Pollution on Plants and Soil," Agr. Res. Counc., pp. 28–29. HM Stationery Office, London.
Webster, G. C., and Varner, J. E. (1955). Aspartate metabolism and asparagine synthesis in plant systems. *J. Biol. Chem.* **215,** 91–93.
Weinstein, L. H. (1961). Effects of atmospheric fluoride on metabolic constituents of tomato and bean leaves. *Contrib. Boyce Thompson Inst.* **21,** 215–231.
Whiteman, T. M., and Schomer, H. A. (1945). Respiration and internal gas content of injured sweet-potato roots. *Plant Physiol.* **20,** 171–182.

Wildman, S. G., and Bonner, J. (1947). The proteins of green leaves. I. Isolation, enzymatic properties and auxin content of spinach cytoplasmic proteins. *Arch. Biochem.* **14,** 381.

Williamson, J. R., Jones E. A., and Azzone, C. F. (1964). Metabolic control in perfused rat heart during fluoroacetate poisoning. *Biochem. Biophys. Res. Commun.* **17,** 696–702.

Woltz, S. S. (1964a). Translocation and metabolic effects of fluorides in gladiolus leaves. *Proc. Fla. State Hort. Soc.* **77,** 511–515.

Woltz, S. S. (1964b). Distinctive effects of root versus leaf acquired fluorides. *Proc. Fla. State Hort. Soc.* **77,** 516–517.

Woltz, S. S., and Leonard, C. D. (1964). Effect of atmospheric fluorides upon certain metabolic processes in valencia orange leaves. *Proc. Fla. State Hort. Soc.* **77,** 9–15.

Yang, S. F., and Miller, G. W. (1963a). Biochemical studies on the effect of fluoride on higher plants. 1. Metabolism of carbohydrates, organic acids and amino acids. *Biochem. J.* **88,** 505–509.

Yang, S. F., and Miller, G. W. (1963b). Biochemical studies on the effect of fluoride on higher plants. 2. The effect of fluroide on sucrose-synthesizing enzymes from higher plants. *Biochem. J.* **88,** 509–516.

Yang, S. F., and Miller, G. W. (1963c). Biochemical studies on the effect of fluoride on higher plants. 3. The effect of fluorine on dark carbon dioxide fixation. *Biochem. J.* **88,** 517–522.

Yu, M. H., and Miller, G. W. (1967). Effect of fluoride on the respiration of leaves from higher plants. *Plant Cell Physiol.* **8,** 483–493.

Yu, M. H., and Miller, G. W. (1970). Gas chromatographic identification of fluoro-organic acids. *Environ. Sci. Technol.* **4,** 492–495.

Zimmerman, P. W., and Hitchcock, A. E. (1946). Fluorine compounds given off by plants. *Amer. J. Bot.* **33,** 233.

Zimmerman, P. W., and Hitchcock, A. E. (1956). Susceptibility of plants to hydrofluoric acid and sulfur dioxide gases. *Contrib. Boyce Thompson Inst.* **18,** 263–279.

5

PEROXYACYL NITRATES

J. B. Mudd

I. Introduction

Field observations of pollutant damage to vegetation revealed in 1956 that certain symptoms could not be accounted for by the chemically characteristic pollutants known at that time (Middleton *et al.*, 1956). An unknown compound was postulated to be the cause of the symptoms; bronzing on the lower surface of leaves. Eventually the symptoms were found to be caused by peroxyacetyl nitrate (PAN) in the ambient air (Stephens *et al.*, 1961). PAN is one of a series of homologs which are both eye irritating and phytotoxic. The most common of these homologs is PAN, $CH_3CO{\cdot}O_2NO_2$, which may rise to concentrations of 0.05 ppm in the Los Angeles basin. Peroxypropionyl nitrate (PPN) has also been detected in the atmosphere and it has been named as the culprit in at least one incident of vegetation damage. Other compounds of the series, peroxybutyryl nitrate (PBN) and peroxybenzoyl nitrate (PBzN) have been synthesized and phytotoxic and eye irritation properties have been studied. The degree of eye irritation increases as the size of the homolog increases, and the phytotoxicity shows the same relationship, i.e., PBN $>$ PPN $>$ PAN (PBzN has not been tested).

The peroxyacyl nitrates have not been studied as much as some other air pollutants, probably because the chemical synthesis and storage have caused numerous explosions (Stephens *et al.*, 1969). This chapter will provide some background on the chemistry of the peroxyacyl nitrates and relate some of these properties to the physiological and biochemical effects on plants.

II. Chemical Properties

A. Formation in the Atmosphere

Peroxyacyl nitrates are secondary pollutants; they are formed by photochemical reactions involving the primary pollutants emitted in automobile exhaust. Leighton's book (1961) is an excellent source of information on the photochemistry of production of secondary pollutants. The basic reaction is

$$NO_2 \xrightarrow{h\nu} NO + O$$

This production of highly reactive oxygen atoms is followed by two other reactions.

$$O + O_2 + M \rightarrow O_3 + M$$
$$O_3 + NO \rightarrow NO_2 + O_2$$

In the absence of hydrocarbons, steady state concentrations of these compounds are reached.

$$NO_2 + O_2 \overset{h\nu}{\rightleftharpoons} NO + O_3$$

When hydrocarbons are present, especially unsaturated hydrocarbons, the hydrocarbon is oxidized, nitric oxide is converted to nitrogen dioxide, and the secondary pollutants, peroxyacyl nitrates, aldehydes, and ketones are produced. In a typical day in Los Angeles, the concentrations of hydrocarbon and nitric oxide rise with the early morning traffic, as these are exposed to sunlight the hydrocarbon concentration drops and the oxidant (ozone plus peroxyacyl nitrates) concentration increases. In the oxidant mixture, ozone concentration is greater than peroxyacyl nitrates by a factor of 10–20.

As the mixture of primary pollutants is irradiated (effective wavelengths are anything less than 430 nm, the limit of the NO_2 absorbance), more oxidant is produced, but when the sun goes down there is no further conversion. The pollutants usually are swept out by winds, but if the air is stagnant, it is found that during the night the ozone concentration declines more rapidly than that of peroxyacyl nitrate.

B. Formation in the Laboratory

Stephens (1969) has presented a number of methods of synthesizing peroxyacyl nitrates in the laboratory. The method which simulates atmospheric conditions with greatest accuracy is the irradiation of an unsaturated hydrocarbon in the presence of NO or NO_2 and air or oxygen. The contents of the reaction mixture are trapped by freezing out and then purified by gas chromatography. In the stages of preparation when products are in liquid form, possibility of explosion is greatest.

The products which can be measured in the photochemical reaction using *trans*-2-butene as the hydrocarbon are various:

$$CH_3CH{=}CHCH_3 + O_2 + NO + NO_2 \xrightarrow{h\nu} CH_3COO_2NO_2 + CH_3CHO + HCHO + CH_3ONO_2 + CO_2 + CO$$

Reactions giving rise to these compounds have been fully discussed by Kerr *et al.* (1972), and possible sequences have been presented for initiation by oxygen atoms or ozone or hydroxyl radicals.

Table I (data of Schuck and Doyle, cited by Leighton, 1961) presents some data on the production of PAN from various hydrocarbons. For the most part, the products reflect the scission of the double bond, but there are some notable exceptions: (1) peroxyformyl nitrate is not formed (or else is not stable); (2) 1-pentene can be fully accounted for without the formation of PBN. Stephens *et al.* (1965), have noted that the preferred method for production of PBN is reaction of *n*-butyraldehyde with NO_2 and O_3. The preferred method for PAN and PPN production in the laboratory is photolysis of alkyl nitrite in oxygen.

TABLE I

PRODUCTS OF PHOTOCHEMICAL REACTION OF OLEFINS AND OXIDES OF NITROGEN[a]

Olefin	Oxide of nitrogen	PAN	HCHO	CH_3-CHO	CH_3-CH_2-CHO	CH_3-$(CH_2)_2$-CHO	CH_2=CH-CHO	Ac_2O	O_3	CO	% Rec.
Ethylene	NO	—	0.63	—	—	—	—	—	0.30	0.56	60
Propylene	NO	0.20	0.68	0.68	—	—	—	—	0.39	—	82
1-Butene	NO_2	0.02	0.47	—	1.00	—	—	—	0.21	0.03	88
trans-2-Butene	NO_2	0.17	—	1.34	—	—	—	—	0.23	0.70	93
1,3-Butadiene	NO_2	0.02	0.88	—	—	—	0.83	—	0.32	0.39	95
1-Pentene	NO	0.08	1.04	—	—	0.78	—	—	0.40	0.53	100
Tetramethylethylene	NO	0.37	0.29	—	—	—	—	1.38	0.23	0.26	90

[a] Data of Schuck and Doyle as presented by Leighton (1961).

C. Detection and Measurement

In the early laboratory studies of peroxyacyl nitrate formation, the product was detected with a long-path infrared spectrophotometer. Stephens (1964) has published the infrared absorptivities of PAN, PPN, and PBN at various wavelengths. The absorptivity of samples of any of these gases can be used for the determination of concentration. In the determination of concentration of peroxyacyl nitrates in ambient air, the sample is injected into a gas chromatograph and PAN, PPN, and PBN can be detected with an electron capture detector. Concentrations of 0.01 ppm can be detected in a 2 ml air sample (Darley *et al.*, 1963).

D. Stability

Peroxyacyl nitrates which have been prepared and purified in the laboratory can be stored in low pressure oxygen tanks pressurized with nitrogen or helium. The compounds can be stored in this fashion, but it is recommended that the concentration not exceed 1000 ppm nor the pressure exceed 100 psi because of the danger of explosion. Provided these precautions are taken, the peroxyacyl nitrates are stable for months. PAN is also stable when dissolved in anhydrous organic solvents (Darnall and Pitts, 1970). If the peroxyacyl nitrate comes in contact with an aqueous solution, however, the breakdown is rapid. While attempting to determine the stoichiometry of the reaction of PAN with reduced nicotinamide adenine dinucleotide (NADH), Mudd and Dugger (1963) found that the most convenient way to determine how much of the PAN bubbled through an aqueous medium had gone into solution was to measure subsequently the inorganic nitrite. Nicksic *et al.* (1967) observed that in alkaline solutions PAN yields stoichiometric amounts of nitrite and acetate, and Stephens (1967) reported that under the same conditions stoichiometric amounts of molecular oxygen were produced. The breakdown of PAN in aqueous solutions can therefore be described by the equation

$$CH_3CO{\cdot}O_2NO_2 + 2OH^- \rightarrow CH_3COO^- + NO_2^- + O_2 + H_2O$$

The half-life of peroxyacyl nitrates is a function of pH and is not the same for PAN, PPN, and PBN (Fig. 1 and Table II). The stability of peroxyacyl nitrates in aqueous media may be related to phytotoxicity although the half-life data do not correlate precisely with the phytotoxicity.

III. Symptoms

Excellent photographs of PAN damage to various plants are presented in "Recognition of Air Pollution Injury to Vegetation: A Pictorial Atlas"

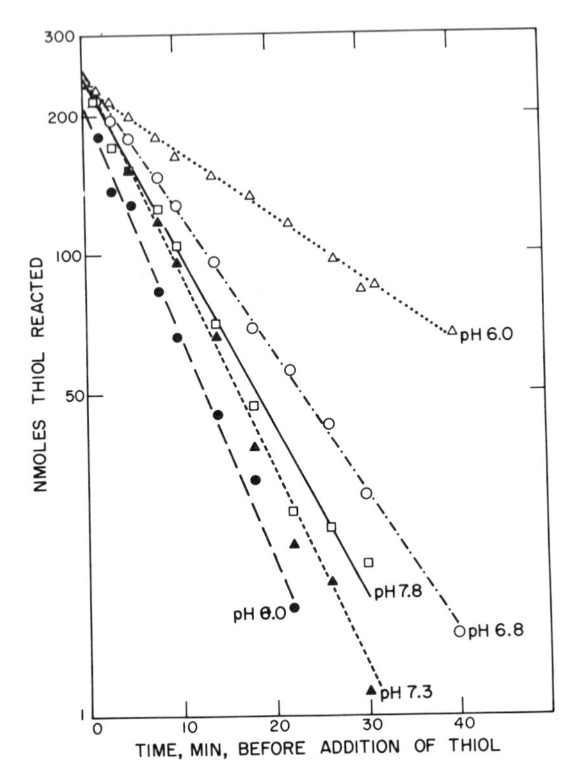

FIG. 1. Half-life of PAN at different pH's. Aliquots of 3.0 ml of 0.033 *M* phosphate buffer of various pH's were gassed at 25° for 30 seconds at 20 ml/minute with 662 ppm PAN (690 nmoles/minute). At the end of the gassing period, the solutions were allowed to stand for varying periods of time and then 1.4 ml aliquots were added to 1.6 ml of 10^{-3} *M* 5-mercapto-2-nitrobenzoic acid in 0.06 *M* phosphate buffer pH 7.8. Samples were read after 45 minutes at 412 nm and the decrease in absorbance used to calculate nmoles oxidized, using the extinction coefficient provided by Ellman (1959). Oxidation is plotted versus time elapsed between gassing and additions of the thiol. Slopes were used to calculate half-lives. (Data obtained in collaboration with T. T. McManus.)

(Jacobson and Hill, 1970). Sensitive species include pinto bean (*Phaseolus vulgaris*), lettuce (*Lactuca sativa*), and oats (*Avena sativa*). Resistant species include cucumber (*Cucumis sativus*), cotton (*Gossypium hirsutum*), and corn (*Zea mays*). Tobacco (*Nicotiana tabacum*) and wheat (*Triticum sativum*) are intermediate. Sensitive species are injured by 20 ppb PAN whereas resistant species show no injury after exposure to 100 ppb for 2–4 hours. The symptoms of damage for PAN, PPN, and PBN cannot be distinguished. The damage is usually a glazing or bronzing of the lower

TABLE II
HALF-LIVES OF PEROXYACYL NITRATES AT DIFFERENT pH's[a]

pH	Half-lives (minutes) of peroxyacyl nitrates		
	PAN	PPN	PBN
6.0	21.5	16	37.5
6.8	10	8	21
7.3	7.5	7	14
7.8	8.5	6	15
8.0	6.5	7	14

[a] Half-lives were determined as described for (Fig. 1). Gas concentrations were PAN, 662 ppm; PPN, 510 ppm; PBN, 318 ppm. Duration of gassing was such that the nmoles of each administered was PAN, 296 nmoles; PPN, 456 nmoles; PBN, 284 nmoles. Thiol reacted at maximum was (PAN) 240 nmoles; (PPN) 144 nmoles; (PBN) 179 nmoles.

surface of the leaves but, in some cases, the upper surface is also affected. Microscopic examination of damaged leaves revealed that the protoplasts of the mesophyll cells in the region of the stomatal cavity were collapsed. The resulting air spaces give rise to the glazed appearance. Development of symptoms of PAN damage is fairly slow: 72 hours may be required for full development.

Taylor *et al.* (1960) compared the symptoms on pinto bean and petunia obtained with (a) ambient air, (b) ozone, (c) reaction products of ozone and 1-hexene, and (d) the photochemical reaction products from NO_2 and 1-hexene. Before the chemical characterization of PAN, the mixture of ozone and hexene was frequently used to simulate Los Angeles smog. The product was phytotoxic but had a short half-life (12 minutes). With the discovery of PAN, which had a half-life comparable to the toxic compounds in the air, interest in "ozonated hexene" declined and nobody determined what the toxic product actually was. In the experiments with petunia, the silvering and bronzing symptoms in ambient air were reproduced only by the irradiated NO_2 + 1-hexene mixture. In the pinto bean, ambient air produced bronzing and glazing on 7-day-old leaves and tissue collapse on the upper surface in 14-day-old leaves. The former symptoms were produced by the photochemical reaction products and the latter by ozone. Heuss and Glasson (1968) have shown that irradiation of NO_2 + 1-hexene gives very eye-irritating products including a small amount of PAN. Carbon recovery data were not given in Heuss and Glasson's paper so the toxic compound could have been one of the higher peroxyacyl nitrates.

IV. Physiological Observations

A. Effect of Leaf Age

It was mentioned in the previous section that pinto bean leaves of different ages show different symptoms. Symptoms of ozone damage are usually not observed until leaves are maximally expanded. In contrast, symptoms of PAN are not observed on very young leaves or old leaves, but only at intermediate ages. This relationship gives rise to banding patterns on leaves with cells of different ages. The compound leaves of tomato provide a good example of the effect of age on susceptibility to PAN. The terminal leaflet of the compound leaf is the oldest tissue. Thus, on the youngest susceptible leaf, the terminal leaflet is damaged. On older leaves, the terminal leaflet is old enough to be resistant, but the lateral leaflets have now moved into the susceptible range.

One would like to know what aspects of physiology and biochemistry dispose the leaves to such a pattern of susceptibility. So far, no attempt has been made to correlate the shape of the curve, age versus susceptibility with chemical composition of the leaf. The chances of either analyzing for the wrong compounds or of finding a correlation with a quite unrelated compound seem equally high.

B. Effect of Illumination

Taylor *et al.* (1961) noted inconsistencies in the elicitation of plant damage by PAN, and during the course of standardizing exposure conditions observed that there is a remarkable requirement for illumination before, during and after exposure to PAN (Fig. 2). If the prefumigation period of 12 hours illumination is followed by only 15 minutes darkness before fumigation, the plants are protected and susceptibility only returns fully after 60 minutes of illumination. After fumigation (30 minutes) the dark period must be 1–2 hours long if damage is to be avoided, and the light period must be 3 hours long in order to achieve maximal damage (Taylor, 1969). Thus, the requirement for darkness before fumigation to afford protection is relatively short, but the effects of light and dark after the fumigation require quite lengthy exposures. These results imply more than a simple switching effect of light, but rather a requirement in addition for maintenance of susceptible compounds in a particular chemical form, so that they are capable of reacting with PAN.

Dugger *et al.* (1963) determined the action spectrum for the light interaction with PAN damage. Maximum quantum responsivity was observed in the 420–480 nm range. There was no evidence for the participation of

FIG. 2. Pinto bean plants fumigated 30 minutes with 1 ppm PAN. DL, 24 hour dark before fumigation; L, 6 hour light before fumigation; LD, 24 hour dark after fumigation; (Photograph obtained from Dr. O. C. Taylor.)

chlorophylls or phytochrome in the sensitization phenomenon. Taken at face value these results should direct our attention to the participation of carotenoids. Reasoning along these lines is hampered, however, by the lack of conclusive evidence of the role of carotenoids in the absence of PAN. Perhaps it is significant that carotenoids are very efficient quenchers of singlet oxygen (Kearns, 1971), for singlet oxygen has been detected in the reaction of PAN with base (Steer *et al.*, 1969). In retrospect, the experiments on the action spectrum may have been oversimplified. The underlying assumption was that there was a single photoreceptor and this need not be the case. Clearly, the testing of a hypothesis in which two light reactions participate (and not necessarily simultaneously) greatly complicates the experimental approach.

C. Uptake

Hill (1971) presented data on the uptake of pollutants by canopies of vegetation (see also chapter 12). HF and SO_2 were readily taken up by the vegetation while CO was not. The data showed that uptake was closely related to water solubility. Ozone did not fit this generalization, presumably because it reacted quickly and so the concentration in solution was always effectively zero. PAN was readily taken up in Hill's study, but the uptake could not be correlated with water solubility because the latter figure is not accurately known.

It is conceivable that differences in toxicity of PAN, PPN, and PBN may be a function of uptake. The alkyl groups may confer different solubility properties particularly in hydrophobic environments. There is plenty of evidence for the importance of this effect in the study of pharmacological compounds (Gould, 1972). We have compared the uptake of PAN and PPN by pinto bean and tomato leaves and found no significant difference (Fig. 3). Thus, the difference in toxicity between PAN and PPN cannot

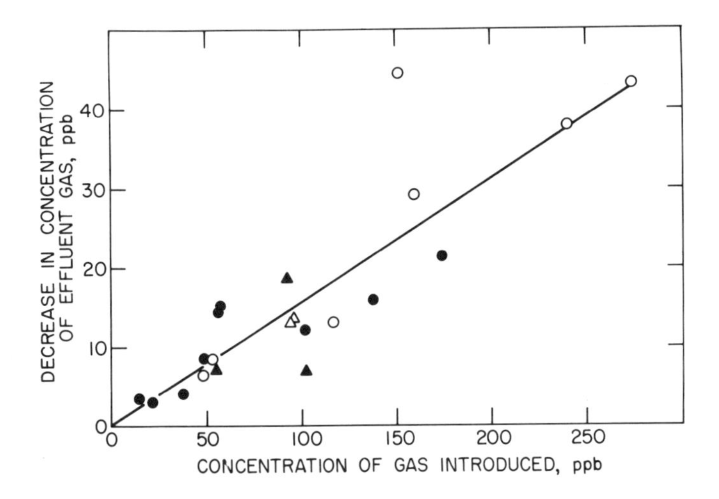

FIG. 3. Uptake of PAN and PPN by leaves. Plants were enclosed in a glass exposure chamber with the stems inserted through the bottom plate and sealed. Nitrogen containing PAN or PPN was introduced into the chamber at a rate of at least 1 liter/minute. The concentration of the inflow was measured and samples of effluent also were analyzed. The concentration in the effluent reached an equilibrium value within 15 minutes. Exposures were for a minimum of 60 minutes. In the absence of plants, the chamber did not lower the gas concentration in the effluent. Carbon dioxide uptake was measured during the experiment: there was some indication of inhibition of CO_2 fixation at the higher PPN concentrations. All of the PPN samples and most of the PAN samples eventually showed visible symptoms. At the end of the experiment the leaf dry weight was determined. Weights were approximately 100 mg. The decrease in concentration of the effluent gas was normalized to 100 mg. dry weight. Peroxyacyl nitrate uptake was proportional to leaf dry weight. Symbols: △, PAN pinto bean; ○, PAN tomato; ▲, PPN pinto bean; ●, PPN tomato. (Data obtained in collaboration with O. C. Taylor and E. A. Cardiff.)

be ascribed to total uptake on the basis of currently available data, and we must seek some other explanation.

In Hill's experiments (1971), exposures were made in the light to ensure that stomata were open. Ozone and chlorine at higher concentrations caused stomatal closure and decreased uptake. The requirement of open stomata for pollutant uptake has been questioned (Dugger *et al.* 1964), but it seems that access to the inner leaf air spaces plays an important role in pollutant uptake and damage. It should be noted that stomatal opening/closure cannot account for the light requirement mentioned in Section IV,B. Protection can be offered by darkness after the exposure period, thus the stomates could close only after the PAN uptake.

V. Biochemical Effects

A. Intact Plants

1. Carbohydrate Metabolism

Ordin (1962) observed that growth of *Avena* coleoptile sections is inhibited by PAN. Since it was well known that growth of coleoptiles is dependent on cell expansion which in turn is dependent on plasticization of the cell wall, a series of investigations was started on the effects of PAN on the metabolism of the cell wall sugars. Incorporation of [^{14}C]glucose into all subcellular fractions was inhibited by PAN, but the cellulose fraction appeared to be particularly susceptible.

Subsequent work has concentrated on the enzymic steps preceding the formation of cellulose.

Cellulase synthetase (enzyme 3 in Fig. 4) was inhibited in coleoptile sections previously exposed to PAN at the rather high concentrations of 35–50 ppm for 4 hours (Ordin and Hall, 1967). In subsequent work of a similar type, Ordin *et al.* (1967) reported that the enzyme phosphoglucomutase (enzyme 1, Fig. 4) obtained from coleoptile sections after expo-

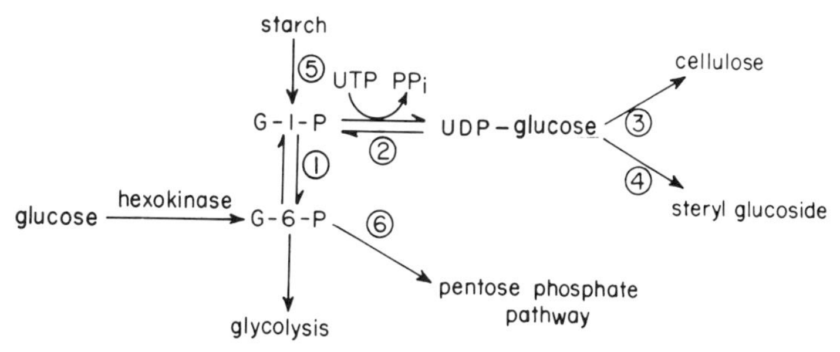

Fig. 4. Effects of PAN on carbohydrate metabolism. See text for detailed explanation.

sure to PAN was inhibited 29–55% depending on the subcellular fraction analyzed. However, direct treatment of a phosphoglucomutase enzyme preparation did not cause marked inhibition. The pyrophosphorylase (enzyme 2, Fig. 4) showed behavior opposite to that of phosphoglucomutase: activity was unaffected by treatment of the coleoptile sections, whereas the isolated enzyme was very susceptible to PAN.

The experiments described above used oat coleoptiles as the biological material. Ordin *et al.* (1971) have applied similar methods to the study of effects of PAN on tobacco leaves and some slightly different results were obtained. The plants were exposed for 1 hour to PAN 0.85–0.97 ppm. Assays were made either immediately after exposure or after 24 hours delay. Measurements were made on the extracts of synthesis from [^{14}C]-UDP-glucose of lipid (enzyme 4, Fig. 4), cellulose synthesis, and synthesis of material soluble in hot 1 *N* NaOH (glucan). The activities of phosphoglucomutase and UDP-glucose pyrophosphorylase were also measured. Immediately after exposure, the synthesis of cellulose and the alkali-soluble glucan were inhibited, but there was no significant effect on lipid synthesis, phosphoglucomutase, or pyrophosphorylase. After 24 hours, all syntheses from UDP-glucose were inhibited, but phosphoglucomutase was unaffected and there was more than a doubling of the activity of the pyrophosphorylase. The latter activity was found as particle bound and soluble forms, and the increase in the latter was responsible for the overall increase in activity. It appears likely that release of the pyrophosphorylase is a consequence of membrane disruption.

Hanson and Stewart (1970) observed that various pollutants, including PAN inhibit the mobilization of starch in darkness. This implies inhibition of the phosphorylase reaction (enzyme 5, Fig. 4). Ordin *et al.* (1967) also noted inhibition of polysaccharide hydrolysis by PAN. The inhibition of starch hydrolysis is an interesting observation that would be worthy of further investigation.

2. Interaction with Indoleacetic Acid (IAA)

Ordin and Propst (1962) reported that both PAN and ozone reacted with indoleacetic acid. Absorption spectra of the products in the two cases showed that they were quite different. In PAN the 280 nm absorbance maximum was eliminated and a new maximum arose at 250 nm. The spectrum of the ozonized product had a maximum at 320 nm probably related to the compound obtained by ozonolysis of the 5-membered ring. Oxidation of IAA by PAN eliminated the hormonal properties. Hall *et al.* (1971) examined the reaction further and found at least five products.

Three of these products were capable of inhibiting coleoptile growth. Only one of the products was identified: 3-hydroxymethyloxindole. In the study of this compound, Hope and Ordin (1971a) have described a new method for its synthesis, photooxidation in the presence of riboflavin. Since the dye-sensitized photooxidations are known to proceed by way of singlet oxygen, Hope's observation gives some circumstantial evidence that the oxidation of IAA by PAN also goes via singlet oxygen.

Since the *in vitro* products of IAA oxidation were inhibitory to coleoptile growth, the possibility existed that exposure of plant to PAN oxidized the endogenous IAA and caused growth inhibition and other symptoms. Hope and Ordin (1971b) attempted to assess this possibility. PAN apparently inhibited the conversion of [^{14}C]IAA to other products, presumed to be conjugates (e.g., with glucose). This interference with IAA metabolism took place even when there was no visible damage.

3. Photosynthesis

Gross and Dugger (1969) examined the effects of PAN on *Chlamydomonas reinhardtii.* Photosynthesis as measured by O_2 evolution was inhibited, but recovery could take place after relatively low doses. During exposure to PAN the cells were irradiated with different quantities of light; subsequent recovery of photosynthesis was measured. Recovery was best when white light was used and worst when 450 nm light had been used. This was in the region of the quality of maximum quantum responsivity in the experiments of Dugger *et al.* (1963). Exposure of the algal cells to PAN also caused a lowering in the free sulfhydryl content of the cells and differential destruction of pigments. Carotenoids were particularly susceptible to destruction and chlorophyll a was more susceptible than chlorophyll b. Dugger and Ting (1968) have related the SH content of leaves to susceptibility to PAN. They have found that illumination increases the amount of SH in bean leaves and that exposure to PAN lowers the amount of SH. The results are consistent with the theory that PAN is toxic because of a reaction with SH groups and the plant is made more susceptible when more SH groups are available for reaction. Dugger and Ting (1968) have related the degree of damage to the formation of SH during the reductive processes of photosynthesis and have attempted to resolve the relative contributions of photosystem I (activated by 700 nm light) and photosystem II (activated by 660 nm light). Damage was greatest when plants were irradiated with 660 nm light, and this effect could be negated by simultaneous exposure to 700 nm light. It was also observed that red light irradiation resulted in higher SH content in bean leaves than for red irradiation.

Even though these results were interpreted in terms of photosynthetic reactions, it should be noted that the wavelengths of light used were very appropriate for phytochrome transformations.

$$\mathrm{Pr} \underset{730\ \mathrm{nm}}{\overset{660\ \mathrm{nm}}{\rightleftharpoons}} \mathrm{Pfr}$$

Pr has an absorbance maximum at 660 nm and Pfr has an absorbance maximum at 730 nm. At 700 nm Pfr has more absorbance than Pr (Mohr, 1972). It may be significant that Pfr reacts with SH reagents whereas Pr does not (Butler *et al.,* 1964). Thus, there is the possibility that PAN can react with Pfr as well as interfere with the processes modulated by Pfr.

Our difficulty in understanding the observations with 660 nm and 700 nm irradiation (Dugger and Ting, 1968) is that neither wavelength appeared important in the action spectrum studies (Dugger *et al.,* 1963). It may be that more than one photoreceptor is involved: (1) a switching mechanism mediated by phytochrome and (2) photooxidative damage with riboflavin or carotenoid as the photoreceptor.

B. Cell-Free Preparations

1. Photosynthesis

Dugger *et al.* (1965) observed that chloroplasts isolated from plants exposed to 0.6 ppm PAN for 30 minutes were capable of normal rates of photophosphorylation but were inhibited as far as O_2 evolution was concerned. Further studies with isolated chloroplasts and rather high concentrations of PAN showed that photophosphorylation could be inhibited, especially if the chloroplasts were illuminated during the fumigation. A similar result was obtained when nicotinamide adenine dinucleotide phosphate (NADP) reduction by chloroplasts was measured: inhibition was most severe when the chloroplasts were illuminated during exposure to PAN. As might be expected from the results on adenosine triphosphate (ATP) and NADPH formations, PAN also inhibited CO_2 fixation, but the effect on CO_2 fixation could also be attributed to diverse effects on the enzymes concerned.

It is not clear that the effect of light in the *in vitro* experiments described above is a true representation of the light effect on PAN damage in intact plants. But the results for phosphorylation can be satisfactorily explained by the observations of McCarty *et al.* (1972) who found that photophosphorylation was inhibited by *N*-ethyl maleimide (NEM) only when chloroplasts were treated in the light. Thus the properties of NEM and PAN are closely comparable: indeed, they are both —SH reagents.

There is further evidence that illumination has its effect on enzymic processes by way of changing the —SH content of the cells. Several enzymes of both the Calvin and the Hatch–Slack pathways of CO_2 fixation are light activated and this activation can be simulated by SH compounds such as dithiothreitol (DTT) (Latzko *et al.*, 1970). Conversely, glucose-6-phosphate (G-6-P) dehydrogenase can be inactivated by light and also by DTT (Johnson, 1972; Anderson *et al.*, 1974). The reactions can be schematically represented in a somewhat conjectural fashion (Fig. 5) and the possible involvement of PAN shows how the regulatory action of light could be sabotaged. The enzymes concerned are located in the stroma of the chloroplast and it is in the stroma of the chloroplast that the first signs of damage detectable with the electron microscope can be seen (Thomson *et al.*, 1965).

2. Enzymes

The response of several enzymes to PAN has been studied. G-6-P dehydrogenase was readily inactivated by PAN, but binding of NADP to the protein afforded protection. The cosubstrate, G-6-P did not prevent the inhibition by PAN. The phenomena of protection and susceptibility of G-6-P dehydrogenase were the same for PAN and reagents such as *p*-chloromercuribenzoate (PMB). This gave rise to the idea that toxicity of PAN depended on its reaction with SH groups (Mudd, 1963).

Enzymes which have no free —SH group should be resistant to PAN and indeed this is the case for pancreatic RNase (Mudd *et al.*, 1966) (reaction 1, Fig. 5). If pancreatic RNase is reduced with SH compounds (re-

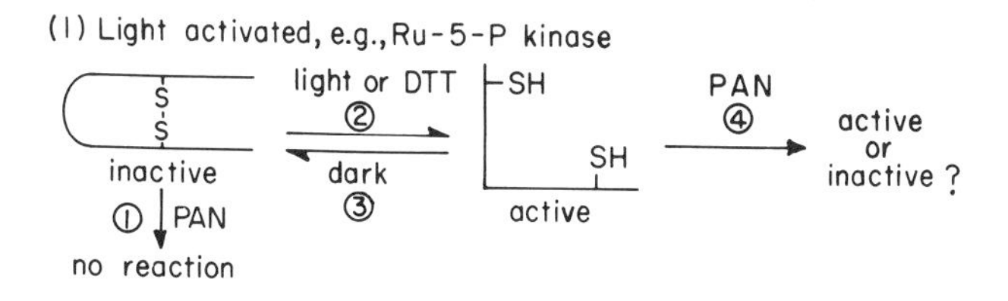

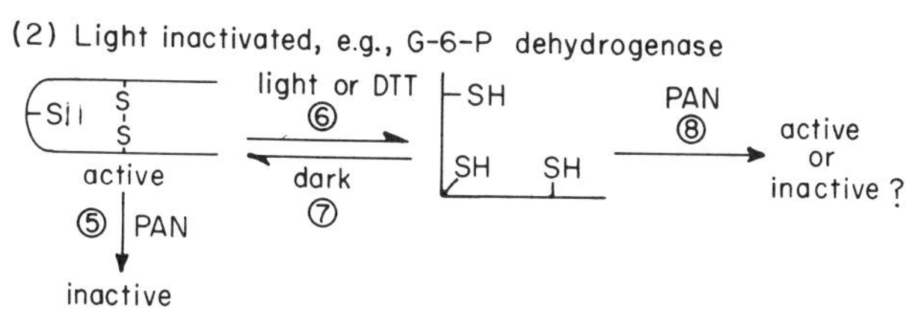

Fig. 5. Effects of light, DTT, and PAN on enzymes of the chloroplast.

action 2, Fig. 5), recovery of the enzyme activity by slow air oxidation (reaction 3, Fig. 5) is prevented by exposure to PAN (reaction 4, Fig. 5). Presumably this is because of reformation of disulfide bonds in unfavorable positions or because of acetylation of SH groups. In the case of papain, evidence has been obtained that dimers are formed (Leh and Mudd, 1974). Thus, the reaction is oxidation rather than acetylation in this case.

The accessibility of protein SH groups to PAN is a consideration which may have biological relevance. Ovalbumin has SH groups which readily react with PMB, but not with PAN (Mudd *et al.*, 1966). Human hemoglobin has 2 SH groups which react with PMB at pH 7, but 6 which react at pH 4.5 (dissociation of tetramer). The reactivity of the SH groups of hemoglobin shows the same pH dependence when PAN is the reagent (Mudd *et al.*, 1966).

One should be cautious in correlating susceptibility to PAN with SH sensitivity. Phosphoglucomutase of oat seedlings is sensitive both *in vivo* and *in vitro* to PAN, and it is also sensitive to 10^{-4} *M* PMB (Hall and Ordin, 1967), but the important amino acid at the active site is serine. There is no evidence that PAN reacts with serine and so one must speculate that the active site is modified in some other fashion. The observations on UDP-glucose pyrophosphorylase are also anomalous: no effect of stimulation *in vivo,* very susceptible *in vitro,* and not affected by 10^{-4} PMB (Hall and Ordin, 1967). Although cysteine and methionine are the only amino acid residues so far known to be affected by PAN, environments of other functional groups in proteins may encourage reaction with PAN. The experiments with pure preparations of enzymes seem to be clear-cut. Perhaps the difficulty with the *in vivo* experiments in the clouding of direct effects on the enzymes by stimulation of protein synthesis, or disruptions of subcellular organizations making enzymes more (or less) accessible to PAN.

3. Fatty Acid Synthesis

Mudd and Dugger (1963) showed that the synthesis of fatty acid from [^{14}C]acetate was inhibited by PAN. The experiment was designed to show whether the oxidation of NADPH by PAN would selectively inhibit fatty acid synthesis. The result, however, is open to a number of interpretations: (1) NADPH was generated by G-6-P dehydrogenase and the inhibition could have been caused by inhibition of this enzyme, (2) the several proteins of the fatty acid-synthesizing system could have been inactivated by PAN. There is no information to indicate whether fatty acid synthesis is inhibited *in vivo.*

C. Specific Biochemicals

1. Nicotinamide Derivatives

Mudd and Dugger (1963) reported that reduced nicotinamide derivatives NADH and NADPH are stoichiometrically oxidized by peroxyacyl nitrates. NAD and NADP were the products and these compounds were unaffected by PAN. The oxidation was more rapid at higher pH's and at higher ionic strengths, although at the higher pH's the total amount of oxidation was less. It became obvious during these experiments that the oxidizing power of PAN dissipated in the absence of an oxidizable substance.

2. Sulfur-Containing Compounds

In addition to the protein thiols already mentioned, PAN reacts with a number of low molecular weight thiols. Results with reduced glutathione showed that the products included both disulfide and *S*-acetylglutathione (Mudd, 1966). Although there have been several reports of oxidation of different mercaptans, there has been no other report of S-acylation by PAN. Nicksic *et al.* (1967) have found, however, that the breakdown of PAN in methanolic solutions gives rise to methyl acetate.

Coenzyme A was one of the thiol compounds tested. The products were separated by ion exchange chromatography and detected by absorbance at 260 nm (the adenine moiety) (Mudd and McManus, 1969). The major product was the coenzyme A disulfide, but there were several minor products which have not been identified. A comparable distribution of products was obtained when coenzyme A was oxidized with hydrogen peroxide. Leh and Mudd (1974) found that PAN readily oxidizes reduced lipoic acid to the cyclic disulfide. The cyclic disulfide is oxidized much more readily than the linear disulfide cystine, and the product from PAN oxidation of lipoic acid disulfide is the monosulfoxide. Methionine is also oxidized by PAN with the production of methionine sulfoxide.

3. Purines and Pyrimidines

Peak and Belser (1969) reported that PAN reacts with bacterial DNA which changes the melting temperature and the viscosity. PAN also lowered the infectivity of bacteriophage. These results were correlated with the susceptibility of purines and pyrimidines. The order of sensitivity was thymine, guanine, uracil, cytosine, and adenine. It should be noted that high concentrations of PAN were used in these experiments (about 1000 ppm) and that no reactions were detectable above pH 5. Thus the reactions are probably of more chemical interest than biological importance.

4. Olefins and Amines

Pitts and co-workers (Darnall and Pitts, 1970; Wendschuh *et al.*, 1973) have dissolved PAN in $CDCl_3$ and reacted it with olefins and amines. Reaction with olefins produced epoxides:

$$\text{PAN} + R_2C{=}CR_2 \longrightarrow R_2C\overset{O}{\frown}CR_2 + \text{MeONO} + \text{MeNO}_2$$

Half-lives of PAN in a series of olefins were measured and they correlated with the reactivity of the same olefins in the reaction with peracids to form epoxides. It would be of interest to find whether such reactions are applicable in biological systems. When PAN reacted with either *cis*- or *trans*-but-2-ene, the epoxide produced was 75% *trans*-butene oxide and excess olefin was isomerized to trans. Such reactions could have important consequences in carotenoid containing systems.

The reaction with amines produced amides:

$$\text{PAN} + RNH_2 \rightarrow CH_3CO{\cdot}NHR + O_2 + HNO_2$$

Since singlet oxygen had been detected in the base catalyzed degradation of PAN (Steer *et al.*, 1969), it was searched for in the reaction with amines, but was not found. Thus the two properties of PAN are shown in the reactions with olefins and amines: oxidation and acylation respectively.

VI. Chemical Basis for Toxicity

A. Eye Irritation

Although the concern of this chapter is with phytotoxicity of peroxyacyl nitrates, discussion of the mechanism of action can include work with animals because the chemical and biochemical basis for toxicity may be the same in the two kingdoms. Eye irritation of humans is the most commonly observed effect of Los Angeles air pollution and the eye irritation has been attributed to aldehydes and peroxycayl nitrates. Eye irritation has received attention from researchers concerned with "riot control agents" and tear gases. Thus, it is possible to compare the chemical properties of these compounds with the peroxyacyl nitrates.

Dixon (1948) and Mackworth (1947) have pointed out the lacrimators are all excellent sulfhydryl reagents and as a consequence are inhibitory to enzymes requiring SH groups for their reactivity. Dixon mentions hexokinase can readily be oxidized in air to an inactive form and in this form it is resistant to ethyl iodoacetate. An analogous situation exists for the reaction of PAN with papain (Mudd *et al.,* 1966). Dixon also cites the case of chloroacetophenone reaction with ovalbumin: the native protein is not alkylated and this is also true for PAN (Mudd *et al.,* 1966). Therefore, the eye irritating properties of PAN are possibly caused by the same type of reactions as those of "riot control agents." One notable difference is that the lacrimators studied by Dixon (1948) did not form disulfide bonds, whereas this is the most common result of exposure to PAN. However, most of the lacrimators studied are SN_2 alkylating agents (Cucinell *et al.,* 1971) rather than oxidizing agents, so a fair comparison cannot be made.

An unfortunate gap in the understanding of lacrimatory compounds in general and peroxyacyl nitrates in particular is the lack of understanding of lacrimation itself. Presumably some important SH groups in the nerve endings are affected, but nothing further is known about potential receptor proteins.

B. Plant Damage

The hypothesis that eye irritation by PAN is caused by reaction with SH groups seems to be satisfactory in the case of lacrimation. Does the same hypothesis apply for plant damage? Evidence presented above indicates that this is the case even though reaction of PAN with other compounds can be demonstrated *in vitro.*

Some observations in plant damage (and eye irritation) are not satisfactorily explained. The increasing toxicity of PAN, PPN, PBN, and PBzN is one of these. In the case of plants, there appear to be three possible explanations: (1) the more toxic homologs are more readily taken up by the plant, having been taken up they may be equally toxic; (2) the homologs are taken up equally well but they have different degrees of persistence, i.e., the half-life of the higher homologs is greater; (3) all homologs are taken up equally well and react equally well with susceptible sites, the difference in toxicity of the homologs is in the ability of the organism to reverse the initial reaction.

1. Uptake

The test with PAN and PPN showed that uptake by vegetation was the same for both homologs. Differences in toxicity, therefore, cannot be atrributed to differences in uptake (Fig. 3).

2. Half-Life in Aqueous Media

Results of experiments presented in Table II and (Fig. 1) show that there is a significant difference in the half-lives of PAN, PPN, and PBN at various pH's. If the toxicity is predicted to depend on the persistence of the compound, we would expect acid pH's to favor damage. This prediction has not been tested: one does not know that the pH of leaf sap is a true indication of the pH at the site of PAN action and the major changes of acidity in the vacuole may not be relevant to PAN action. The data in Table II also show that PBN has the longest half-life at all pH's thus its durability is consistent with its toxicity. Unfortunately, the stabilities of PAN and PPN are not in the order predicted. The lower toxicity would have caused us to expect PAN's half-life to be shortest. Nevertheless, this seems the most likely explanation of the different toxicities, and the experiments on half-life should be tested under different conditions.

3. Reversal of Toxic Reactions

One of the possible toxic reactions is the acylation of SH groups as demonstrated in the case of glutathione and also possible in the case of proteins. The hypothesis in this case would predict that the *S*-acetyl compound would be hydrolyzed most readily, and the *S*-butyryl derivative poorly. These predictions have been tested in spinach leaves by determining the Michaelis constants and maximum velocities for the enzymic hydrolysis of *S*-acetyl, *S*-propionyl, and *S*-butyryl glutathione (Winberry and Mudd, 1974). The results were opposite to those predicted by the hypothesis, that is, the butyryl derivative was most readily hydrolyzed, and so one cannot explain the differences in toxicity of PAN, PPN, and PBN by reversal of the acylation reaction. Reversal of the disulfide formation would not be relevant, since the various peroxyacyl nitrates would produce the same disulfides.

References

Anderson, L. E., Ng, T.-C. L., and Park, K.-E. Y. (1974). Inactivation of pea leaf chloroplastic and cytoplasmic glucose-6-phosphate dehydrogenase by light and dithiothreitol. *Plant Physiol.* **53**, 835–839.

Butler, W. L., Siegelman, H. W., and Miller, C. O. (1964). Denaturation of phytochrome. *Biochemistry* **3**, 851–857.

Cucinell, S. A., Swentzel, K. C., Biskup, R., Snodgrass, H., Lorre, S., Stark, W., Feinsilver, L., and Vocci, F. (1971). Biochemical interactions and metabolic fate of riot control agents. *Fed. Proc., Fed. Amer. Soc. Exp. Biol.* **30**, 86–91.

Darley, E. F., Kettner, K. A., and Stephens, E. R. (1963). Analysis of peroxyacyl nitrates by gas chromatography with electron capture detection. *Anal. Chem.* **35**, 589–591.

Darnall, K. R., and Pitts, J. N., Jr. (1970). Peroxyacetyl nitrate. A novel reagent for oxidation of organic compounds. *Chem. Commun.* pp. 1305–1306.

Dixon, M. (1948). Reactions of lachrymators with enzymes and proteins. *Biochem. Soc. Symp.* **2,** 39–49.

Dugger, W. M., and Ting, I. P. (1968). The effect of peroxyacetyl nitrate on plants: Photoreductive reactions and susceptibility of bean plants to PAN. *Phytopathology* **58,** 1102–1107.

Dugger, W. M., Taylor, O. C., Cardiff, E. A., and Thompson, C. R. (1962). Stomatal action in plants as related to damage from photochemical oxidants. *Plant. Physiol.* **37,** 487–491.

Dugger, W. M., Taylor, O. C., Klein, W. H., and Shropshire, W. (1963). Action spectrum of peroxyacetyl nitrate damage to bean plants. *Nature (London)* **198,** 75–76.

Dugger, W. M., Mudd, J. B., and Koukol, J. (1965). Effect of PAN on certain photosynthetic reactions. *Arch. Environ. Health* **10,** 195–200.

Ellman, G. L. (1959). Tissue sulfhydryl groups. *Arch. Biochem. Biophys.* **82,** 70–77.

Gould, R. F., ed. (1972). "Biological Correlations—The Hansch Approach," Advan. Chem. Ser. No. 114. Amer. Chem. Soc., Washington, D.C.

Gross, R. E., and Dugger, W. M. (1969). Responses of *Chlamydomonas reinhardtii* to Peroxyacetyl nitrate. *Environ. Res.* **2,** 256–266.

Hall, M. A., and Ordin, L. (1967). Subcellular location of phosphoglucomutase and UDP-glucose pyrophosphorylase in *Avena* Coleoptiles. *Physiol. Plant.* **20,** 624–633.

Hall, M. A., Brown, R. L., and Ordin, L. (1971). Inhibitory products of the action of peroxyacetyl nitrate upon indole-3-acetic acid. *Phytochemistry* **10,** 1233–1238.

Hanson, G. P., and Stewart, W. S. (1970). Photochemical oxidants: Effect on starch hydrolysis in leaves. *Science* **168,** 1223–1224.

Heuss, J. M., and Glasson, W. A. (1968). Hydrocarbon reactivity and eye irritation. *Environ. Sci. Technol.* **2,** 1109–1116.

Hill, A. C. (1971). Vegetation: A sink for atmospheric pollutants. *J. Air. Pollut. Contr. Ass.* **21,** 341–356.

Hope, J. J., and Ordin, L. (1971a). An Improved Procedure for the synthesis of oxindole-3-carbinol (Hydroxymethyl Oxindole). *Phytochemistry* **10,** 1551–1553.

Hope, H. J., and Ordin, L. (1971b). Metabolism of 3-indoleacetic acid in tobacco exposed to the air pollutant peroyacetyl nitrate. *Plant Cell Physiol.* **12,** 849–857.

Jacobson, J. S., and Hill, A. C., eds. (1970). "Recognition of Air Pollution Injury to Vegetation: A Pictorial Atlas." Air Pollut. Contr. Ass., Pittsburgh, Pennsylvania.

Johnson, H. S. (1972). Dithiothreitol: An inhibitor of glucose-6-phosphate-dehydrogenase activity in leaf extracts and isolated chloroplasts. *Planta* **106,** 273–277.

Kearns, D. R. (1971). Physical and chemical properties of singlet molecular oxygen. *Chem. Rev.* **71,** 395–427.

Kerr, J. A., Calvert, J. G., and Demerjian, K. L. (1972). The mechanism of photochemical smog formation. *Chem. Brit.* **8,** 252–257.

Latzko, E., von Ganier, R., and Gibbs, M. (1970). Effect of photosynthesis, photosynthetic inhibitors and oxygen on the activity of ribulose-5-phosphate kinase. *Biochem. Biophys. Res. Commun.* **39,** 1140–1144.

Leh, F., and Mudd, J. B. (1974). Reaction of peroxyacetyl nitrate with cysteine, cystine, methionine, lipoic acid, papain, and lysozyme. *Arch. Biochem. Biophys.* **161,** 216–221.

Leighton, P. A. (1961). "Photochemistry of Air Pollution." Academic Press, New York.

McCarty, R. E., Pittman, P. R., and Tsuchiya, Y. (1972). Light-dependent inhibition of photophosphorylation by *N*-ethyl maleimide. *J. Biol. Chem.* **247,** 3048–3051.

Mackworth, J. F. (1947). The inhibition of thiol enzymes by lachrymators. *Biochem. J.* **43,** 82–90.

Middleton, J. T., Crafts, A. S., Brewer, R. F., and Taylor, O. C. (1956). Plant damage by air pollution. *Calif. Agr.* June 1956, 9–12.

Mohr, H. (1972). "Lectures on Photomorphogenesis." Springer-Verlag, New York and Berlin.

Mudd, J. B. (1963). Enzyme inactivation by peroxyacetyl nitrate. *Arch. Biochem. Biophys.* **102,** 59–65.

Mudd, J. B. (1966). Reaction of peroxyacetyl nitrate with glutathione. *J. Biol. Chem.* **241,** 4077–4080.

Mudd, J. B., and Dugger, W. M. (1963). The oxidation of reduced pyridine nucleotides by peroxyacetyl nitrates. *Arch. Biochem. Biophys.* **102,** 52–58.

Mudd, J. B., and McManus, T. T. (1969). Products of the reaction of peroxyacetyl nitrate with sulfhydryl compounds. *Arch. Biochem. Biophys.* **132,** 237–241.

Mudd, J. B., Leavitt, R., and Kersey, W. H. (1966). Reaction of peroxyacetyl nitrate with sulfhydryl groups of proteins. *J. Biol. Chem.* **241,** 4081–4085.

Nicksic, S. W., Harkins, J., and Mueller, P. K. (1967). Some analyses for PAN and studies of its structure. *Atmos. Environ.* **1,** 11–18.

Ordin, L. (1962). Effect of peroxyacetyl nitrate on growth and cell wall metabolism of *Avena* Coleoptile sections. *Plant Physiol.* **37,** 603–608.

Ordin, L., and Hall, M. A. (1967). Studies on cellulose synthesis of a cell-free oat coleoptile enzyme system: Inactivation by air pollutant oxidants. *Plant Physiol.* **42,** 205–212.

Ordin, L., and Propst, B. (1962). Effect of air-borne oxidants on biological activity of indoleacetic acid. *Bot. Gaz.* (*Chicago*) **123,** 170–175.

Ordin, L., Hall, M. A., and Katz, M. (1967). Peroxyacetyl nitrate-induced inhibition of cell wall metabolism. *J. Air Pollut. Contr. Ass.* **17,** 811–815.

Ordin, L., Garber, M. J., Kindinger, J. I., Whitmore, S. A., Greve, C., and Taylor, O. C. (1971). Effect of peroxyacetyl nitrate (PAN) in vivo on tobacco leaf polysaccharide synthetic pathway enzymes. *Environ. Sci. Technol.* **5,** 621–626.

Peak, M. J., and Belser, W. L. (1969). Some effects of the air pollutant peroxyacetyl nitrate, upon deoxyribonucleic acid and upon nucleic acid bases. *Atmospheric Environment* (1969) **43,** 385–397.

Steer, R. P., Darnall, K. R., and Pitts, J. N., Jr. (1969). The base-induced decomposition of peroxyacetyl nitrate. *Tetrahedron Lett.* **43,** 3765–3767.

Stephens, E. R. (1964). Absorptivities for infrared determination of peroxyacyl nitrates. *Anal. Chem.* **36,** 928–929.

Stephens, E. R. (1967). The formation of molecular oxygen by alkaline hydrolysis of peroxyacetyl nitrate. *Atmos. Environ.* **1,** 19–20.

Stephens, E. R. (1969). The formation, reactions and properties of peroxyacyl nitrates (PANS) in photochemical air pollution. *Advan. Environ. Sci. Technol.* **1,** 119–146.

Stephens, E. R., Darley, E. F., Taylor, O. C., and Scott, W. E. (1961). Photochemical reaction products in air pollution. *Int. J. Air Water Pollut.* **4,** 79–100.

Stephens, E. R., Burleson, F. R., and Holtzclaw, K. M. (1969). A damaging explosion of peroxyacetyl nitrate. *J. Air Pollut. Contr. Ass.* **19,** 261–264.

Taylor, O. C. (1969). Importance of peroxyacetyl nitrate (PAN) as a phytotoxic air pollutant. *J. Air Pollut. Contr. Ass.* **19,** 347–351.

Taylor, O. C., Stephens, E. R., Darley, E. F., and Cardiff, E. A. (1960). Effect of air-borne oxidants on leaves of pinto bean and petunia. *Proc. Amer. Soc. Hort. Sci.* **75,** 435–444.

Taylor, O. C., Dugger, W. M., Cardiff, E. A., and Darley, E. F. (1961). Interaction of light and atmospheric photochemical products (SMOG) within plants. *Nature (London)* **192,** 814–816.

Thomson, W. W., Dugger, W. M., and Palmer, R. L. (1965). Effects of peroxyacetyl nitrate on ultrastructure of chloroplasts. *Bot. Gaz. (Chicago)* **126,** 66–72.

Wendschuh, P. M., Fuhr, H., Gaffney, J. S., and Pitts, J. N., Jr. (1973). Reaction of peroxyacetyl nitrate with amines. *Chem. Commun.* pp. 74–75.

Winberry, L. K., and Mudd, J. B. (1974). *S*-Acyl glutathione thioesterase of plant tissue. *Plant Physiol.* **53,** 216–219.

6

OXIDES OF NITROGEN

O. C. Taylor, C. R. Thompson, D. T. Tingey, and R. A. Reinert

Of the several oxides of nitrogen which may be found in the atmosphere, the most important as air pollutants are nitric oxide (NO) and nitrogen dioxide (NO_2). According to Robinson and Robbins (1970) the major component of worldwide atmospheric nitrogen oxides is biologically produced NO. These, primarily bacteria produced oxides, amount to about 50×10^7 tons per year, and by comparison man-made sources emit about 5×10^7 tons per year. The biological sources are widely distributed over the face of the earth, while man-made sources are concentrated in a relatively few densely populated areas. Consequently, natural sources are usually not considered seriously when an inventory of sources of oxides of nitrogen is being compiled.

I. Formation of Nitrogen Oxides

During combustion some of the nitrogen in the air is oxidized to NO and a comparatively small amount of NO_2. The rate of NO formation increases in proportion to the temperature of combustion. Rate of NO decomposition decreases rapidly as temperature of the gaseous by-products of combustion falls. Both NO and NO_2 are formed at combustion temperatures above 2000°F, but NO_2 accounts for less than 1% of the total nitrogen oxides in these gases. Within a few seconds after the combustion gases are ejected into the atmosphere they are diluted, and part of the NO, perhaps as much as 10%, is oxidized to NO_2. Once the NO has been diluted to about one part per million (ppm) parts of air it no longer reacts readily with oxygen to produce NO_2.

During daylight, atmospheric NO may be quantitatively converted to NO_2 by photochemical reactions involving the absorption of sunlight and interaction with hydrocarbons and oxygen. A detailed description of the nitrogen dioxide photolytic cycle and photochemical reactions is not included here but may be obtained from such sources as the Air Quality Criteria for Photochemical Oxidants, Air Pollution Control Office Publication No. AP-63, U.S. Government Printing Office.

Natural scavenging processes prevent excessive buildup of the bacteria produced nitrogen oxides which are spread over broad nonurban areas. These scavenging processes also effectively decompose nitrogen oxides emitted by man's activities, but under most conditions the emissions exceed the capabilities of the scavenging processes on frequent occasions.

One of the important scavenging processes is the photochemical reaction through which NO is oxidized to NO_2, and NO_2 is subsequently consumed in the production of ozone, peroxyacyl nitrates (PAN's), and other components of oxidant smog. Characteristically, the atmosphere near urbanized and industrialized complexes contains considerably more NO and NO_2 during early morning hours than at any other time of day. Diurnal variations of NO and NO_2 have been illustrated graphically as concentration profiles. Typically, in such areas as the South Coast Air Basin of California, the NO concentration will reach a peak coinciding with or shortly after the hours of peak traffic. During weekday mornings NO concentrations typically increase rapidly to a peak about 7:00 or 8:00 AM. After sunrise, NO is rapidly oxidized to NO_2 and by about 9:00 or 10:00 AM NO_2 usually reaches a peak concentration. As the NO concentration approaches zero, ozone begins to accumulate rapidly and NO_2 concentration declines to a low level for the remainder of the day. The monitoring data summarized in the Air Quality Criteria for Nitrogen Oxides (Anonymous, 1971) indicates that NO and NO_2 levels near major metropolitan areas are higher during the winter months. The seasonal variation for NO_2 is less distinct than for NO, but there is a definite trend toward higher concentrations in winter than in summer months.

Robinson and Robbins (1970) concluded that from an air pollutant standpoint the natural scavenging processes are probably not rapid enough to materially affect hour-to-hour concentrations of nitrogen oxides in urban areas. But there appears to be little reason to expect a long-term buildup of phytotoxic nitrogen compounds in the global environment. Adverse, direct effects of nitrogen oxides on plant life are therefore limited to areas in close proximity to urban and industrial developments where the emissions are concentrated.

Attempts to define "acceptable" limits of NO_2 in the atmosphere to protect all vegetation are beset by the same problems found with other phytotoxic agents, i.e., a wide range of responses related to stage of growth and conditions of light, temperature, humidity and/or water stress, and fertilization at the time of exposure. Acceptable limits of NO_2 may be defined as levels which fail to meet three criteria. These criteria are to induce the following:

1. Acute responses are manifested after 1–2 hours exposure and result in intercostal bifacial necrotic collapse of leaf tissue similar to the effects of SO_2. These foliar markings first appear as water-soaked lesions and later turn white, tan, or bronze in color. Lesions may also be marginal and tend to be near the apex (Fig. 1). An overall waxy appearance is shown by some weeds.
2. Chronic effects which may be an enhancement of green color followed by chlorosis and extensive leaf drop.
3. Physiological effects which result in altered photosynthesis, stunting, spindly growth, reduced fruit set, or reduced yield.

Nitrogen dioxide, a brown gas with a characteristic acrid odor, produces little eye irritation at exposures of 1 ppm without the addition of organic substances. The threshold for odor is 1–15 ppm depending upon the sensitivity of the individual. In addition to the widespread biological sources, NO_2 is produced by some chemical plants, including manufacture of explosives and nitrogenous fertilizers, but the greatest source of NO and subsequently of NO_2 is the high temperature combusion of fossil fuels. In some cities, such as Los Angeles, motor vehicle emissions are the dominant source, but in other areas industry and space heating contribute the greatest amount.

II. Factors Affecting Injury

Several factors, such as concentration of pollutant, length of exposure, species of plant, stage of plant development, plant environment (tempera-

FIG. 1. Collapsed, necrotic lesions on *Impatiens* leaves exposed for 2 hours to 15 ppm NO_2 under full sun conditions.

ture, light, humidity, soil moisture, mineral nutrition), and resistance of species (variety or clone), influence the degree of injury suffered by vegetation from air pollutants. Depending upon the kind of plant and its environment, one factor may be of much greater importance than others during any given exposure.

Detection of injury from pollutants often requires the measurement of subtle responses, such as photosynthesis, transpiration, rate of enzymatic processes, accumulation of the toxicant, and unusual physiological changes. Unfortunately, a minimal amount of these measurements has been made with NO_2. Generally, investigations have relied on obvious outward symptoms, such as leaf lesions, color changes, or reduction in growth and production.

III. Mechanism Causing Plant Injury

Mechanism(s) by which nitrogen dioxide causes injury to plants has received little attention in biochemical and histological studies. When NO_2 reacts with water it forms a stoichiometric mixture of nitrous and nitric acids. This probably occurs as the gas reaches the wet surfaces of the spongy parenchyma in the leaves of plants and when the acid exceeds a given threshold the tissues are injured.

In vitro studies by Weill and Caldwell (1945) on the effect of 1.0 *M* nitrous acid on β-amylase from barley (*Hordeum vulgare*) showed the enzyme to be slowly inactivated. They concluded that easily oxidized groups such as sulfhydryls could be affected. Di Carlo and Redfern (1947) conducted similar, more detailed studies with α-amylase obtained brom *Bacillus subtilis* and concluded that the nitrite reacted with an essential amino group of proteins. It reacts with two oxidation states of catalase, a hemoprotein occurring in mammalian tissues and with a peroxidase obtained from horseradish (*Armoracia rusticana*) Boyer *et al.,* (1959). The enzymatic studies used concentrations of 1.0 *M* nitrite. No enzymatic studies at the low concentrations of nitrite and nitrate which would occur in plant tissues from exposure to a few parts per million of NO_2 in air have been made. Matsushima (1972) fumigated citrus seedlings with NO_2 and SO_2 and compared the increased synthesis of various amino acids and organic acids from introduced $^{14}CO_2$. Nitrogen dioxide at 40 ppm for 16 hours stimulated about twice as much amino acid synthesis as SO_2, but produced less increase in organic acids.

Wellburn *et al.* (1972) compared the effects of SO_2 and NO_2 on the ultrastructure of chloroplasts in bean (*Phaseolus vulgaris*) tissue. Fumigation with 1, 2, or 3 ppm NO_2 for 1 hour caused thylakoid swelling which was reversible if the NO_2 containing atmosphere was replaced with pure air. No extra chloroplastic damage occurred at the test levels. SO_2 caused a similar effect, but at lower concentrations. Berge (1963) described NO_2 injury as causing plasmolysis of palisade tissue cells, disappearance of starch granules, and browning of cell walls. Ehrlich and Miller (1972) tested the effect of 10 ppm NO_2 on spores of *Bacillus subtilis,* but found no reduction in viability.

IV. Effects on Higher Plants

It has often been observed that vegetation in the vicinity of nitric acid factories is injured by gases from this manufacturing activity (Thomas, 1951). Leaves show brown or black spots, especially on the margins. Levels of exposure at which these effects become apparent are unknown, but probably are extremely variable with high peak concentrations.

Haagen-Smit (1951) recognized early the importance of NO_2 as a combustion by-product in generation of photochemical smog effects on vegetation. He found that inclusion of NO_2 in experimental fumigation mixtures caused plant injury similar to that caused by ozone, but he gave no information on the direct effects on plants. Subsequently Haagen-Smit *et al.* (1952) tested NO_2 at 0.4 ppm on five species, but observed no injury. Middleton (1958), Middleton *et al.* (1958), and Thomas (1961) all recognized the presence of NO_2 in photochemically polluted atmospheres as a product of combustion but likewise postulated that levels were, and would continue to be, too low to cause injury to vegetation.

In fumigation studies of NO_2 on vegetation, Benedict and Breen (1955) selected 10 annual or perennial weeds which were judged by 25 state agricultural experiment stations to be most common in the United States. These were tested to serve as indicators of plant injury caused by air pollutants. They exposed these species for 4 hours each at midday to a mixture of NO and NO_2 in outdoor transparent fumigation chambers. Nitric oxide from a cylinder was released into the incoming airstream of the chambers and allowed to react for 5 minutes before final injection. The levels of dispensing were adjusted to provide exposure levels of 50 or 20 ppm of NO_2. The amount of unreacted NO remaining was not measured.

Two types of leaf markings developed: (1) a discoloration associated with cell collapse and necrosis, and (2) a general overall waxy appearance of the leaf. The individual plants were rated by counting all leaves which were discolored and dividing this value by the total number of leaves on the plant to obtain a percentage (Table I).

With broad-leaved plants, the collapsed irregular-shaped necrotic markings were between the veins but nearer the margins of the leaf. Similar irregular-shaped, collapsed lesions were produced on many broad-leaved species at the Statewide Air Pollution Research Center at Riverside (Fig. 1). Middle-aged leaves developed the most markings but with sunflower (*Helianthus annuus*), annual bluegrass (*Poa annua*), and nettleleaf goosefoot (*Chenopodium* sp.) both the middle-aged and oldest leaves were equally susceptible. With mustard (*Brassica* sp.) the oldest leaves were most susceptible. In all species the young leaves were marked the least. Pigweed (*Chenopodium* sp.) showed no injury. The data indicated that moist exposure conditions caused several times as much injury as dry.

MacLean *et al.* (1968) exposed 14 ornamental species and 6 citrus varieties to extremely high concentrations of NO_2 for short periods. NO_2 levels of 10–250 ppm and exposure times of 0.2–8 hours were used. The fumigations caused nonspecific marginal and intercostal necrosis which was often visible within 1 hour. Tissue collapse appeared on the upper leaf surface, but later spread through the leaf making spots.

Citrus responded by wilting and defoliation of young leaves. Older leaves became necrotic. Exposure to 200 ppm NO_2 for 4 or 8 hours or 250 ppm for 1 hour caused young shoot necrosis. Varieties listed in the order of decreasing sensitivity were: Marsh seedless grapefruit, pineapple, orange, Valencia orange, tangelo orange, Hamlin orange, and Temple orange.

Ornamentals represented a wide range of susceptibility. The relative susceptibility is shown in Table II. Azaleas showed rapid leaf tissue collapse

TABLE I
PERCENTAGE OF LEAF AREA OF COMMON WEEDS MARKED BY OXIDES OF NITROGEN (4-HOUR FUMIGATIONS)

Weed	Concentration of oxides of nitrogen and age and soil conditions of the plants						Mean[a]
	50 ppm			20 ppm			
	3 weeks moist	6 weeks moist	6 weeks dry	3 weeks moist	6 weeks moist	6 weeks dry	
Mustard (*Brassica arvensis*)	32	54	4	9	26	1	21
Sunflower (*Helianthus annuus*)	21	23	3	5	6	7	11
Annual bluegrass (*Poa annua*)	7	18	1	5	6	1	6
Dandelion (*Taraxicum officinale*)	2	10	4	1	4	3	4
Cheeseweed (*Malva parviflora*)	12	8	1	1	1	1	4
Kentucky bluegrass (*Poa pratensis*)	4	6	1	2	2	1	3
Chickweed (*Stellaria media*)	2	3	1	4	2	1	2
Nettleleaf goosefoot (*Chenopodium murale*)	4	3	1	1	1	1	2
Lamb's quarters (*Chenopodium album*)	1	2	1	1	1	1	1
Pigweed (*Amaranthus retroflexus*)	0	1	0	0	1	0	<1
Mean[a]	9	13	2	3	5	2	

[a] Difference required for significance between means at $P < 0.05$ for species = 3%.

TABLE II
RELATIVE SENSITIVITY OF 14 ORNAMENTAL SPECIES TO NITROGEN DIOXIDE

Very sensitive	Moderately sensitive	Relatively resistant
Azalea (*Rhododendron canescens*)	Pittosporum (*P. tobira*)	Carissa (*C. carandas*)
Oleander (*Nerium oleander*)	Melaleuca (*M. leucadendra*)	Croton (*Codiaem variegatum*)
Bougainvillea (*B. spectabilis*)	Ligustrum (*L. lucidum*)	Shore juniper (*Juniperus coniferta*)
Pyracantha (*P. coccinea*)	Ixora (*I. coccinea*)	
Hibiscus (*H. rosasinensis*)	Cape Jasmine (*Gardenia radicans*)	
	Gardenia (*G. jasminoides*)	

and defoliation. Croton (*Codiacum* sp.) plants given 15 ppm for 4 hours showed slight intercostal necrosis, but Carissa (*Carissa carandas*) showed no observable effects. Although necrosis and defoliation were complete in sensitive plants, all species survived and axillary buds were produced in 2–6 weeks.

Middleton *et al.* (1958) found pinto beans to be much more susceptible to NO_2 than the weeds tested by Benedict and Breen (1955). Threshold concentrations for visible damage were 3–4 ppm for an 8-hour fumigation. The symptoms were similar to those produced by SO_2 but 3–4 times as much NO_2 was required to produce the same degree of injury. Two-hour fumigations with 30 ppm injured the leaves completely.

Czech and Nothdurft (1952) fumigated agricultural and horticultural crops with NO_2 in the laboratory and in small greenhouses. Rape (*Brassica napus*), wheat (*Triticum sativum*), oats (*Avena sativa*), peas (*Pisum* sp.), potatoes (*Solanum tuberosum*), and beans showed little or no injury from 30 ppm NO_2 for 1 hour. Alfalfa (*Medicago sativa*), sugar beets (*Beta vulgaris*), winter rye (*Secale cereale*), and lettuce (*Lactuca sativa*) showed some effects. A very significant finding was that nighttime fumigation caused more severe injury than treatment during the day.

Taylor (1968) recognized the increased sensitivity of plants to NO_2 caused by low light intensity and reported that 3 ppm NO_2 in darkness caused as much damage as 6 ppm in light. In published comments on this report Heck noted that a fumigation of 2 ppm NO_2 for 4 hours did not injure tobacco (*Nicotiana tabacum*), nor did 0.7 ppm SO_2 for 4 hours. However, 0.1 ppm NO_2 + 0.1 ppm SO_2 for the same period caused moderate injury to Bel W3 tobacco. Also a mixture of 0.05 ppm of O_3 + 0.5 ppm SO_2 injured sensitive tobacco considerably.

Thompson *et al.* (1970) fumigated navel orange trees continuously with five levels of NO_2 at 1.00, 0.5, 0.25, 0.12, and 0.06 ppm. After 35 days the two highest levels had caused chlorosis of leaves and extensive defoliation.

Van Haut and Strattman (1967) fumigated 60 species of plants with a 1 : 1 mixture of NO and NO_2 and compared the degree of acute leaf injury to that caused by SO_2. They reported the early symptoms of acute injury to leaves as "gray-green or slightly brownish" spots which later became necrotic. After induction of acute symptoms, by fumigation with 2.5 to 10 ppm NO + NO_2 for 4 to 8 hours, they removed plants from the fumigator and allowed symptoms to develop. They found the leaf injury to be similar to that caused by SO_2. NO_2 was 1.2–5 times less toxic than SO_2 to plant foliage. They classified plants as sensitive, intermediate, or resistant (Table III). Specific data showed that 6 ppm NO_2 for 4, 7, and 8 hours, respectively, gave moderately severe necrosis on peas, bush beans, and alfalfa (H. Strattman, personal communication, 1969). Spierings (1971) studied effects of NO_2 on growth and yield of tomato (*Lycopersicon esculentum*) and showed that 0.4–0.5 ppm for 21–45 days gave longer stems but smaller leaves and petioles than in controls. Use of 0.25 ppm during the entire growth period decreased crop yield 22%.

In studies of the effect of the time of day on injury, during some periods 4–6 times as much leaf damage occurred as at other times. Rye plants were most sensitive at 12–2 PM, but oats had a bimodal sensitivity. This species showed a sensitivity at 12–2 AM which was twice that at 6–8 AM.

Korth *et al.* (1964) fumigated beans, wrapper tobacco, and petunias (*Petunia* sp.) with 1.0 ppm NO_2 and observed no injury after 2 hours. Kaendler and Ulbrich (1964) found that carotene both *in vitro* and in green leaves was decomposed rapidly by nitrate ion or NO_2 gas to form a gray pheophytinlike substance. This effect did not occur with SO_2 fumigation, heat, or formaldehyde exposure.

Heck (1964) fumigated cotton (*Gossypium hirsutum*), pinto beans (*Phaseolus vulgaris*), and endive (*Cicorium endivia*) under controlled conditions with 1.0 ppm NO_2 for 48 hours and observed slight but definite spotting of leaves. In another study the same species were fumigated with 0.5, 2.0, and 3.5 ppm NO_2 for 21 hours. At exposures of 3.5 ppm NO_2 mild necrotic spots appeared on cotton and beans, but the leaves on endive were completely necrotic. No injury was produced at 1.0 ppm NO_2 in 12 hours.

W. W. Heck and D. T. Tingey (personal communication, 1969) extended these findings to include fumigation of 10 field and vegetable crops

TABLE III

Empirical Resistance to NO_2 as Measured by Leaf Sensitivity

Resistance Group I: Sensitive
Field and Horticultural Crops
Spring vetch (*Vicia sativum*)
Garden peas (*Pisum sativa*)
Lucerne (*Medicago sativa*)
Crimson or Italian clover (*Trifolium incarnatum*)
Red clover (*Trifolium pratense*)
Carrots (*Daucus carota*)
Common lettuce (*Lactuca sativa*)
Common tobacco plant (*Nicotiana tabacum*)
White mustard (*Sinapis alba*)
Lupine (*Lupinus augustrifolius*)
Common oats (*Avena sativa*)
Parsley (*Petroselinum hortense*)
Leek (*Allium porrum*)
Viper's grass (*Scorzonera hispanica*)
Barley (*Hordeum distichon*)
Rhubarb (*Rheum rhubarbarum*)
Ornamental Plants
Great snapdragon (*Antirrhinum majus*)
Tuberous-rooted begonia (*Begonia multiflora*)
Rose (*Rosa* sp.)
Sweet pea (*Lathyrus odoratus*)
China aster (*Callistephus chinensis*)
Coniferous Trees
Larch (*Larix europea*)
Japanese larch (*Larix leptolepis*)
Deciduous Trees
Weeping birch (*Betula pendula*)
Showy apple (*Malus* sp.)
Wild pear tree (*Pyrus* sp.)
Resistance Group II: Medium Sensitive
Deciduous Trees
Norway maple (*Acer platanoides*)
Fan maple (*Acer palmatum*)
Winter lime (*Tilia parvifolia*)
Summer lime (*Tilia grandiflora*)
Coniferous Trees
Blue spruce (*Picea pungens glauca*)
White spruce (*Picea alba*)
Lawson's cypress (*Chamaecyparis lawsoniana*)
Nikko or Japanese fir (*Abies homolepis*)
Common silver fir (*Abies pectinata*)

TABLE III (*Continued*)

Resistance Group II: Medium Sensitive (*Continued*)
Ornamental Plants
Fuchsia (*Fuchsia hybrida*)
Petunia (*Petunia multiflora*)
Rhododendron (*Rhododendron catawbiense*)
Dahlia (*Dahlia variabilis*)
Field and Horticultural Crops
Rye (*Secale cereale*)
Celery (*Apium graveolens* var. *rapaceum*)
Maize (*Zea mays*)
Common wheat (*Triticum sativum*)
Tomato (*Solanum lycopersicum*)
Potato (*Solanum tuberosum*)
Pine strawberry (*Fragaria chiloensis* var. *grandiflora*)
Resistance Group III: Relatively Insensitive
Deciduous Trees
Black locust (*Robinia pseudoacacia*)
Hornbeam (*Carpinus betulus*)
Common beech (*Fagus sylvatica*)
Common elder (*Sambucus nigra*)
Gingko tree (*Ginkgo biloba*)
Mountain elm (*Ulmus montana*)
Purple-leaved beech (*Fagus sylvatica atropurpurea*)
Common oak (*Quercus pendunculata*)
Coniferous Trees
Common yew tree (*Taxus baccata*)
Black pine (*Pinus austriaca*)
Knee pine or dwarf mountain pine (*Pinus montana mughus*)
Field and Horticultural Crops
Kohlrabi (*Brassica oleracea* var. *gongylodes*)
Onion (*Allium cepa*)
White cabbage (*Brassica oleracea* var. *capitata alba*)
Kale (*Brassica oleracea acephala*)
Red cabbage (*Brassica oleracea* var. *capitata rubra*)
Ornamental Plants
Oxeye daisy (*Chrysanthemum leucanthemum*)
Lily of the Valley (*Convallaria majalis*)
Common gladiolus (*Gladiolus communis*)
Plantain lily or Funkia (*Hosta* sp.)

in one series and 22 crops in another. The 10 crops were subjected to approximately 8, 16, and 32 ppm NO_2 for 1 hour. Traces of injury occurred at the lowest level on tomato and bromegrass (*Bromus inermis*), but at 32 ppm all crops of tobacco, tomato, bromegrass, soybean (*Glycine max*), beet, cotton, orchard grass (*Dactylis glomerata*), wheat, and swiss

TABLE IV

PERCENT OF LEAF AREA INJURED BY DESIGNATED DOSAGE (CONCENTRATION (ppm) X LENGTH OF EXPOSURE (HOUR)) OF NITROGEN DIOXIDE[a]

Experiment A									
Dosage (ppm × hour)	2	4	6	8	13	14	16	20	25
(ppm)	4	4	3	16	13	2	8	5	5
(hr)	0.5	1	2	0.5	1	7	2	4	7
% Injury[b]									
Begonia (*Begonia* sp.)	0	1	0	26	35	0	49	4	5
Periwinkle (*Vinca* sp.)	0	0	0	12	20	0	23	1	1
Chrysanthemum (*Chrysanthemum marifolium*)	1	1	1	34	41	0	25	4	1
Azalea (*Rhododendron canescens*)	0	0	0	0	1	0	0	0	0
Experiment B									
Dosage (ppm × hour)	2.5	3	6	11	14	16	18	24	42
(ppm)	5	3	3	22	2	16	9	6	6
(hr)	0.5	1	2	0.5	7	1	2	4	7
% Injury[c]									
Tobacco (*Nicotiana tabacum* Bel. W3)	0	0	6	15	0	2	0	0	0
Cotton (*Gossypium hirsutum* var. Paymaster)	0	0	6	50	0	27	2	2	1
Tobacco (*N. tabacum* var. White Gold)	0	0	1	18	0	6	0	0	0
Tobacco (*N. tabacum* Bel B.)	0	0	3	18	0	17	0	0	0
Cotton (*G. hirsutum* var. Acala 4–42)	0	0	0	28	0	28	1	1	1
Tobacco (*N. tabacum* var. Burley 21)	0	0	0	8	0	0	0	0	0
Experiment C									
Dosage (ppm × hour)	2.5	3	4	10	10	14	16	20	28
(ppm)	5	3	2	10	20	2	8	5	4
(hr)	0.5	1	2	1	0.5	7	2	4	7
% Injury[c]									
Corn (*Zea mays* var. Pioneer)	1	0	0	1	1	0	0	0	0
Sorghum (*Sorghum vulgare*)	0	0	0	0	0	0	0	0	0
Radish (*Raphanus sativus*)	15	0	0	67	95	0	30	1	2
Oats (*Avena sativa* var. Clintland 64)	14	0	0	63	80	2	40	0	21

TABLE IV (*Continued*)

Experiment C (*Continued*)									
Corn (*Zea mays* var. Golden Cross)	0	0	0	0	0	0	0	0	2
Bromegrass (*Bromis inermis*)	8	0	0	37	69	0	26	1	0
Experiment D									
Dosage (ppm × hour)	2.5	4	4	7	8	13	18	20	21
(ppm)	5	4	2	1	16	13	9	5	3
(hr)	0.5	1	2	7	0.5	1	2	4	7
% Injury[c]									
Wheat (*Triticum sativum*)	3	2	1	3	31	34	2	3	1
Oats (*Avena sativa*)	2	2	1	1	32	18	14	9	14
Broccoli (*Brassica oleracea*)	0	0	0	0	18	21	0	0	0
Cucumber (*Cucumus sativus*)	0	0	0	0	0	0	0	0	0
Oats (*A. sativa* var. *Pendek*)	1	2	0	0	39	2	2	1	2

[a] From W. W. Heck and D. T. Tingey (personal communication, 1969).
[b] % area of all leaves affected by NO_2.
[c] % area of three most sensitive leaves affected by NO_2.

chard (*Beta chilensis*) showed some or very extensive injury. The 22 crops series were given nine different time and concentration treatments, varying the times from 0.5–7 hours and NO_2 concentrations from 2–20 ppm (Table IV). The greatest injury occurred when gas concentrations were high even for short periods.

Taylor and Eaton (1966) exposed tobacco (*Nicotiana glutinosa*), pinto bean, and tomato in experimental chambers to controlled levels of NO_2 and comparisons were made with tissue injury caused by ozone. In preliminary work with tobacco, one strain of *N. glutinosa* was injured severely by exposure to 2.3 ppm NO_2 for 8⅔ hours. A 5-hour exposure caused no effect. Pinto beans were more resistant, requiring 10.0 ppm NO_2 for 5 hours before injury occurred. In low level exposures (0.30 ppm) for 10–19 days, pinto beans declined in dry weight and increased in chlorophyll content per unit weight. In similar studies with tomatoes at 0.15–0.26 ppm NO_2 for 10–22 days, dry weight and leaf area were decreased. NO_2 fumigation also caused a darker green color and a downward curvature of the leaves (Fig. 2). H. Strattman (personal communication, 1969) also reported sig-

FIG. 2. Pinto bean plants. Left: grown in filtered air. Right: exposed to about 0.4 ppm NO_2 for 7 hours a day for 10 days. NO_2 induced curved leaves, dark green color, and suppressed growth.

nificant growth depression in bush beans after exposure to 1.0 ppm NO_2 for 14 days. The lower levels, continued for 8 months, gave increased leaf drop and lowered yield from the trees.

O. C. Taylor and E. A. Cardiff (unpublished data, 1969) fumigated several agricultural and ornamental crops in small sunlit chambers with successively higher levels of NO_2 to determine how much was required to cause overt leaf damage within a short period. They found a fairly sharp threshold level for all species tested, somewhere between 10 and 15 ppm. With tomato, an increase in levels from 10–15 ppm NO_2 for 90 minutes fumigation caused an increase from 0 to 90% injury of the leaves.

Hill and Bennett (1970) compared the effects of NO and NO_2 on the rate of apparent photosynthesis of alfalfa and oats. A threshold concentration of about 0.6 ppm of each gas was required to reduce CO_2 assimilation. Combining the two gases gave an additive effect. NO produced a more rapid reduction in apparent photosynthesis than NO_2 and recovery was more rapid when fumigation was stopped. O. C. Taylor and E. A. Cardiff (unpublished data, 1969) found that NO at concentrations below 10 ppm induced an immediate decline in the rate of carbon dioxide (CO_2) absorption by pinto bean and tomato plants. Absorption of CO_2 was reduced by approximately 10% when plants were exposed to 4 ppm NO and by as much as 45% when they were exposed to 10 ppm. There was no evidence of permanent injury, and CO_2 absorption immediately returned to the original rate when NO was removed from the atmosphere. NO caused a greater reduction per unit concentration after about 2 hours of treatment. Concentrations which caused 50% reduction in apparent photosynthesis after 1–2 hours did not cause injury to foliage.

The present ambient levels of NO_2 occurring in areas of the country with the highest emissions and poorest dispersion, such as Los Angeles, are surely severalfold less than that required to cause acute damage symptoms on vegetation. With summer mean hourly averages 8–15 parts per hundred million (pphm) NO_x and 24 pphm mean value in winter and with one-half or less of these compounds being NO_2, present data would indicate no effects. About 1.0 ppm NO_2 seems to be required for 1 day to produce overt leaf damage on sensitive plants. H. Strattman (personal communication, 1969) gave a "first approximation" level required for leaf damage as 0.4 ppm.

The effects of exposure to low levels, i.e., chronic effects, for long periods are less clear. Recently completed studies suggest that 25 pphm *or less* of NO_2 supplied continuously for 8 months will cause increased leaf drop and reduced yield of navel oranges.

A study in which ambient and twice ambient levels of NO_2 at Upland, California were added to carbon-filtered air supplied to navel oranges showed no effects on leaf drop or yield (Thompson *et al.,* 1971).

V. Combined Effects of NO_2 and SO_2

A limited number of studies with combinations of NO_2 and SO_2 have been made because both are present in ambient polluted air in many regions. Electrical generating plants and industries which burn sulfur-containing petroleum fuels may emit significant quantites of both, and in other instances, the sources are separate but occupy the same air basin. Tingey *et al.* (1971) reported that below 200 pphm of NO_2 or 50 pphm of SO_2 no leaf injury occurred, but plants exposed to mixtures of 5 to 25 pphm of both bases were injured, indicating a synergistic or potentiating effect. They observed that the levels of NO_2 and SO_2 which occur in urban areas may result in yield losses to plants grown under field conditions. Fugiwara (1973) fumigated green peas with 10 pphm $SO_2 + 20$ pphm NO_2 and observed only additive effects, but 10 pphm of both gases did give the potentiating effect. Hill *et al.* (1974) in experiments on 87 native cold desert species failed to observe a synergistic response with a mixture containing 28% as much NO_2 as SO_2.

VI. Evaluation of Air Quality Standards for Vegetation Effects

An evaluation of results from studies previously conducted indicate the following conclusions: (1) The Federal Air Quality Standards of 1973 for 0.05 ppm NO_2 on an annual basis, maintained continuously in the absence of other air pollutants, might induce small chronic reduction in growth and production of very susceptible plants. (2) This standard could occur for a few hours each day with little detectable effect and under no environmental conditions would leaf symptoms be seen. (3) The 1973 California standards of 0.25 ppm for 1 hour might induce some chronic injury only if the 1 hour periods occurred very frequently and no leaf symptoms would develop.

Acceptable levels of NO_2 as a single pollutant represent a time–concentration relationship best expressed by (Fig. 3) and Table V. Plants are

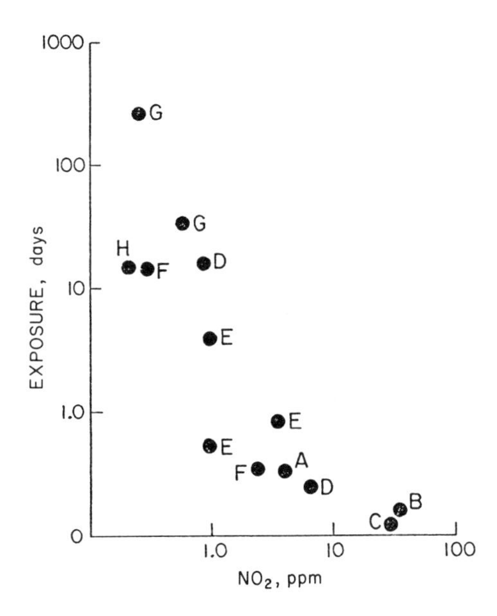

FIG. 3. Summary of effects of NO_2 on vegetation. The points describe a dosage line above which injury was detected. Individual points were obtained from the following references: (A) Middleton *et al.* (1958), (B) Hill *et al.* (1974), (C) Czeck and Nothdurft (1952), (D) H. Strattman (personal communication, 1969), (E) Heck (1964), (F) Taylor and Eaton (1966), (G) Thompson *et al.* (1970), and (H) Matsushima (1972).

TABLE V

TABULAR SUMMARY OF EFFECTS OF NITROGEN DIOXIDE ON VEGETATION

No.	Effect	Concentration ppm	Length of exposure	Reference
1	Decrease in dry wt and leaf area of tomatoes, darker green color and downward curvature of leaves	0.15–0.26	10–22 days	Taylor and Eaton (1966)
2	Increased leaf drop and reduced yield of fruit in navel oranges	0.25	8.5 months	Thompson *et al.* (1970)
3	Decrease in dry wt and increased chlorophyll per unit wt of pinto beans	0.30	10–19 days	Taylor and Eaton (1966)
4	Leaf chlorosis and leaf drop in navel oranges	0.5	35 days	Thompson *et al.* (1970)
5	Leaf injury to Bel #3 tobacco	0.1[a]		Taylor (1968)
6	Growth depression	1.0	14 days	H. Strattman (1969)[b]
7	Necrotic leaf spots on cotton, beans, and endive	1.0	48 hours	Heck (1964)
8	No leaf injury on cotton, beans, and endive	1.0	12 hours	Heck (1964)
9	Severe leaf necrosis on tobacco	2.3	8.6 hours	Taylor and Eaton (1966)
10	Necrotic spots on cotton and beans but complete leaf necrosis on endive	3.5	21 hours	Heck (1964)
11	Leaf lesions on beans	3–4	8.0 hours	Middleton *et al.* (1958)
12	Leaf lesions on beans, peas and alfalfa	6.0	4–8 hours	H. Strattman (1969)[b]
13	Leaf discoloration and waxy appearance on leaves of 10 common weeds: mustard very sensitive, pigweed resistant	20–50	4.0 hours	Hill *et al.* (1974)
14	Some leaf damage to alfalfa, beets, rye, and lettuce	30	1.0 hour	Czeck and Nothdurft (1952)
15	Marginal and intercostal leaf necrosis, plus shoot dieback on 14 ornamentals, 6 citrus species	10–250	0.2–8.0 hours	MacLean *et al.* (1968)

[a] Plus 0.10 ppm SO_2.

[b] Personal communication.

uninjured by many parts per million if exposures are for a few minutes. Continuous exposure to 5–10 pphm will result in reduced performance, but this effect will be very difficult to detect unless careful comparisons are made.

References

Anonymous. (1971). "Air Quality Criteria for Nitrogen Oxides," Air Pollut. Contr. Office Publ. No. AP-84. U.S. Govt. Printing Office, Washington, D.C.

Benedict, H. M., and Breen, W. H. (1955). The use of weeds as a means of evaluating vegetation damage caused by air pollution. *Proc. Nat. Air Pollut. Symp., 3rd, 1955,* pp. 117–190.

Berge, H. (1963). "Phytotoxische immissionen." Parey, Berlin.

Boyer, P. D., Lardy, H., and Myrbäck, K., eds. (1959). "The Enzymes," 2nd ed., Vol. 1, pp. 180–181.

Czech, M., and Nothdurft, W. (1952). Investigations of the damage to field and horticultural crops by chlorine, nitrous and sulfur dioxide gases. *Landwirt. Forsch.* **4,** 1–36.

Di Carlo, F. J., and Redfern, S. (1947). α-amylase from *B. subtilis* II. Essential groups. *Arch. Biochem.* **15,** 343–350.

Ehrlich, R., and Miller, S. (1972). Effect of NO_2 on airborne Venezuelan equine encephalomeyelitis virus. *Appl. Microbiol.* **23,** 481–484.

Fugiwara, T. (1973). Effects of nitrogen oxides in the atmosphere on vegetation. *J. Pollut. Contr.* **9,** 253–257.

Haagen-Smit, A. J. (1951). What is smog? *Calif. Inst. Tech., Res. Bull.* **15.**

Haagen-Smit, A. J., Darley, E. F., Zaitlin, M., Hull, H., and Noble, W. (1952). Investigation on injury to plants from air pollution in the Los Angeles area. *Plant Physiol.* **27,** 18–34.

Heck, W. W. (1964). Plant injury induced by photochemical reaction products of propylene-nitrogen dioxide mixtures. *J. Air Pollut. Contr. Ass.* **14,** 255–261.

Hill, A. C., and Bennett, J. H. (1970). Inhibition of apparent photosynthesis by nitrogen oxides. *Atmos. Environ.* **4,** 341–348.

Hill, A. C., Hill, S., Lamb, C., and Barrett, T. W. (1974). Sensitivity of native desert vegetation to SO_2 and NO_2 combined. *J. Air Pollut. Contr. Ass.* **24,** 153–157.

Kaendler, U., and Ulbrich, H. (1964). Detection of NO_2 damage to leaves. *Naturwissenschaften* **51,** 518.

Korth, M. W., Rose, A. H., and Stahman, R. C. (1964). Effects of hydrocarbon to oxides of nitrogen ratios on irradiated auto exhaust. Part I. *J. Air Pollut. Countr. Ass.* **14,** 168–175.

MacLean, D. C., McCune, D. C., Weinstein, L. H., Mandl, R. H., and Woodruff, G. N. (1968). Effects of acute hydrogen fluroide and nitrogen dioxide exposures on citrus and ornamental plants of central Florida. *Environ. Sci. Technol.* **2,** 444–449.

Matsushima, J. (1972). Influence of SO_2 and NO_2 on assimilation of amino acids, organic acids, and saccharoid in *Citrus natsudaidai* seedlings. *Bull. Fac. Agr., Mie Univ.* **44,** 131–139.

Middleton, J. T. (1958). Clean air essential for good citrus. *West. Fruit Grower* **1,** 6–9.

Middleton, J. T., Darley, E. F., and Brewer, R. F. (1958). Damage to vegetation from polluted atmospheres. *J. Air Pollut. Contr. Ass.* **8,** 9–15.

Robinson, E., and Robbins, R. C. (1970). Gaseous nitrogen compound pollutants from urban and natural sources. *J. Air Pollut. Contr. Ass.* **20,** 303–306.

Spierings, F. H. F. G. (1971). Influence of fumigations with NO_2 on growth and yield of tomato plants. *Neth. J. Plant Pathol.* **77,** 194–200.

Taylor, O. C. (1968). Effects of oxidant air pollutants. *J. Occup. Med.* **10,** 485–499.

Taylor, O. C., and Eaton, F. M. (1966). Suppression of plant growth by nitrogen dioxide. *Plant Physiol.* **41,** 132–135.

Thomas, M. D. (1951). Gas damage to plants. *Annu. Rev. Plant Physiol.* **2,** 293–322.

Thomas, M. D. (1961). Effects of air pollution on plants. *World Health Organ. Monog. Ser.* **46,** 233–278.

Thompson, C. R., Hensel, E. G., Kats, G., and Taylor, O. C. (1970). Effects of continuous exposure of navel oranges to NO_2. *Atmos. Environ.* **4,** 349–355.

Thompson, C. R., Kats, G., and Hensel, E. G. (1971). Effects of ambient levels of NO_2 on navel oranges. *Environ. Sci. Technol.* **5,** 1017–1019.

Tingey, D. T., Reinert, R. A., Dunning, J. A., and Heck, W. W. (1971). Vegetation injury from the interaction of NO_2 and SO_2. *Phytopathology* **61,** 5106–1511.

Van Haut, H., and Strattman, H. (1967). Experimental investigations of the effect of nitrogen dioxide on plants. *Trans. Land Inst. Pollut. Contr. Soil Conserv. Land North Rhine-Westphalia, Essen* **7,** 50–70.

Weil, E. C., and Caldwell, M. L. (1945). A study of the essential groups of β-amylase. *J. Amer. Chem. Soc.* **67,** 212–214.

Wellburn, A. R., Majernik, O., and Wellburn, A. M. (1972). Effects of SO_2 and NO_2 polluted air upon the ultrastructure of chloroplasts. Environ. *Pollut.* **3,** 37–49.

7

PARTICULATES

Shimshon L. Lerman and Ellis F. Darley

I. Introduction

In comparison to gaseous air pollutants, many of which are readily recognized as being the cause of injury to various types of vegetation, relatively little is known and limited studies have been carried out on the effects of particulate air pollutants on vegetation. Published experimental results are confined principally to settleable dusts emitted from the kilns of cement plants. There are a few reports on effects of fluoride dusts, soot, sulfuric acid mist, lead particles, and particulate matter from various types of metal processing. Most of the research is related to the direct effects of dusts on leaves, twigs, and flowers, as opposed to indirect effects from dust accumulation in the soil.

Because of the dearth of experimental results, the tenor of many reports is directed to the question of whether particles in fact have deleterious effects on plants rather than discussions of the actual extent of plant injury. The pollutant called particulate matter is a conglomerate of chemically heterogeneous substances. Therefore, it is reasonable to anticipate some disagreement on its impact on vegetation. The various types of particles and their effects on vegetation will be discussed.

II. Effects of Specific Particulate Matter on Vegetation

A. Cement-Kiln Dust

Cement-kiln dust is the dust contained in waste gases from the kilns and is not derived directly from processing of cement. It is apparent from some reports, however, that the composition of wastes from different kilns operating at different efficiencies varies considerably, and at times the effluents may contain cementitious materials that more properly belong in the finished product. Another important factor to consider is that literature reports describing effects of dust deposited on various plants in the field relate to kiln-stack materials, whereas experimental dusts applied in laboratory or field studies were taken from various collectors in the waste-gas system between the kiln and the stack. Differences in results that may be due to this factor have not been reconciled.

Fallout levels, as well as the physical and chemical properties of kiln dusts, are determined by factors such as the nature of raw material, cement manufacturing processes, and the type of equipment employed for the control of particulate matter emissions. Lerman (1972) recorded dust deposits of 1.5 g per m^2 per day in the vicinity of a cement plant in California. The rotary kilns in the plant were equipped with a multiple cyclone emission control device. According to reports from Germany, the maximum amounts of dust that might be deposited in the vicinity of cement factories vary from 1.5 g per m^2 per day (Pajenkamp, 1961) to 3.8 g per m^2 per day (Bohne, 1963). The diversity in chemical composition and pH reactions of cement-kiln dust samples from various sources are presented in Table I.

1. Direct Effects

a. Observations in the Field and the Nature of Dust Deposition. Most of the reports concerning harmful effects of cement-kiln dust on plants stress the fact that crusts form on leaves, twigs, and

TABLE I
CHEMICAL COMPOSITION (%) OF SOME CEMENT-KILN DUSTS

Sample	Na	Ca	K	Mg	Al	Fe	pH of water-dust suspension
Cal-1	0.40	31.20	2.16	0.40	0.40	1.15	12.3
Ariz-3	0.13	23.00	8.00	1.00	0.32	0.55	10.8
G-1	2.80	17.56	13.00	1.08	0.94	1.10	10.4
G-2	7.60	8.00	32.50	0.56	0.48	0.60	7.3

flowers. Pierce (1910) and Parish (1910) noted in California that settled dust in combination with mist or light rain formed a relatively thick crust on upper leaf surfaces of affected plants. The crust would not wash off and could be removed only with force. The central theme about which Czaja (1961a,b, 1962, 1966) builds his case for harmful effects is the crust formation in the presence of free moisture. He states that crust is formed because some portion of the settling dust consists of the calcium silicates which are typical of the clinker (burned limestone) from which cement is made. When this dust is hydrated on the leaf surface, a gelatinous calcium silicate hydrate is formed, which later crystallizes and solidifies to a hard crust. When the crust is removed, a replica of the leaf surface is often found, indicating intimate contact of dust with the leaf. The relatively thick crust formed from continuous deposition is confined to the upper leaf surface of deciduous species but completely encloses needles of conifers. Prolonged dry periods during the time dust is deposited provide no opportunity for hydration, and crusts are not formed. Dust deposits which are not crusted are readily removed by wind or hard rain.

Photographs taken in Germany (by the authors) (Figs. 1 and 2) show incrustations of cement-kiln dust on branches of fir trees. Atmospheric levels of dust in this area were probably in excess of 85 tons/mile month (1.0 g/m^2 day). Incrustations built up on the older twigs (Fig. 1) and caused needles to fall prematurely. Some of the twigs were dead and incrustations were forming on the newest needles. The net effect was a shortening of each succeeding year's flush of growth. A dead tree had heavy incrustations on the branches (Fig. 2).

Bohne (1963) reported a marked reduction of growth of poplar trees located about one mile from a cement plant after production in the plant was more than doubled. The change in growth rate was determined by the width of annual rings in the wood. Darley (1966) observed a reduction of spring growth elongation on conifers in Germany, where the oldest needles were incrusted.

FIG. 1. Cement-kiln dust on fir branches. Incrustation has built up on the older twigs of a fir tree exposed to cement-kiln dust particles in the vicinity of cement plant. Needles have fallen prematurely.

Anderson (1914) observed in New York that cherry fruit set was reduced on the side of the tree nearest a cement plant. He demonstrated that the dust on the stigma prevented pollen germination.

Pajenkamp (1961) reviewed the unpublished work of several German investigators, some of whom had applied dust artificially to test plants, and stated that he was opposed to the view that dusts are harmful to plants. He concluded that depositions of from 0.75 to 1.5 g/m^2 day (the latter amount representing the maximum that might be found in the vicinity of a cement factory) had no harmful effect on plants.

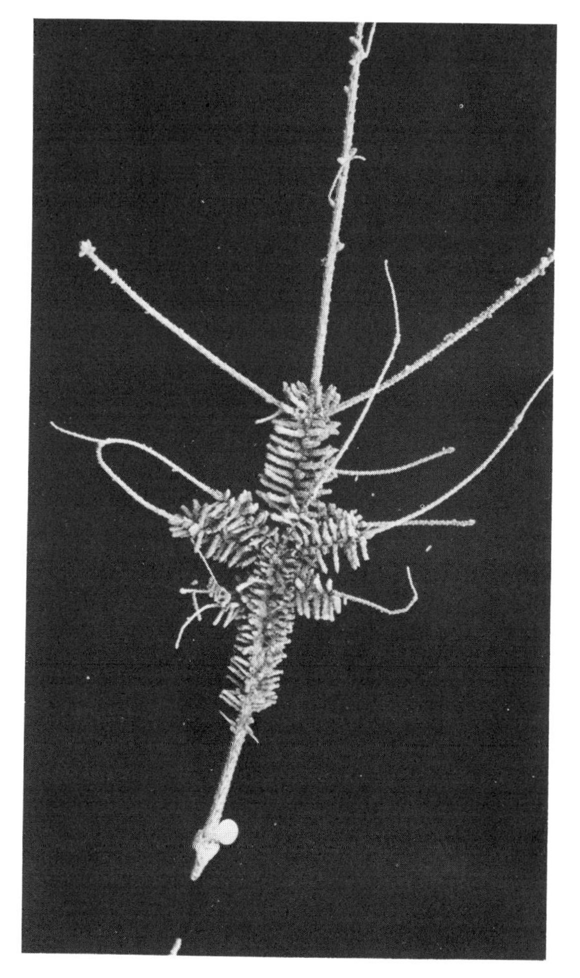

FIG. 2. Cement-kiln dust on fir tree branches. Very heavy incrustation on a branch of a dead fir tree exposed to cement-kiln dust particles in the vicinity of cement plant.

Raymond and Nussbaum (1966) also stated that cement dusts have little effect on wild plants. On the other hand, Guderian (1961) and Wentzel (1962) disagreed with Pajenkamp and stated that the limited evidence at best presented a contradictory picture and that Pajenkamp had not cited Czaja's earlier work (1961a,b, 1962, 1966). They also pointed out that a deposit of 1.5 g/m^2 day was not maximum, since other workers had found up to 2.5 g/m^2 day, and Bohne (1963) has since reported weekly averages of up to 3.8 g/m^2 day.

b. Physical Effects. Peirce (1910) demonstrated that incrustations of cement-kiln dust on citrus leaves interfered with light required for photosynthesis and reduced starch formation. This was later confirmed by Czaja (1962) and Bohne (1963) in a variety of plants. More recently, Steinhubel (1962) compared starch reserve changes in undusted common holly leaves and those dusted with foundry dust. He concluded that the critical factor in starch formation was the light absorption by the dust layer, and that the influence on transpiration or overheating of leaf tissue was of minor significance. Lecenier and Piquer (see Czaja, 1962) attributed the reduced yields from dusted tomato and bean plants to interference with light imposed by the dust layer. Darley (1966) demonstrated that dust deposited on bean leaves in the presence of free moisture interfered with the rate of carbon dioxide exchange, but no measurements of starch were made.

Czaja (1966) has presented good histological evidence that stomata of conifers may be plugged by dust, preventing normal gas exchange by the leaf tissue. Uninhibited exchange of carbon dioxide and oxygen by leaf tissue is necessary for normal growth and development.

Lerman (1972) demonstrated limited clogging of stomatal openings on bean leaves which were heavily dusted with dry dust. Scanning electron micrographs of upper and lower surfaces of dusted bean leaf are shown in Figs. 3–4. Only a few dust particles can be observed on the lower surface of the leaf (Fig. 3), on which most of the stomata are located. Many stomatal openings on the upper leaf surface (Fig. 4) are clear of dust particles in spite of the fact that the leaf was dusted with 6.64 g/m^2, a dust load which would normally cause severe damage to bean plants in the presence of free moisture.

c. Chemical Effects. Darley (1966) applied kiln dusts of particle size less than 10 μm in diameter at rates of 0.5 to 3.8 g/m^2 day to leaves for 2–3 days in the laboratory. Water mist was applied several times each day. Even though the dust adhered to the leaf in a uniform layer, it did not appear to be crustlike, probably because the experiments were of short duration. Reduction in CO_2 uptake was reported in these experiments.

Leaves of bean plants dusted for 2 days with cement-kiln dust of 8 to 20 μm size at the rate of 4.7 g/m^2 day and then exposed to naturally occurring dew were moderately damaged. The inury (Fig. 5) appeared as a rolling of the margins of the leaves, and some interveinal tissues were killed. Leaves which were dusted but kept dry were not injured.

In short-term experiments of 2–3 days, Darley (1966) dusted the primary leaves of bean plants with fractionated precipitator dust obtained from Germany. The dust contained relatively high amounts of potassium

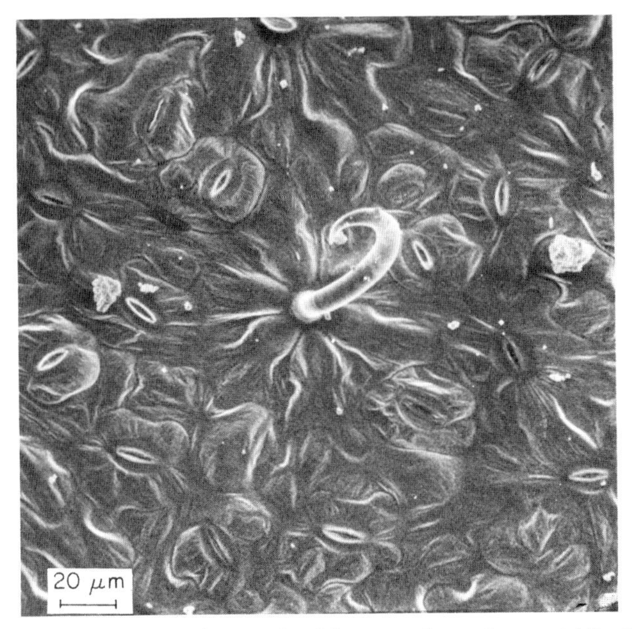

FIG. 3. Scanning electron micrograph of lower surface of cement-kiln dusted bean leaf. Only a few dust particles can be observed on the lower surface of a leaf which was dusted with a total of 6.64 g/m². × 300.

chloride, KCl. When a fine mist was applied to dusted leaves, a portion of the leaf tissue was killed (up to 29%) and it was presumed that the action was due to KCl. In later experiments other fractions of the same dust containing very little KCl caused an almost equivalent amount of injury, this indicating that KCl was apparently not the only factor involved.

Detail on the injury to be expected from certain cement-kiln precipitator dusts was given by Czaja (1966). His work is based on comparisons of chemical composition of dusts and resultant injury to leaf cells of a sensitive moss plant, *Mnium punctatum*. A cut leaf was mounted in water on a microscope slide, and dust was placed in contact with the water at the edge of the cover slip. Any effect of the resultant solution on leaf cells could be observed directly. Eighteen of the dusts tested in this way fell into the following categories:

1. No permanent injury to living cells, but some plasmolysis from the concentration effect of the solution.
2. Slight injury to readily accessible cells, disruption of the cytoplasm, and displacement of chloroplasts.
3. Severe injury to all cells of the leaflet.

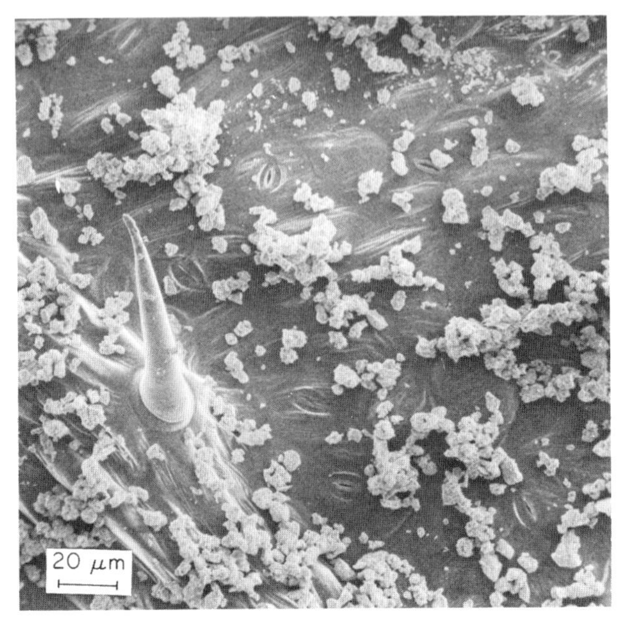

FIG. 4. Scanning electron micrograph of upper surface of cement-kiln dusted bean leaf. Many stomatal openings of the upper leaf surface are clear of dust particles in spite of the fact that the leaf was dusted with 6.64 g/m^2. × 300.

FIG. 5. The effect of cement-kiln dust on bean plants. The plants were dusted with particles 8–20 μm in diameter at the rate of 4.7 g/m^2 day. (a) Control plant. (b) Dusted but kept dry, no injury symptoms. (c) Dust-dew treated leaf. The injury appeared as a rolling of leaf margins and interveinal necrosis.

Dusts were further described as follows: group 1, pH 9.5–11.5, relatively high rate of carbonation, an intermediate amount (19–29%) of clinker phase (calcium silicates), and a high (36–79%) amount of secondary salts; group 2 pH about 11, a high rate of carbonation, a lower (13–16%) clinker phase and a significantly high (81–85%) proportion of raw feed; group 3, pH 11–12, a very slow carbonation rate and a significantly high (17–49%) clinker phase. The greater injury was thus related to the larger amounts of clinker phase, which in turn resulted in higher and prolonged alkalinity. But Czaja also pointed out that the composition of dusts within the three groups was not consistent and that, although not yet demonstrated, the constituents of a given dust undoubtedly influence one another.

Czaja (1962) stated that the hydration process of crust formation released calcium hydroxide. The hydrated crusts gave solutions of pH 10–12. Severe injury of naturally dusted leaves, including killing of palisade and parenchyma cells, was revealed by microscopic examination. The alkaline solutions penetrated stomata on the upper leaf surface, particularly the rows of exposed stomata on needles of conifer species, and injured the cells beneath. Czaja (1962) stated that on broad-leaved species with stomata only on the lower leaf surface, the alkaline solutions first saponified the protective cuticle on the upper surface, permitting migration of the solution through the epidermis to the palisade and parenchyma tissues. Typical alkaline precipitation reaction with tannins, especially in leaves of rose and strawberry, was evidence that calcium hydroxide penetrated the leaf tissue. Bohne (1963) described similar "corrosion" of tissues under the crust formed on oak leaves.

Observations made by Lerman (1972) with a scanning electron microscope revealed disorganization of the cuticle on the surface of dust-dew treated bean plants (Fig. 6). The breakdown of the layer appeared first in the form of cracks which later developed into a mass of dust particles surrounded by peeled-off cuticular strips. Figure 7 shows smooth upper surface of control leaf. Chemical analysis of extracts from the leachate of the treated leaves revealed relatively large quantities of free hexadecanoic acid, and small amounts of free tetradecanoic, dodecanoic, and nonanoic acid. These fatty acids are among the cuticular wax constituents. Extracts from the leachate of control plants did not have any detectable amounts of these fatty acids.

Studies in our laboratory revealed that bean plants respond in various ways to dust-dew treatment using dusts from different sources. Dust from California (Cal-1) caused moderate damage: the pH of the dust-water suspension reached 12.3 (Table I). Ariz-3 dust caused severe damage but the pH of the suspension was only 10.8. Dust from G-1 cement plant had a relatively low content of calcium with a pH reaction of 10.4; G-1 dust

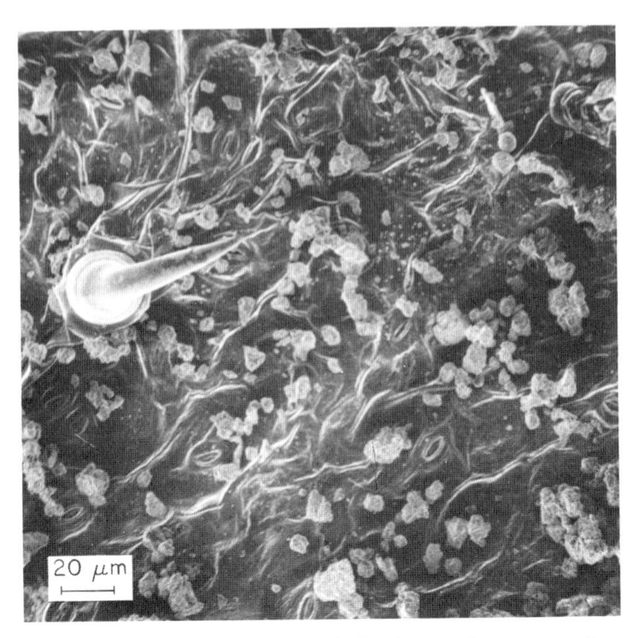

FIG. 6. Scanning electron micrograph of the damaged upper surface of cement-kiln dust-dew treated leaf. Dust deposit 5.0 g/m^2, an advanced stage in the breakdown of the cuticle, marked by the mass of dust particles surrounded by peeled off cuticle strips. × 300.

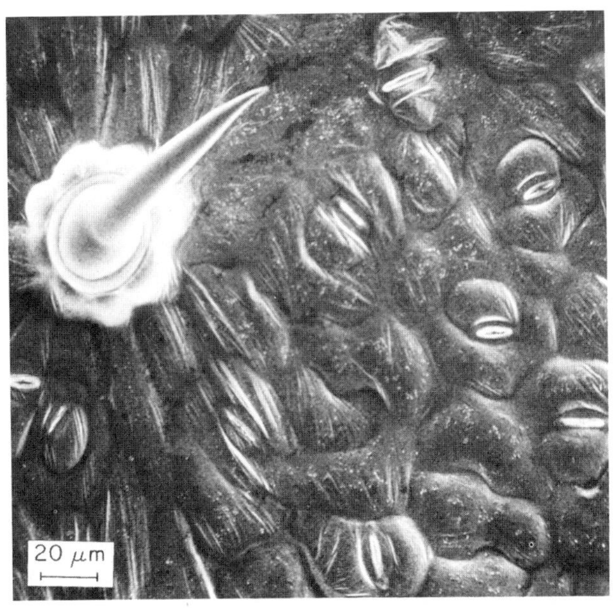

FIG. 7. Scanning electron micrograph of the upper surface of primary bean leaf. Control undusted leaf. × 300.

had no harmful effects on plants. The level of potassium was very high in the dust sample from G-2 and the pH reaction was near neutral, yet bean plants dusted with the latter showed damage symptoms and their ability to fix CO_2 was reduced. This high KCl dust could act on the plant as airborne salinity. Most of the calcium in the dust samples was in the form of calcium oxide, calcium hydroxide, and calcium carbonate. The high pH reactions of the dust-water suspensions were positively correlated with a high content of calcium hydroxide, as was demonstrated by X-ray powder diffractometry. These results indicate that the response of plants to dust-dew treatment is directly related to the chemical composition of the dust, and that the physical effects of the dust are negligible since dust treatment without dew had no detectable harmful effects on plants.

2. Indirect Effects

Pajenkamp (1961) reported on unpublished work by Scheffer in Germany during two growing seasons, indicating that even considerable quantities of precipitator dust applied to the soil surface brought about no harmful effects and no other lasting effects on growth or crop yield of oats, rye grass, red clover, and turnips. The dust had a content of about 29.3% limestone (analyzed as lime, CaO) and 3.1% potassium oxide. The maximum rate of deposit was 1.5 g/m^2 day. Discontinuous dustings were made at 2.5 g/m^2 day to give an average of 0.75 g/m^2 day. In 1 year, the yield of red clover and the weight of turnips were higher in the dusted plots, although the yield of leaves in the latter crop was reduced. Acid manuring of the soil appeared to increase yield but the interaction of dusting and manuring was not understood.

While Scheffer *et al.* (1961) found no direct injury to plants, they indicated that there might be indirect effects through changes in soil reaction, which in time might impair yield.

Stratmann and van Haut (1956) dusted plants with quantities of dust ranging from 1.0 to 48.0 g/m^2 day; dust falling on the soil caused a shift in pH to the alkaline side, which was unfavorable to oats but favorable to pasture grass.

Schonbeck (1960) treated a field planting of sugar beets biweekly with 2.5 g/m^2 of dust and observed that infection by leaf-spotting fungus, *Cercospora beticola,* was significantly greater than in nondusted plots. He postulated that the physiological balance was altered by dust, increasing susceptibility to infection.

Darley (1966) noted that plants were stunted and had few leaves in the heavily dusted portions of an alfalfa field downwind from a cement plant in California. Plants appeared normal in another part of the field where there was no visible dust deposit. The dusted plants were also heavily infested with aphids, and it was not clear whether the poor growth was due to the aphids feeding or a direct effect of the dust. Entomologists suggested that the primary effect of the dust may have been to eliminate aphid predators, thus encouraging high aphid populations, which in turn cause poor plant growth because of their feeding.

In an attempt to assess the long term effects of limestone dust on vegetation under natural environmental conditions, Brandt and Rhoades (1972) compared dusted and nondusted forest communities in the vicinity of limestone quarries and processing plants. The experimental site which was subjected to a heavy dust fall showed significant changes in structure and composition of the seedling-shrub, sapling, and tree strata. In a comparative study the same investigators (1973) demonstrated a reduction of at least 18% in the lateral growth of *Acer rubrum, Quercus prinus,* and *Q. rubra.* However, lateral growth of *Liriodendron tulipifera* increased by 76%. Changes in soil reaction and nutrient availability were considered as possible factors in both the increase in lateral growth of *L. tulipifera* and in reductions of growth of dominant trees in the dusty site.

B. Fluorides

Particles containing fluoride appear to be much less injurious than gaseous fluorides to vegetation. Pack *et al.* (1959) reported that 15% of gladiolus leaf was killed when plants were exposed 4 weeks to 0.79 $\mu g/m^3$ fluoride as HF, but no necrosis developed when plants were exposed to fluoride aerosol averaging 1.9 $\mu g/m^3$ fluoride. Inasmuch as the material was collected from a gas stream which was treated with limestone and hydrated lime, the aerosol was probably calcium fluoride. Moreover, when the accumulated levels of fluoride in leaf tissues were about the same, whether from gas or particulate, injury from the latter was much less.

McCune *et al.* (1965) reported an increase of only 4 mm tipburn on gladiolus exposed to cryolite (sodium aluminum fluoride dust). The washed leaf tissue from this treatment showed an accumulation of 29 ppm fluoride. A 70 mm increase in tipburn would have been expected if a similar accumulation had occurred from exposure to HF. Except for the slight tipburn noted above, these authors found that cryolite produced no visible effects on a variety of plants nor did it reduce growth or yield.

It is evident from the work of McCune *et al.* (1965) that fluoride in plant tissue is accumulated from cryolite treatment, but the rate of accumulation is much slower than would be expected from a comparable treatment with HF. For example, when comparing washed leaf samples, exposure to gladiolus to HF for 3 days at 1.01 $\mu g/m^3$ fluoride resulted in an accumulation of 26.4 ppm fluoride, whereas only 34 ppm was accumulated from an exposure to cryolite for 40 days at 1.7 $\mu g/m^3$ fluoride. Pack *et al.* (1959) reported only one-third as much fluoride accumulated from particulate matter as from gaseous forms.

Both the investigations cited above indicate that much of the particulate matter remains on the surface of the leaf and can be washed off, although what remains after washing is not necessarily internal fluoride. Reduced phytotoxicity of particulate fluoride is ascribed in part to the inability of the material to penetrate the leaf tissue. In addition, McCune *et al.* (1965) suggest that inactivity of particles may be due to their inability to penetrate the leaf in a physiologically active form.

C. Lead Particles

Published results of several studies (Cannon and Bowles, 1962; Kloke and Riebartsch, 1964; Leh, 1966; Motto *et al.,* 1970; Page and Ganje, 1970; Schuck and Locke, 1970; Warren and Delavault, 1962) show that lead accumulation by vegetation near highways varies with motor vehicle traffic densities and generally declines with distance from heavily traveled roads. Page *et al.* (1971) demonstrated that lead accumulations in and on plants next to highways in southern California were caused principally by aerial deposition and not to a great extent by absorption from lead-contaminated soils. Although lead concentrations close to highways have been high, there are no known reports of injury symptoms on vegetation.

D. Soot

Jennings (1934) noted the suggestion that soot may clog stomata and prevent normal gas exchange but that most investigations tend to discount this effect. Microscopic examination failed to show enough clogging of stomata on leaves of shade trees (broad-leaved species) to be significant. He further states that interference with light can be more serious, but he offers no data from critical experiments to substantiate this theory.

A well-illustrated report by Berge (1965) showed plugged stomata on conifers growing near Cologne, Germany. He also stated that growth was adversely affected.

Necrotic spotting was observed on leaves of several plants where soot from a nearby smokestack had entered a greenhouse (Miller and Rich, 1967). The necrosis was attributed to acidity of the soot particles. Plants outside the greenhouse were not damaged, possibly because the particles had been removed by rain before severe injury could occur.

E. Magnesium Oxide

The possible indirect effect on vegetation of magnesium oxide falling on agricultural soils was reported by Sievers (1924). He noted poor growth in the vicinity of a magnesite processing plant in Washington. Experiments were designed to grow plants in soil collected at various distances from the processing plant, in normal soil, and in soil to which magnesium oxide was added. Suppression of plant growth was demonstrated with the high levels of magnesium. After the processing plant ceased operation, injury to crops in the area became less pronounced, indicating that the injurious effect was not a permanent one.

F. Iron Oxide

Berge (1966), in Germany, dusted experimental plots with iron oxide at the rate of 0.15 mg/cm^2 day over 1- to 10-day intervals through the growing season for 6 years. The plots were planted with cereal grains or turnips, and effects of treatment on the primary product, on straw, and on leaves were noted. No harmful effect of the dust was detected on either crop. There was a tendency for improvement of yields of grain and turnip roots, but this was not statistically significant.

G. Foundry Dusts

Changes in starch reserves were compared in common holly leaves, untreated and treated with dusts emitted from foundry operations (Steinhubel, 1962). The critical factor was the amount of light absorbed by the dust layer and not the effects of dust particles on transpiration or temperature of the leaf. These observations agree with some of those reported above on the range of effects of cement-kiln dust on vegetation.

H. Sulfuric Acid Aerosols

Sulfuric acid mist may form in the air as a result of sulfur dioxide oxidation to sulfur trioxide. The latter rapidly hydrates to form aerosols. Middleton *et al.* (1958) observed necrotic spots on the upper leaf surface of vegetation in the Los Angeles area following periods of heavy air pollution accompanied by fog. Thomas *et al.* (1952) exposed plants to sulfuric acid mists in concentrations ranging from 108 to 2160 mg/m^3. Necrotic spots were developed only on the upper surface of wet leaves.

III. Conclusions

There has been relatively little research on the effects of particulate matter on vegetation, and most of the experiments done to date have dealt with specific kinds of dusts rather than the conglomerate mixture normally encountered in the atmosphere.

The significance of dusts as phytotoxicants is not yet entirely clear but there is considerable evidence that certain fractions of cement-kiln dusts adversely affect plants when naturally deposited on moist leaf surfaces. Dry cement-kiln dusts appear to have little deleterious effect, but in the presence of moisture, the dust solidifies into a hard adherent crust which can damage plant tissue and inhibit growth. Moderate damage has been observed on the leaves of bean plants dusted at the rate of about 0.47 mg/cm^2 day (400 $tons/mile^2$ month) for 2 days and followed by exposure to naturally occurring dew. Similarly, a marked reduction in the growth of poplar trees 1 mile from a cement plant was observed after cement production was more than doubled.

At levels in excess of 1.0 gm^2 day (85 $tons/mile^2$ month), incrustations, premature needle drop, and shortening of each succeeding year's flush growth have been observed on the branches of fir trees. Although the mechanism by which injury occurs is not entirely understood, there is evidence to support theories such as screening of light, plugging of stomatal openings, and direct injury to the plant tissue by the chemical reaction of the dust on the leaf surface.

It should be noted, however, that the harmful effect of cement dusts on vegetation is not fully substantiated and has been questioned by some workers. The controversy that surrounds this subject is not surprising in view of the limited research to date. In addition, not all studies have been carried out under identical conditions or with dusts of the same composition. Studies of the effects of cement-kiln dusts deposited on the soil also raise questions. Some investigators report no harmful effects at levels from 1.5 to 7.5 g/m^2 day (130 to 640 $tons/mile^2$ month), while others report that concentrations from 1.0 to 48 g/m^2 day (86 to 4000 $ton/mile^2$ month) cause shifts in the soil alkalinity which may be favorable to one crop but harmful to another.

The great disparity between experimental results and the conclusions drawn by many investigators is due to the fact that the pollutant called "cement-kiln dust" is actually a heterogeneous substance whose constituents and amounts vary with time and location. No general conclusions can be drawn about the effects of cement-kiln dust until each dust source is classified and studied separately.

Fluorides in particulate form are less damaging to vegetation than gaseous fluorides. Fluoride may be absorbed from depositions of soluble fluoride on lead surfaces. However, the amount absorbed is small in relation to that entering the plant in gaseous form. The fluoride from particulates apparently has great difficulty penetrating the leaf tissue.

The research evidence suggests that few, if any, effects occur on vegetation at fluoride particulate concentrations below 2 $\mu g/m^3$. Concentrations of this magnitude can be found in the immediate vicinity of sources of fluoride particulate pollution, but they are rarely found in urban atmospheres. Fluorides absorbed or deposited on plants may be detrimental to animal health. Fluorosis in animals has been reported due to the ingestion of vegetation covered with particulates containing fluorides. In a similar manner arsenic poisoning of cattle and sheep has occurred from ingestion of arsenic-containing particles that had settled on vegetation.

Soot may clog stomata and may produce necrotic spotting if it carries with it a soluble toxicant, such as one with excess acidity. Magnesium oxide deposits on soils have been shown to reduce plant growth, while iron oxide deposits appear to have no harmful effects and may be beneficial. Sulfuric acid aerosols will cause leaf spotting. The levels at which these materials may produce a toxic response are not well defined.

Although lead concentrations and accumulations by plants in the vicinity of highways have been high, there are no known reports of injury to vegetation.

Acknowledgment

Portions of the research reported in this chapter were supported in part by the Environment Protection Agency under Research Grant AP 00272.

References

Anderson, P. J. (1914). The effect of dust from cement mills on the setting of fruit. *Plant World* **17,** 57–68.

Berge, H. (1965). Luftverunreinigungen in Raume Koln. *Allg. Forstztg.* **51–52,** 834–838.

Berge, H. (1966). Emissionsbedingte Eisenstaube und ihre Auswirkungen auf Wachstum and Ertrag landwirtschaftlicher Kulturen. *Z. Luftvereinigung (Dusseldorf)* **2,** 1–7.

Bohne, H. (1963). Schadlichkeit von Staub aus Zementwerken fur Waldbestande. *Allg. Forstz.* **18,** 107–111.

Brandt, C. J., and Rhoades, R. W. (1972). Effects of limestone dust accumulation on composition of a forest community. *Environ. Pollut.* **3,** 217–225.

Brandt, C. J., and Rhoades, R. W. (1973). Effects of limestone dust accumulation on lateral growth of forest trees. *Environ. Pollut.* **4,** 207–213.

Cannon, H. L., and Bowles, J. M. (1962). Contamination of vegetation by tetraethyl lead. *Science* **137,** 765–766.

Czaja, A. T. (1961a). Die Wirkung von verstaubtem Kalk und Zement auf Pflanzen. *Qual. Plant. Mater. Veg.* **8,** 184–212.

Czaja, A. T. (1961b). Zementstaubwirkungen auf Pflanzen: Die Entstehung der Zementkrusten. *Qual. Plant. Mater. Veg.* **8,** 201–238.

Czaja, A. T. (1962). Uber das Problem der Zementstaubwirkungen auf Pflanzen. *Staub* **22,** 228–232.

Czaja, A. T. (1966). Uber die Einwirkung von Stauben, speziell von Zementofenstaub auf Pflanzen. *Angew. Bot.* **40,** 106–120.

Darley, E. F. (1966). Studies on the effect of cement-kiln dust on vegetation. *J. Air Pollut. Contr. Ass.* **16,** 145–150.

Guderian, R. (1961). Kurzberichte: H. Pajenkamp. Einwirkung des Zementofenstaubes auf Pflanzen und Tiere. *Staub* **21,** 518–519.

Jennings, O. E. (1934). Smoke injury to shade trees. *Proc. Nat. Shade Tree Conf., 10th, 1934,* pp. 44–48.

Kloke, A., and Riebartsch, K. (1964). The contamination of plants with lead from motor exhaust gases. *Naturwissenschaften* **51,** 367–368.

Leh, H. O. (1966). Contamination of crop plants with lead from motor vehicle exhaust gases. *Gesunde Pflanz.* **18,** 21–24.

Lerman, S. (1972). Cement-kiln dust and the bean plant (*Phaseolus vulgaris* L. Black Valentine var.); in-depth investigations into plant morphology, physiology and pathology. Ph.D. Dissertation, University of California, Riverside.

McCune, D. C., Hitchcock, A. E., Jacobson, J. S., and Weinstein, L. H. (1965). Fluoride accumulation and growth of plants exposed to particulate cryolite in the atmosphere. *Contrib. Boyce Thompson Inst.* **23,** 1–22.

Middleton, J. T., Darley, E. F., and Brewer, R. F. (1958). Damage to vegetation from polluted atmosphere. *J. Air Pollut. Contr. Ass.* **8,** 7-15.

Miller, P. M., and Rich, S. (1967). Soot damage to greenhouse plants. *Plant Dis. Rep.* **51,** 712.

Motto, H. L., Daines, R. H., Chilko, D. M., and Motto, C. K. (1970). Lead in soils and plants: Its relationship to traffic volume and proximity to highways. *Environ. Sci. Technol.* **4,** 231–237.

Pack, M. R., Hull, A. C., Thomas, M. D., and Transtrum, L. G. (1959). Determination of gaseous and particulate inorganic fluorides in the atmosphere. *Amer. Soc. Test. Mater., Spec. Tech. Publ.* **281,** 27–44.

Page, A. L., and Ganje, T. J. (1970). Accumulation of lead in soils for regions of high and low motor vehicle traffic density. *Environ. Sci. Technol.* **4,** 140–142.

Page, A. L., Ganje, T. J., and Joshi, M. S. (1971). Lead quantities in plants, soil, and air near some major highways in southern California. *Hilgardia* **41,** No. 1.

Pajenkamp, H. (1961). Einwirkung des Zementofenstaubes auf Pflanze and Tiere. *Zem.-Kalk-Gips* **14,** 88–95.

Parish, S. B. (1910). The effects of cement dust on citrus trees. *Plant World* **13,** 288–291.

Peirce, G. J. (1910). An effect of cement dust on orange trees. *Plant World* **13,** 283–288.

Raymond, V., and Nussbaum, R. (1966). A propos des poussières de cimentéries et leurs effets sur l'homme, les plants, et les animaux. *Pollut. Atmos.* **3,** 284–294.

Scheffer, F., Przemeck, E., and Wilms, W. (1961). Untersuchungen uber den Einfluss von Zementofen-Flugstaub auf Boden and Pflanze. *Staub* **21,** 251–254.

Schonbeck, H. (1960). Beobachtungen zur Frage des Einflusses von industriellen Immissionen auf die Krankbereitschaft der Pflanze. *Ber. Landesanstalt Bodennutzungsschutz (Bochum)* **1,** 89–98.

Schuck, E. A., and Locke, J. K. (1970). Relationship of automotive lead particulates to certain consumer crops. *Environ. Sci. Technol.* **4,** 324–330.

Sievers, F. J. (1924). Crop injury resulting from magnesium oxide dust. *Phytopathology* **14,** 108–113.

Steinhubel, G. (1962). Zmeny v skrobovych rezervach listov cezminy po umelom znecisteni pevnym popraskom. *Biologia (Bratislava)* 18, 23–33; Abstract, "Veranderungen in den Starkereserven der Blatter der gemeinen Stechpalme nach einer kunstlichen Verunreinigung durch Staub," pp. 32–33, (in German).

Stratmann, H., and van Haut, H. (1956). "Vegetations versuche mit Zementflugstaub" (unpublished investigations of the Kohlen-Stoff-biologischen Forschungsstation). Essen, Germany.

Thomas, M. D., Hendricks, R. H., and Hill, G. R., Jr. (1952). Some impurities in the air and their effects on plants. *In* "Air Pollution" (L. C. McCabe, ed.), pp. 41–46. McGraw-Hill, New York.

Warren, H. V., and Delavault, R. E. (1962). Lead in some food crops and trees. *J. Sci. Food Agr.* **13,** 96–98.

Wentzel, K. F. (1962). Literaturberichte: H. Pajenkamp. Einwirking des Zementofenstaubes auf Pflanzen und Tiere. *Z. Pflanzenkr. (Pflanzenpathol.) Pflanzenschutz* **69,** 478.

8

PLANT RESPONSES TO POLLUTANT COMBINATIONS

R. A. Reinert, A. S. Heagle and W. W. Heck

I. Introduction

Air pollutants, responsible for vegetation injury and crop yield losses, are causing increased concern (Fujii, 1973; Heck and Brandt, 1975; Heggestad and Heck, 1971; Jacobson and Hill, 1970; Treshow, 1971). Pollutants rarely exist alone; instead, the air environment consists of a complex mixture of phytotoxic gases. Several reports in the 1950's suggested that the amount and type of foliar injury differed when plants were exposed to more than one pollutant. From these early studies, initial concepts were developed of how air pollutants in combination may affect plant response.

A. Early Considerations of Pollutant Combinations

Thomas *et al.* (1952) first observed that one pollutant might influence the effects of another pollutant on plants. They observed increased ambient oxidant injury to bean (*Phaseolus vulgaris* L. cv Pinto) grown in greenhouses equipped with water-spray scrubbers to remove sulfur dioxide (SO_2). This suggested that SO_2 lessened the effect of oxidants in causing foliar injury in *P. vulgaris*. However, the increased humidity in the greenhouse provided by the water-spray scrubbers might also have increased sensitivity of *P. vulgaris* to oxidants. Middleton *et al.* (1958) exposed *P. vulgaris* to SO_2 and ozone (O_3) separately and in mixtures of different ratios. Foliar injury was observed after 2 hours of exposure to 1.5 ppm SO_2 or 0.25 ppm O_3 separately. When *P. vulgaris* was exposed to a combination of 1.5 ppm SO_2 and 0.3 ppm O_3, a 5:1 ratio SO_2:O_3, SO_2 did not interfere with the amount of foliar injury caused by O_3. If the ratio was increased to 6:1, foliar symptoms of injury from both gases appeared; but when the ratio was decreased to 4:1, O_3 appeared to interfere with the expected injury from SO_2. Hitchcock *et al.* (1962) found no interacting effects on foliar injury to gladiolus (*Gladiolus hortulanus*) from combinations of SO_2 and hydrogen fluoride (HF) or SO_2 and hydrocarbons. They did observe, however, that hydrocarbons decreased the effects of fluorosilicic acid (H_2SiF_6) on *G. hortulanus*. Heck (1964) investigated foliar injury response to three hydrocarbons (ethylene, propylene, and acetylene) and to mixtures of these with products of irradiated propylene-nitrogen dioxide (NO_2) mixtures. He concluded that various combinations of these hydrocarbon gases did not interfere with the development of foliar injury.

After O_3 was identified as the cause of weather fleck in tobacco (*Nicotiana tabacum* L.) (Heggestad and Middleton, 1959), Menser and Heggestad (1966) reported that exposure to SO_2 (0.24 ppm) or O_3 (0.03 ppm) for 2 to 4 hours did not injure sensitive *N. tabacum* (cv Bel W_3). However, the same concentrations of SO_2 and O_3 together caused 23–48% foliar injury on four *N. tabacum* cultivars. This study encouraged plant scientists to review the potential effects of pollutant combinations on plants and stimulated new research on this subject.

B. Concepts and Terminology

Environmental and plant variables that affect the response of plants to single pollutants are discussed in previous chapters and these same variables probably influence the response of plants to pollutant combinations. Three additional variables may affect plant response to pollutant combinations; (1) the concentration of each gas in the combination exposures with respect to the injury thresholds of the separate pollutants; (2) the ratio of the concentration of each gas to the other; and (3) whether the combined pollutant stress is applied simultaneously, sequentially, and/or intermittently. In most studies of pollutant combinations, plants have been exposed to *simultaneous* pollutant additions or, mixtures. In several studies plants were exposed to one pollutant and followed by a second pollutant in a *sequential* exposure. The term *reciprocal* exposure has been used (Matsushima, 1971; Matsushima and Brewer, 1972) when the order of a *sequential* exposure has been reversed. However, the term appears redundant. Exposures are termed *intermittent* when a time period without pollutants is inserted between exposures to pollutants.

Tingey and Reinert (1975) have discussed terminology used to describe plant response to pollutant combinations. The effects of pollutant combinations can be less than the additive effects of the single pollutants (antagonistic, interference), greater than the additive effects of the single pollutants (synergistic, potentiate), or equal to the additive effects of the single pollutants. The term *predisposition* has been used when a stress factor predisposes a plant to a greater effect than expected from a given level of the same or another stress. Studies involving gaseous pollutant effects on plants have used all the above terms and concepts, but confusion may exist when these terms are used to describe a variety of plant responses (i.e., foliar injury, plant growth and yield, and/or biochemical changes). Before recommending the use of terms it is important to understand dose–response phenomena, physiological responses to pollutants, effects of pollutants on sites of entry, and the mechanism of pollutant action. In this chapter we will use the statements less than additive, additive, and greater than additive to describe the response of plants to pollutant combinations.

II. Plant Responses to Pollutant Combinations

Research on pollutant combinations has generally involved only two pollutants. Only one report concerns pollutant mixtures in combination with a biotic stress (Weber *et. al.,* 1974). Initially, foliar injury was the plant response measured, but recently, physiological, biomass, and yield responses have been reported. This section will discuss the response of plants to pollutant combinations.

A. Sulfur Dioxide and Ozone

The first pollutant combinations studied involved SO_2 with oxidants and subsequently with O_3 (Haagen-Smit *et al.,* 1952; Middleton *et al.,* 1958;

Thomas *et al.,* 1952). The greater than additive or "synergistic" effects first noted on *N. tabacum* (Menser and Heggestad, 1966) and the widespread occurrence of both SO_2 and O_3 have stimulated special interest in both pollutants. Concentrations of SO_2 and O_3 that might cause foliar injury exist mainly within and around major metropolitan areas of eastern United States. However high oxidant values do occur in rural areas of eastern United States and in large metropolitan areas throughout the world. The advent of large "mine-mouth" power plants burning high sulfur coals has increased the concern for air pollution effects in the rural east (United States). Hourly concentrations of 0.1 ppm SO_2 and 0.1 ppm O_3 singly and in mixtures may injure the foliage of a wide range of plant species, and multiple exposures may reduce productivity.

Generally, foliar injury caused by O_3 is easily distinguished from that caused by SO_2. Mixtures of these two gases at concentrations below the injury threshold for SO_2 and at or below the threshold for O_3 produce symptoms similar to those caused by O_3. Symptoms of SO_2 injury are seldom observed unless the concentration of SO_2 in the mixture is well above the injury threshold. Results of studies where plants were exposed simultaneously to SO_2 and O_3 are summarized in Tables I–III.

As mentioned earlier, one variable that may affect plant response to pollutant mixtures is the ratio of the concentration of one gas to another. Tingey *et al.* (1973a) exposed alfalfa (*Medicago sativa* L. cv Vernal), broccoli (*Brassica oleracea* var. *botrytis* L. cv Calabrese), cabbage *B. oleracea* var. *capitata* L. cv All Season, radish (*Raphanus sativus* L. cv Cherry Belle), tomato (*Lycopersicon esculentum* Mill. cv Roma VF) and tobacco (*Nicotiana tabacum* L. cv Bel W_3) to SO_2 and O_3 mixed in different ratios. Pollutant concentrations were 0.1, 0.25 and 0.5 ppm SO_2 and/or 0.05 or 0.1 ppm O_3 for 4 hours. Foliar injury response of the 6 species is summarized as additive, greater than additive or less than additive (Table IV). *Nicotiana tabacum, R. sativus,* and *M. sativa* had greater than additive foliar injury at certain ratios but not at others. There was no general trend indicating how ratios of pollutant concentrations influenced foliar injury response. Visual injury due to the SO_2/O_3 combinations frequently appeared on both leaf surfaces. The undersurface injury was often a silvering and a collapse of epidermal tissue. The upper surface injury was usually an interveinal necrotic fleck, stipple, and/or pigment accumulation.

Eastern white pine (*Pinus strobus* L.) has been reported to be sensitive to both SO_2 and O_3. Young pine needles on different branchlets of the same tree were independently injured by either 0.05–0.15 ppm SO_2 or 0.3 ppm O_3 for 2 hours (Costonis, 1970). In later studies, Costonis (1973) observed the greatest amount of injury to white pine needles from a sequential

TABLE I

RESPONSE OF HORTICULTURAL CROPS TO SULFUR DIOXIDE AND OZONE MIXTURES[a,b]

Plant species	SO_2/O_3 (ppm)	Exposure duration (hours)	Plant response[c]	Mixture response[d]	Plant age during exposure (weeks)
Phaseolus vulgaris L.	1.7/0.19	0.5	24	+	3
	1.6/0.42	0.33	74	*	3
Phaseolus limensis Macf.	0.25/0.05	4	0	0	3
	1.0/0.1	4	0	0	3
Brassica oleracea var. *botrytis* L.	0.25/0.05	4	3	0	3
	1.0/0.1	4	32	+	3
	0.5/0.05	4	17	+	3
	0.1–0.5/0.1	4	11–34	+,0	3
Brassica oleracea var. *capitata* L.	0.25/0.05	4	0	0	3
	1.0/0.1	4	28	0	3
	0.5/0.05	4	4	0	3
	0.1–0.5/0.1	4	14–54	0,+	3
Allium cepa L.	0.25/0.05	4	0	0	3
	1.0/0.1	4	3	0	3
Spinacea oleracea L.	0.25/0.05	4	0	*	3
	1.0/0.1	4	1	0	3
Lycopersicon esculentum Mill.	0.25/0.05	4	3	0	4
	1.0/0.1	4	2	0	4
	0.5/0.05	4	1	0	4
	0.1–0.5/0.1	4	10–50	–,0	4
Raphanus sativus L.	0.25/0.05	4	7	0	3
	1.0/0.1	4	45	+	3
	0.5/0.05	4	7	0	3
	0.1–0.5/0.1	4	22–50	+	3
R. sativus	0.05/0.05	8/day, 5 days/week, 5 weeks	10 TDW	0	1–5
			55 RDW	—	
R. sativus[e]	0.45/0.45	2	16 TDW	0	2
	0.45/0.45		64 RDW	0	
	0.45/0.45	4	16 TDW	0	2
	0.45/0.45		70 RDW	0	

[a] Plants were exposed in greenhouse chambers.

[b] Data taken from Matsushima, 1971; Tingey *et al.*, 1973a; Tingey *et al.*, 1971a; Tingey and Reinert, 1975.

[c] The values listed are percent injury or percent reductions from control for the indicated growth measures. Top dry weight, TDW; root dry weight, RDW.

[d] The mixture responses are +, greater than additive; 0, additive; –, less than additive; *, not defined.

[e] The plants were harvested at 4 weeks from seeding.

TABLE II
RESPONSE OF FIELD CROPS AND CERTAIN CONIFER SPECIES TO SULFUR DIOXIDE AND OZONE MIXTURES[a]

Plant species	Exposure chamber[b]	SO_2/O_3 (ppm)	Exposure duration (hours)	Plant response[c]	Mixture response[d]	Plant age during exposure (weeks)
Pinus strobus L.	GH	0.1/0.1	8 hours/day, 5 days/week, 8 weeks	16	+	[e]
P. strobus	GH	0.025/0.05	6	–	*	[e]
Medicago sativa L.	GH	0.05/0.05	8 hours/day, 5 days/week, 12 weeks	18 TDW 24 RDW	– –	1–12
M. sativa	GH	0.25/0.05	4	0	*	3
		1.0/0.1	4	3	0	3
		0.5/0.05	4	2	0	3
		0.1–0.5/0.1	4	21–60	+	3
Bromus inermis Leyss.	GH	0.25/0.05	4	0	*	3
		1.0/0.1	4	4	0	3
Glycine max (L.) Merr.	GH	0.25/0.05	4	0	*	3
		1.0/0.1	4	1	0	3
G. max (2 cv)	GH	0.5/0.5	7 hours/day, 5 days/week, 3 weeks	24 RDW	+	0–3
G. max	F	0.1/0.1	6 hours/day, 5 days/week	52 TDW 63 seed wt	0 0	2–19
G. max (8 cv)	F	1.0/0.3	6	13–38	–	8
Gossypium hirsutum L. (12 cv)	F	1.0/0.3	6	10–24	–	9

[a] Data taken from Dochinger *et al.*, 1970; Houston, 1970; Tingey *et al.*, 1973a; Tingey *et al.*, 1973b; Tingey and Reinert, 1975; Heagle *et al.*, 1974; Heagle and Neely, personal communication.

[b] GH, greenhouse; F, field.

[c] The values listed are percent injury or percent reductions from control for the indicated growth and yield measures. Top dry weight, TDW, root dry weight, RDW.

[d] +, Greater than additive; 0, additive; –, less than additive; *, not defined.

[e] The white pine ramets were grafted onto 1–2-year-old seedling rootstocks and the injury evaluated on current year needles.

TABLE III

RESPONSE OF *Nicotiana* sp. TO SULFUR DIOXIDE AND OZONE MIXTURES[a]

Nicotiana type or species[b]	Exposure chamber[c]	SO_2/O_3 (ppm)	Exposure duration (hours)	Plant response[d]	Mixture response[e]	Plant age during exposure (weeks)
Samsun	CE	0.45/0.03	2–4	2	+	10–12
Samsun NN	CE	0.45/0.03	2–4	7	+	10–12
Xanthi	CE	0.45/0.03	2–4	4	+	10–12
N. glutinosa L.	CE	0.45/0.03	2–4	32	+	10–12
N. rustica L. var. *brasilia*	CE	0.45/0.03	2–4	34	+	10–12
Havana (4 cv)	CE	0.45/0.03	2–4	8–15	+	9–10
Flue-cured	CE	1.0/0.5	½–10	0–60	+	5
(2 cv)	CE	0.2–1.0/0.2	½–15	0–30	+	5
Burley (9 cv)	CE	0.4/0.03	2–4	4–27	+	10–12
Maryland (6 cv)	CE	0.5/0.1	2	–	+	
Cigar wrapper	GH	0.25/0.05	4	17	+	5
Cigar wrapper	GH	1.0/0.1	4	48	+	5
Cigar wrapper	GH	0.5/0.05	4	60	+	5
Cigar wrapper	GH	0.1–0.5/0.1	4	88–96	+,0	5
Flue-cured	GH	0.25/0.05	4	1	0	5
Flue-cured	GH	1.0/0.1	4	3	0	5
Cigar wrapper	GH	0.05/0.05	7 hours/day, 5 days/week, 4 weeks	32 TDW 49 RDW	0 0	3–7

[a] Data taken from Grosso *et al.*, 1971; Hodges *et al.*, 1971; Macdowall and Cole, 1971; Menser and Hodges, 1970; Tingey *et al.*, 1973a; Tingey and Reinert, 1975.

[b] Types listed are from *N. tabacum* L.

[c] CE, Controlled environment; GH, greenhouse.

[d] The values listed are percent injury or percent reductions from control for the indicated growth measures. Top dry weight, TDW; root dry weight, RDW.

[e] +, Greater than additive; 0, additive.

exposure to 0.05 ppm O_3 for 2 hours and then 0.05 ppm SO_2 for 2 hours, followed 24 hours later by a 2-hour exposure to a mixture of 0.05 ppm SO_2 and 0.05 ppm O_3. A mixture of 0.05 ppm SO_2 and 0.05 ppm O_3 produced less needle injury than exposure to SO_2 or O_3 alone. The exposure durations and concentrations of the pollutant mixture reported by Costonis to be injurious were less than those reported from other studies. The results from Costonis may reflect extremely sensitive white pine ramets, sensitive exposure conditions, or both.

The foliar injury response of various *N. tabacum* types—havana (Hodges *et al.,* 1971), burley (Menser and Hodges, 1970), Maryland

TABLE IV
SUMMARY EFFECTS OF SULFUR DIOXIDE AND OZONE MIXTURES ON FOLIAR INJURY[a]

Plant species	Sulfur dioxide/ozone concentrations (ppm)			
	0.5/0.05	0.1/0.1	0.25/0.1	0.5/0.1
Medicago sativa L.	–[b]	+	+	+
Brassica oleracea var. *botrytis* L.	+	+	0	0
B. oleracea var. *capitata* L.	0	0	0	+
Raphanus sativus L.	0	+	+	+
Lycopersicon esculentum Mill.	0	–	0	0
Nicotiana tabacum L. Bel W_3	+	0	+	+

[a] Data taken from Tingey *et al.*, 1973a. Exposure duration was 4 hours.
[b] +, Injury from the mix was greater than the additive injury from the individual pollutants (greater than additive); 0, injury from the mix was equal to the additive injury from the individual pollutants (additive); –, injury from the mix was less than the additive injury from the individual pollutants (less than additive).

(Menser *et al.,* 1973), and cigar wrapper (Menser and Heggestad, 1966)—to SO_2 and O_3 alone and in a mixture has been studied. In addition, various *Nicotiana* species and cultivars within the various *N. tabacum* types have been evaluated for foliar injury from exposure to pollutant mixtures (Grosso *et al.,* 1971). In some instances, relative sensitivity of the cultivars in the mixture and single pollutant exposures was compared with relative sensitivity to photochemical oxidants under field conditions. The relative sensitivity among certain tobacco cultivars exposed to O_3 was similar to sensitivity of the same cultivars exposed to a SO_2–O_3 mixture, or ambient air. In many of the other cultivars tested, there was very little similarity. Heagle and Neely (personal communication) compared relative foliar injury among 12 cultivars of cotton (*Gossypium hirsutum* L.) and 8 cultivars of soybean [*Glycine max* (L.) Merr.] exposed in the field to SO_2 and O_3 alone and as a mixture, as well as ambient air. Comparable rankings among the cultivars exposed to the single pollutants and mixtures were rare.

Most research on the effects of pollutant combinations on plant development and productivity has concerned SO_2 and O_3. Several studies reported reduced growth, whereas others show an inhibition in some other aspect of plant development or productivity from pollutant combinations. Ozone and SO_2 have been suspect in the chlorotic dwarf decline, a growth disorder of *P. strobus*. Evidence that air pollution was the primary cause of chlorotic dwarf was obtained when dwarfed *P. strobus* recovered and grew normally in chambers with charcoal-filtered air (Dochinger, 1968).

Dochinger *et al.* (1970) showed that SO_2 and O_3 in a mixture could shorten and mottle needles and dwarf plants. In these studies sensitive and resistant white pine ramets were exposed in greenhouse chambers to 0.1 pm SO_2 and 0.1 ppm O_3 alone and in a mixture for 8 hours/day, 5 days/week for 4–8 weeks. Control plants were also in charocal-filtered air chambers within the greenhouse. No chlorotic dwarf symptoms were observed on the exposed resistant white pine ramets or on sensitive control plants. The percent mottling of current needles and premature drop of the older needles exposed to the mixture was approximately 16%. Exposure to each gas, separately, produced a 3–4% mottling and premature drop. It was suggested that the breakdown of chlorophyll and loss of needles from these chronic exposures could reduce growth through the inability of the needles to supply the normal photosynthetic needs of white pine.

A series of studies have attempted to demonstrate that SO_2 and O_3 acting alone or in a mixture at concentrations corresponding to those recorded adjacent to urban areas (Reinert *et al.,* 1970) significantly influenced growth and yield in several crops. Tingey *et al.* (1971a) exposed *R. sativus* to 0.05 ppm SO_2 or 0.05 ppm O_3 alone and in a mixture. Exposures began 3–4 days after seeding and lasted 8 hours/day, 5 days/week until harvest at 5 weeks. All treatments reduced top and root growth when compared with control plants grown in charcoal-filtered air. The inhibition of top growth in the pollutant mixture was additive, but the inhibition of root growth was significantly less than additive.

In studies similar to the *R. sativus* study, Reinert *et al.* (1969) investigated the early growth and development of *N. tabacum* Bel W_3 and Tingey *et al.* (1973b) studied the early growth of *G. max.* Tingey and Reinert (1975) summarized the chronic effects of SO_2 and O_3 acting alone or in various mixtures on these three crops, as well as *M. sativa* (Table V). Exposures to a mixture of 0.05 ppm SO_2 and 0.05 ppm O_3 (7–8 hours/day, 5 days/week/3–5 weeks) caused greater than additive inhibition of the early root growth of *G. max,* additive inhibition of *N. tabacum* growth, and less than additive inhibition of *M. sativa* growth. Exposures to 0.45 ppm SO_2 and 0.45 ppm O_3 alone and in a mixture caused inhibition of *R. sativus* growth that was additive for both foliage and roots.

A few studies involving SO_2 and O_3 have been done in portable field exposure chambers, about 8 ft^3, which has enabled measurement of yields of mature plants. *G. max* exposed to a mixture of 0.1 ppm SO_2 and 0.1 ppm O_3 (6 hours/day. 19 weeks) showed growth and yield decreases that were somewhat greater than the additive effects of the single gases, but the differences between the mixture and O_3 treatments were not significant (Heagle et al., 1974). The effects of acute doses of a SO_2–O_3 mixture on yield of peanut (*Arachis hypogaea* L.) was somewhat more

TABLE V
SUMMARY EFFECTS OF SULFUR DIOXIDE AND OZONE MIXTURES ON PLANT GROWTH[a,b]

Plant species	Foliage growth	Root growth
Medicago sativa L.	−	−
Raphanus sativus L.[c]	0	0
R. sativus[d]	0	−
Glycine max[e] L. Morr.	0	+
Nicotiana tabacum L. Bel W_3[f]	0	0

[a] Data taken from Tingey and Reinert, 1975.

[b] +, Growth reduction from the mix was greater than the additive reduction from the individual pollutants (greater than additive); 0, growth reduction from the mix was equal to the additive reduction from the individual pollutants (additive); −, growth reduction from the mix was less than the additive reduction from the individual pollutants (less than additive).

[c] Radishes received a single exposure for 2 hours at 14 days of age to 0.45 ppm SO_2 and/or 0.45 ppm O_3.

[d] Data from Tingey *et al.*, 1971a.

[e] Data from Tingey *et al.*, 1973b.

[f] Data from Reinert *et al.*, 1969.

but not significantly different from the additive effects of each gas alone (Heagle and Trent, personal communication).

There is evidence that microbial activity in the rhizosphere can be altered by mixtures of SO_2 and O_3. *Rhizobium* nodulation of soybean was decreased throughout a 79-day growth period by exposure to 0.25 ppm O_3 and to a mixture of 0.25 ppm SO_2 and 0.25 ppm O_3 (4 hours/day, 3 day/week) (Weber *et al.*, 1974). Sulfur dioxide at 0.25 ppm did not decrease nodulation. The SO_2–O_3 mixture also affected the feeding habits of four species of nematodes parasitizing soybean. *Heterodera glycines,* a sedentary semi-endoparasite, was inhibited by O_3 and by the mixture of SO_2–O_3 as measured by the number of cysts and males per plant. The reproduction of *Trichodorus christiei* was suppressed by O_3 and by the mixture of SO_2–O_3, but reproduction in *Belonolaimus longicaudatus* was unaffected. *Pratylenchus penetrans* a migratory endoparasite, was stimulated by SO_2 and the SO_2–O_3 mix. Studies of this type are important in understanding how pollutant stress affects parasitism of plants by microorganisms both in cultivated and natural plant ecosystems.

B. Sulfur Dioxide and Nitrogen Dioxide

Foliar injury response to a SO_2–NO_2 mixture was first reported on *N. tabacum* Bel W_3 (Heck, 1968; Tingey *et al.*, 1971b). This pollutant combination may be as important as combinations of SO_2 and O_3, because SO_2 and NO_2 are components of fossil fuel combustion (primarily coal for power plants) and thus normally occur together. In large metropolitan areas the NO_2 from automobiles is added to the concentrations emitted from power plants. Foliar injury to SO_2–NO_2 mixtures are summarized in Table VI.

Foliar injury on six plant species exposed to mixtures of SO_2 and NO_2 at different ratios has been described by Tingey *et al.* (1971b). The concentrations of SO_2 and NO_2 were below the foliar injury threshold for the individual gases and ranged from 0.05–0.25 ppm for a 4-hour duration. The six plant species in order of decreasing sensitivity to the SO_2–NO_2 mixtures were *G. max* Hark, *R. sativus* Cherry Belle, *N. tabacum* Bel W_3, *P. vulgaris* Pinto, oat (*Avena sativa,* L. cv Clintland 64) and *L. esculentum* Roma VF. Upper surface injury usually was typical of symptoms caused by O_3. Lower surface injury was either a reddish pigmentation or a silvering that was not usually related spatially to upper surface injury.

Matsushima (1971) exposed *P. vulgaris* Pinto, *L. esculentum, A. sativa,* cucumber (*Cucumis sativus* L.), and pepper (*Capsicum frutescens* L.) to combinations of SO_2 (1.5–2.3 ppm) and NO_2 (12–15 ppm) for 40–70 minutes. Foliar injury occurred on all five species. At these high concentrations the individual gases often caused injury. The SO_2–NO_2 mixture caused both greater than and less than additive injury, depending on the plant species. For example, 1-hour exposures to 2.3 ppm SO_2 or 13 ppm NO_2 caused 10–18% injury on the third leaf of *L. esculentum;* the mixture caused 78% injury. On the second leaf of *A. sativa,* the effects were less than additive; 2.4 ppm SO_2 caused 58% injury, 13 ppm NO_2 caused none, and the mixture caused 39% injury.

Bennett *et al.* (1974) exposed several crop species to SO_2, NO_2, and mixtures of the two gases for 1 and 3 hours. Concentrations of the two pollutants ranged from 0.1–1.0 ppm and the species exposed included *A. sativa, P. vulgaris* Pinto, *R. sativus,* L. *esculentum,* Swiss chard (*Beta vulgaris* var. *cicla* L.), and sweet pea (*Lathyrus odoratus* L.). Visible foliar injury was observed in *R. sativus* exposed to a mixture of 0.5 ppm SO_2 and 0.5 ppm NO_2. Between 0.75 and 1.0 ppm of each pollutant was required in the mixture to injure *B. vulgaris* var. *cicla, A. sativa,* and *L. odoratus.* The individual gases did not produce injury at the concentrations used.

Symptoms from exposure to SO_2–NO_2 combinations often resemble those caused by O_3, especially when concentrations of the combined pollutants are near or below the injury threshold of the pollutants individually. This mimicking of O_3 symptoms, coupled with the fact that O_3 injury is widespread throughout the United States, makes identification of the cause of

TABLE VI

PLANT RESPONSE TO SULFUR DIOXIDE AND NITROGEN DIOXIDE MIXTURES[a]

Plant species	Exposure chamber[b,d]	SO_2/NO_2 (ppm)	Exposure duration (hours)	Plant response (% injury)	Mixture response[c,d]	Plant age (weeks)
Avena sativa L.	CE	0.75/0.75	1 or 3	0–5	+	4–5
Beta vulgaris var. *cicla* L.	CE	0.75/0.75	1 or 3	0–5	+	4–5
Lathyrus odoratus L.	CE	0.75/0.75	1 or 3	0–5	+	4–5
Raphanus sativus L.	CE	0.75/0.75	1 or 3	5–8	+	4–5
A. sativa	CE	0.15–0.25/0.1–0.2	4	0	*	2–3
R. sativus	CE	0.15–0.25/0.1–0.2	4	0	*	2–3
Phaseolus vulgaris L.	CE	0.15–0.25/0.1–0.2	4	0	*	2–3
R. sativus	CE	0.5/0.5	1 or 3	0–5	+	4–5
Nicotiana tabacum L.	GH	0.1/0.1	4	0–10	+	7–8
Oryzopsis hymendoides (R&S.) Ricker	F	0.5–0.7/0.15–0.21	2	16	0	*
Populus tremuloides Michx.	F	0.5–0.7/0.15–0.21	2	1	0	*
Sphaeralcea munroana Spach.	F	0.5–0.7/0.15–0.21	2	31	0	*
P. vulgaris	*	1.5/15	1.17	70–75	+	3–4
Lycopersicon esculentum Mill.	*	2.3/13	1	35–85	+	3–5
Cucumis sativus L.	*	2.3/12	0.67	50–100	+	3–4
A. sativa	*	2.4/13	1	40–75	−	3–4
Capsicum frutescens L.	*	2.4/15	1	10–58	+	5–6
P. vulgaris Pinto	GH	0.05–0.25/0.05–0.25	4	0–24	+	3–4
A. sativa	GH	0.05–0.25/0.05–0.25	4	0–27	+	3–4
R. sativus	GH	0.05–0.25/0.05–0.25	4	0–27	+	3–4
Glycine max (L.) Merr.	GH	0.05–0.25/0.05–0.25	4	0–35	+	3–4
N. tabacum	GH	0.05–0.25/0.05–0.25	4	0–18	+	7–8
L. esculentum	GH	0.05–0.25/0.05–0.25	4	0–17	+	5–6

[a] References used: Bennett *et al.*, 1975; Heck, 1968; Hill *et al.*, 1974; Matsushima, 1971; Tingey *et al.*, 1971b.

[b] CE, Controlled environment; GH, greenhouse; F, field.

[c] +, Greater than additive; 0, additive; −, less than additive.

[d] * Not defined.

pollutant injury more difficult. Foliar injury heretofore assumed to be caused by O_3 may, in fact, be caused by combinations of SO_2 and NO_2.

There are no known reports of plant growth responses to combinations of SO_2 and NO_2. Skelly *et al.* (1972) observed that growth of *P. strobus* was limited near a source of SO_2 and NO_2 where the levels of each pollutant often exceeded 0.3 and 0.5 ppm, respectively. There are several reports of growth effects from the individual pollutants (NO_2 and SO_2) at concentrations below 1.0 ppm. Thus, the impact of combinations of SO_2 and NO_2 below concentrations of 1.0 ppm may have pronounced effects on plant growth and development.

C. Sulfur Dioxide and Hydrogen Fluoride

Both SO_2 and HF are emitted from various industrial sources. Foliar symptoms and sensitivity of many species to exposure to each gas singly are well known. Both gases can disrupt metabolic processes in plants, which may reduce growth and yield. All reported effects of SO_2 and HF combinations are summarized in Table VII.

Matsushima and Brewer (1972) studied the influence of SO_2 and HF singly and in combination on Koethen sweet orange (*Citrus sinensis*) and Satsuma orange (*Citrus nobilis* var. *unshiu*). No visible injury or leaf abscission were observed on *C. sinensis*. However, the mixture of 0.8 ppm SO_2 and 2.3 to 17.1 ppb HF in a continuous 23-day exposure additively decreased linear growth of branches and reduced leaf area. With *C. nobilis* var. *unshiu,* exposure to the same mixture for 15 days caused foliar injury that was less than additive and did not affect growth.

Mandl *et al.* (1975) reported the effects of SO_2 and HF on foliar injury of *P. vulgaris* Pinto, barley (*Hordeum vulgare* L.), and sweet corn (*Zea mays* L.). A 7-day exposure to a mixture of 0.15 ppm SO_2 and 0.60 ppb HF caused a similar amount of injury on *H. vulgare* and *Z. mays* as did SO_2 alone; HF alone caused no injury. When the concentration of SO_2 in the mixture was decreased to 0.08 ppm and the exposure time increased to 21 days, the amount of foliar injury on *Z. mays* and *H. vulgare* was greater than additive.

D. Other Pollutant Combinations

Combinations of SO_2 and peroxyacetyl nitrate (PAN) may be important, but little is known of the potential effects. PAN is a major phytotoxic component of the photochemical oxidant complex in the Los Angeles basin and may occur with SO_2 in local areas. We do not know whether significant levels of PAN occur in the eastern United States where SO_2 pollution is

TABLE VII

PLANT RESPONSE TO SULFUR DIOXIDE AND HYDROGEN FLUORIDE MIXTURES[a]

Plant species	Exposure chamber[b,g]	SO_2/HF (ppm/ppb)	Exposure duration	Plant response[c]	Mixture response[d,g]	Plant age during exposure (weeks)
Citrus sinensis, Osbeck	GH	0.8/2.5–13	23 days, 10-day interruption[h]	52 SL 36 LA	0 0	[e]
Citrus nobilis var. *unshiu*, Swingle	GH	0.8/2.5–13	15 days, 10-day interruption[h]	4 2 abs	– –	[e]
Phaseolus vulgaris L.	CE	0.08/0.6	27 days	0	*	1–4
Hordeum vulgare L.	CE	0.08/0.8	27 days	60	+	1–4
Zea mays L.	CE	0.08/0.6	27 days	3019#	+	3–7
Lycopersicon esculentum Mill.	*	0.5/5	4 hr, 2 day/wk	[f]	*	2
P. vulgaris L.	*	0.5/5	4 hr, 2 day/wk	[f]	*	2

[a] Data taken from Matsushima and Brewer, 1972; Mandl *et al.*, 1975; Solberg and Adams, 1956.
[b] GH, Greenhouse; CE, controlled environment.
[c] The values listed are percent injury or percent reduction from control for the indicated response measures (SL, stem length; LA, leaf area; abs, abscission; #, number of lesions).
[d] +, Greater than additive; 0, additive; –, less than additive; *, not defined.
[e] Experiments were initiated on 2-year-old seedlings.
[f] Microscopic injury based on various microtechnique methods.
[g] *, Not defined.
[h] The interruption occurred following the first 12 days of exposure.

fairly common. Matsushima (1971) found that a 30- to 45-minute exposure to a mixture of 1.5 to 2.1 ppm SO_2 and 0.27 to 0.4 ppm PAN caused foliar injury on *P. vulgaris* Pinto, *C. frutescens,* and *L. esculentum*. Foliar injury observed from exposure to the two pollutants alone was usually less than injury observed from exposure to the mixture.

Ozone and NO_2 occur together in photochemical air pollution, but little is known of how such a combination might affect plants. Matsushima (1971) found less than additive foliar injury on *L. esculentum* and *C. frutescens* from a 50-minute to both 0.4 ppm O_3 and 15 ppm NO_2.

Ozone and PAN exist together in photochemical oxidant pollution in the West and possibly in eastern United States. Except for observations of plant injury in the field where such mixtures occur, we know very little about their combined effects on vegetation. Kress (1972) studied the response of hybrid poplar (*Populus maximowiczii* X *trichocarpa*) exposed sequentially to O_3 and PAN. In many of the exposures he found that the total foliar injury was greater than the additive injury from the separate exposures. Kohut (1972) exposed hybrid poplar to O_3 and PAN singly and to a mixture of both gases simultaneously and found that the amount of foliar injury from the mixture varied from less than additive to greater than additive compared with foliar injury from separate exposures to the two pollutants.

Thompson and Taylor (1969) undertook a major study of the effects of photochemical oxidants and fluoride on growth and yield of several *Citrus* sp. They were not studying pollutant combinations per se but their methodology included several mixtures of HF, O_3, PAN, and NO_2. However, the nature of their experimental design does not permit an evaluation of the effects of pollutant combinations. They did report as much as 50% reduction of fruit yield in these experiments.

III. Implications of Pollutant Combinations

Plant scientists should be challenged to define the effects of pollutant combinations so that the impacts on the production of quality food, feed, and fiber can be understood. We now know that some plant species may respond differently to pollutant combinations than to single pollutants. However, our knowledge is fragmentary and many questions remain unanswered. For example: (1) Does the ratio of one pollutant to another influence plant response? (2) Does one pollutant predispose plant systems to the effects of another? (3) What are the dose–response relationships for pollutant combinations? (4) Does response of different cultivars within plant species differ for different pollutant combinations? (5) Are plant development, yield, quality influenced by pollutant combinations in ways different from single pollutant responses? Mechanisms of action of individual pollutants are poorly understood; obviously much less is known for combined pollutants. We know very little of physiological or biochemical changes in plants exposed to pollutant combinations. One report indicates a greater than additive inhibitory effect on photosynthesis in alfalfa by a mixture of SO_2 and NO_2 (White *et al.*, 1974).

Although pollutant combinations may cause less than, greater than, or additive responses in a number of plant species, scientists have not begun to consider the long-term importance of pollutant combinations on plant communities. Cultivated and native ecosystems are subject to a vast list of selection pressures to which adaptation is a prerequisite for high yield or survival. Air pollutants both individually and in combination offer new components to the complexity of environmental stresses influencing plant populations and communities. Thus we suggest that: (1) Air pollutants may cause irreversible changes within natural ecosystems that influence either productivity or plant succession (*P. strobus* and *P. ponderosa* decline may be indicative of these impacts on an ecosystem). (2) Plant ecosystems at great distances from pollutant sources may be affected. There is strong evidence that species diversity and frequency do change around industrial sources of SO_2 and HF (Treshow, 1968; Ferry *et al.*, 1973). Yet even here, pollutants do not exist alone (i.e., in the combustion of coal, fluoride, sulfur oxides, and oxides of nitrogen are all released into the atmosphere).

Meteorological factors affecting pollutant distribution must be adequately understood. We know that photochemical pollution is a dynamic mixture and that primary pollutants from point sources, such as smelters and power plants, are slowly transformed via oxidation and hydration into components that may also be important toxicants in an ecosystem. This dynamic air environment and the many environmental and biotic factors that impinge on plant species and affect their response to pollutants and pollutant combinations must be understood. Our lack of understanding as to how these factors interact makes it difficult to accurately predict plant response.

Some of the present concepts of air pollution control relate to the development of defensible air quality standards for given pollutants. It may be impossible to develop a standard that protects all plants from foliar injury or other deleterious effects without being economically harmful. Some plants may be injured under some conditions with our present standards. However the specific conditions may occur so infrequently that even sensitive plants will normally show no effects. It is possible then, that standards allowing higher pollutant concentrations could be tolerated in some areas even though some plant injury might occur. These arguments are valid only when individual pollutants are considered. The few available results suggest that the greater than additive, less than additive, and additive effects of pollutant combinations can make any attempts to set reasonable standards for individual toxicants very difficult. It is critical that we learn more concerning plant response to pollutant combinations.

We have attempted to identify most of the implications and known effects of pollutant combinations on plants. Considering the knowledge base of 5–10 years ago, we are slowly moving toward expanding this base. This chapter has attempted to stress the importance of understanding pollutant combinations. Hopefully it has provided inspiration to acquire the necessary knowledge as quickly as possible.

References

Bennett, J. H., Hill, A. C., Soleimani, A., and Edwards, W. H. (1975). Acute effects of combinations of sulfur dioxide and nitrogen dioxide on plants. *Environ. Pollut.* (in press).

Costonis, A. C. (1970). Acute foliar injury of eastern white pine induced by sulfur dioxide and ozone. *Phytopathology* **60,** 994–999.

Costonis, A. C. (1973). Injury of eastern white pine by sulfur dioxide and ozone alone and in mixtures. *Eur. J. Forest Pathol.* **3,** 50–55.

Dochinger, L. S. (1968). The impact of air pollution on eastern white pine: The chlorotic dwarf disease. *J. Air Pollut. Contr. Ass.* **18,** 814–816.

Dochinger, L. S., Bender, F. W., Fox, F. O., and Heck, W. W. (1970). Chlorotic dwarf of eastern white pine caused by ozone and sulfur dioxide interaction. *Nature* **225,** 476.

Ferry, B. W., Baddeley, M. S., and Hawksworth, D. L., eds. (1973). "Air Pollution and Lichens." Oxford Univ. Press (Athlone), London and New York.

Fujii, S. (1973). The current state of plant damage by air pollution in Okayama Perfecture *Shokubutsu Boeki.* **27,** 249–252.

Grosso, J. J., Menser, H. A., Hodges, G. H., and McKinney, H. H. (1971). Effects of air pollutants on Nicotiana cultivars and species used for virus studies. *Phytopathology* **61,** 945–950.

Haagen-Smit, A. J., Darley, E. F., Zaitlin, M., Hull, H., and Noble, W. (1952). Investigation on injury to plants from air Pollution in the Los Angeles area. *Plant Physiol.* **27,** 18–34.

Heagle, A. S., Body, D. E. and Neely, G. E. (1974). Injury and yield responses of soybean to chronic doses of ozone and sulfur dioxide in the field *Phytopathology* **64,** 132–136.

Heck, W. W. (1964). Plant injury induced by photochemical reaction products of propylene-nitrogen dioxide mixtures. *J. Air Pollut. Contr. Ass.* **14,** 255–261.

Heck, W. W. (1968). Effects of oxidant and pollutants (discussion of A. C. Taylor, 1968). *J. Occup. Med.* **10,** 496–499.

Heck, W. W., and Brandt, C. S. (1975). *In* "Air Pollution" (A. C. Stern, ed.), 3rd ed., Vol. 1, Chapter 14, (in press). Academic Press, New York.

Heggestad, H. E., and Heck, W. W. (1971). Nature, extent and variation of plant response to air pollutants. *Advan. Agron.* **23,** 111–145.

Heggestad, H. E., and Middleton, J. T. (1959). Ozone in high concentrations as cause of tobacco leaf injury. *Science* **129,** 208–210.

Hill, A. C., Hill, S., Lamb, C., and Barrett, T. W. (1974). Sensitivity of native desert vegetation to SO_2 and to SO_2 and NO_2 combined. *J. Air Pollut. Contr. Ass.* **24,** 153–157.

Hitchcock, A. E., Zimmerman, P. W., and Coe, R. R. (1962). Results of ten year's work (1951–1960) on the effect of fluorides on Gladiolus. *Contrib. Boyce Thompson Inst.* **21,** 303–344.

Hodges, G. H., Menser, H. A., Jr., and Ogden, W. B. (1971). Susceptibility of Wiscousin Havana tobacco cultivars. *Agron. J.* **63,** 107–111.

Houston, D. B. (1970). Physiological and genetic response of *Pinus strobus,* L. clones to sulfur dioxide and ozone exposures. Ph.D. Thesis, University of Wisconsin, Madison.

Jacobson, J. S., and Hill, A. C., eds. (1970). "Recognition of Air Pollution Injury to Vegetation: A Pictorial Atlas." Air Pollut. Contr., Ass., Pittsburgh, Pennsylvania.

Kress, L. W. (1972). Response of Hybrid Poplar to Sequential Exposures of Ozone and PAN, MS Thesis, CAES Publ. No. 259–72. Cent. Air Environ. Stud., Penn. State University, State College, Pa.

Kohut, R. J. (1972). Response of hybrid poplar to simultaneous exposure to ozone and PAN. MS Thesis, CAES Publ. No. 288–72 Cent. Air Environ. Stud., Penn. State University, State College, Pa.

Macdowall, F. D. H., and Cole, A. F. W. (1971). Threshold and synergistic damage to tobacco by ozone and sulfur dioxide. *Atmos. Environ.* **5,** 553–559.

Mandl, R. H., Weinstein, L. H., and Keveny, M. (1975). Effects of hydrogen fluoride and sulfur dioxide alone and in combination on several species of plants. *Environ. Pollut.* (in press).

Matsushima, J. (1971). On composite harm to plants by sulfurous acid gas and oxidant. *Sangyo Kogai* **7,** 218–224.

Matsushima, J., and Brewer, R. F. (1972). Influence of sulfur dioxide and hydrogen fluoride as a mix or reciprocal exposure on citrus growth and development. *J. Air Pollut. Contr. Ass.* **22,** 710–713.

Menser, H. A., and Heggestad, H. E. (1966). Ozone and sulfur dioxide synergism; injury to tobacco plants. *Science* **153,** 424–425 .

Menser, H. A., and Hodges, G. H. (1970). Effects of air pollutants on burley tobacco cultivars. *Agron. J.* **62,** 265–269.

Menser, H. A., Hodges, H. A., and McKee, C. G. (1973). Effects of air pollution on Maryland type 32 tobacco. *J. Eviron. Qual.* **2,** 253–258.

Middleton, J. T., Darley, E. F., and Brewer, R. F. (1958). Damage to vegetation from polluted atmospheres. *J. Air Pollut. Contr. Ass.* **8,** 9–15.

Reinert, R. A., Tingey, D. T., Heck, W. W., and Wickliff, C. (1969). Tobacco growth influenced by low concentrations of sulfur dioxide and ozone. *Argon. Abstr.* **61,** 34.

Reinert, R. A., Heagle, A. S., Miller, J. R., and Geckeler, W. R. (1970). Field studies of air pollution injury to vegetation in Cincinnati, Ohio. *Plant Dis. Rep.* **54,** 8–11.

Skelly, J. M., Moore, L. D., and Stone, L. L. (1972). Symptom expression of eastern white pine located near a source of oxides of nitrogen and sulfur dioxide. *Plant Dis. Rep.* **56,** 3–6.

Solberg, R. A., and Adams, D. F. (1956). Histological responses of some plant leaves to hydrogen fluoride and sulfur dioxide. *Amer. J. Bot.* **43,** 755–760.

Taylor, O. C. (1968). Effects of oxidant air pollutants. *J. Occup. Med.* **10,** 485–492.

Thomas, M. D., Hendricks, R. H., and Hill, G. R. (1952). *In* "Air Pollution" (L. C. McCabe, ed.), pp. 41–47. McGraw-Hill, New York.

Thompson, C. R., and Taylor, O. C. (1969). Effects of air pollutants on growth, leaf drop, fruit drop and yield of citrus trees. *Environ. Sci. Technol.* **3,** 934–940.

Tingey, D. T., Heck, W. W., and Reinert, R. A. (1971a). Effect of low concentrations of ozone and sulfur dioxide on foliage, growth and yield of radish. *J. Amer. Soc. Hort. Sci.* **96,** 369–371.

Tingey, D. T., and Reinert, R. A. (1975). The effect of ozone and sulfur dioxide singly and in combination on plant growth. *Environ. Pollut.* (in press).

Tingey, D. T., Reinert, R. A., Dunning, J. A., and Heck, W. W. (1971b). Vegetation injury from the interaction of nitrogen dioxide and sulfur dioxide. *Phytopathology* **61,** 1506–1511.

Tingey, D. T., Reinert, R. A., Dunning, J. A., and Heck, W. W. (1973a). Foliar injury responses of eleven plant species to ozone/sulfur dioxide mixtures. *Atmos. Environ.* **7,** 201–208.

Tingey, D. T., Reinert, R. A., Wickliff, C., and Heck, W. W. (1973b). Chronic ozone or sulfur dioxide exposures, or both affect the early vegetative growth of soybean. *Can. J. Plant Sci.* **53,** 875–879.

Treshow, M. (1968). Impact of air pollutants on plant populations. *Phytopathology* **58,** 1108–1113.

Treshow, M. (1971). Fluorides as air pollutants affecting plants. *Annu. Rev. Phytopathol.* **9,** 21–44.

Weber, D. E., Reinert, R. A., and Barker, K. R. (1974). The effects of ozone and sulfur dioxide on selected species of plant parasitic nematodes. *Proc. Amer. Phytopath. Soc.* **1,** 113.

White, K. L., Hill, A. C., and Bennett, J. H. (1974). Synergistic inhibition of apparent photosynthesis rate of alfalfa by combinations of sulfur dioxide and nitrogen dioxide. *Environ. Sci. Technol.* **8,** 574–576.

9

EFFECTS OF AIR POLLUTANTS ON PLANT ULTRASTRUCTURE

William W. Thomson

I. Introduction

Two factors limit the scope and extent of this chapter. First, as circumscribed by the title, this treatment will be limited mainly to the effects of air pollutants on plant ultrastructure. Other chapters in this book will be concerned with the effects of the various toxicants on biochemical, physiological, and ecological aspects. Second, although there are several well recognized pollutants, only a few have been studied with emphasis on their effects on fine structure of plant cells and the number of these studies have been quite limited. These include the oxidants, ozone and peroxyacetyl nitrate (PAN), nitrogen dioxide (NO_2), Hydrogen fluoride (HF), sulfur dioxide (SO_2), and ethylene.

II. Ozone and Peroxacetyl Nitrate

These compounds are relatively recently recognized air pollutants which accumulate to toxic levels in the atmosphere through photochemical reactions with unsaturated hydrocarbons and nitrous oxide released through auto emissions and other sources (Dugger and Ting, 1970a,b; Hill *et al.*, 1970). As pointed out by Hill *et al.* (1970), ozone is probably the most serious phytotoxicant in the United States of the air pollutants. In electron microscopic studies on bean (*Phaseolus*), Thomson *et al.* (1966) found that after fumigation with 0.6–1.0 ppm of ozone for a half hour, there were alterations in the ultrastructure of the palisade parenchyma cells prior to and considerably earlier than the appearance of visual damage to the leaves. The first observed changes were an increased granulation and electron density of the chloroplast stroma (Fig. 1) and in some chloroplasts the appearance of clusters and order arrays of fibrils (Figs. 2 and 3). The fibrils measured approximately 85 nm in width. Subsequently the plasmalemma, tonoplast, and the chloroplast envelope ruptured and broke down; the mitochondria became swollen; and electron-dense material accumulated within these organelles. The final stages were marked by a collapse of cellular contents into a large mass in the center of the cell. Only the ordered arrays of fibrils and the granal membranes were clearly identifiable within this mass. This sequence and the alterations in ultrastructure were almost identical to those earlier observed by the same investigators (Thomson *et al.*, 1965) in a study on the effects of peroxyacetyl nitrate on the fine structure of bean leaves. In more recent studies on cotton (*Gossypium*), Thomson and Swanson (1972) also observed the presence of crystalline structures in the stroma early after fumigation with ozone. In similar studies on tobacco (*Nicotiana*), Swanson *et al.* (1973) did not observe the early formation of crystalline arrays in the stroma. However, they found that the chloroplasts became irregular in outline, such as illustrated in (Fig. 1), and there was an increase in density of the stroma. Determinations of the axial ratios of the chloroplasts indicated there was a general decrease in their volume. The absence of the crystalline arrays is a bit puzzling since we have observed them in chloroplasts of ozone fumigated tobacco leaves of both sensitive and insensitive varieties. Nevertheless, all the studies indicate a significant change in the chloroplast stroma after fumigation. Since the stroma is the repository of the enzymes involved in CO_2 fixation (Trebst *et al.*, 1958; Park and Pon, 1961), these changes are probably directly related to the observed inhibition of CO_2 fixation by these oxidant pollutants (Dugger *et al.*, 1963; Dugger and Ting, 1970a,b).

Crystalline arrays of fibrils in chloroplasts have been observed a number

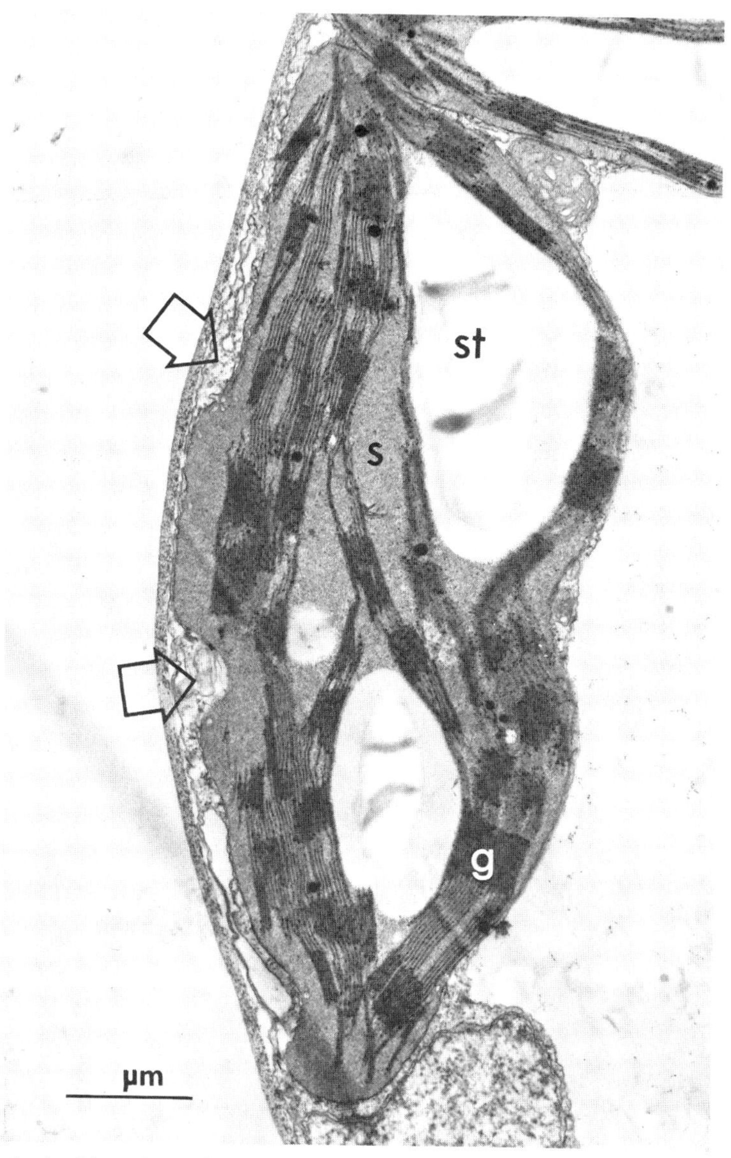

FIG. 1. A chloroplast of an ozone fumigated *Phaseolus* leaf. Note the irregular outline of chloroplast (arrows) and the general density of the stroma (s). Starch grains (st) and the grana (g) are clearly evident. ×20,000.

of times. Similar arrays have also been observed in chloroplasts fumigated with other air pollutants (see later discussion). As pointed out by De Greef and Verbelen (1973), they generally seem to form when the plant is chal-

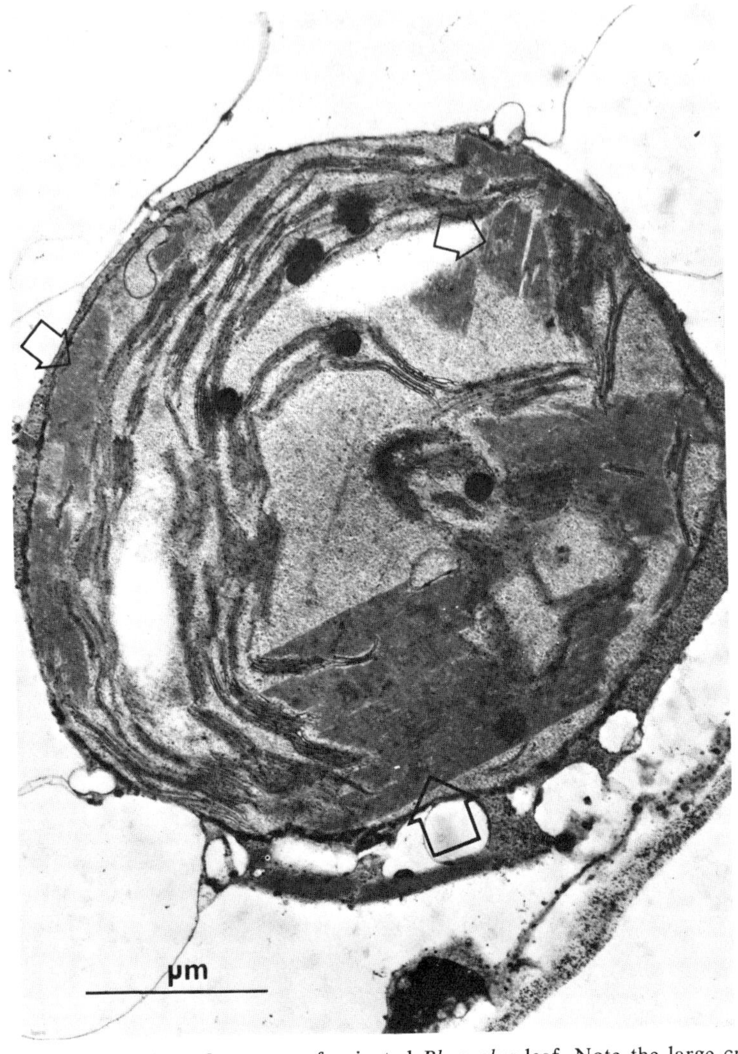

FIG. 2. A chloroplast of an ozone fumigated *Phaseolus* leaf. Note the large crystalloids in the stroma (arrows). × 29,000.

lenged with stress conditions, such as starvation (Ragetli *et al.,* 1970), removal of roots (de Greef and Verbelen, 1973), and under conditions of water stress (Perner, 1962), 1963; Shumway *et al.,* 1967; Gunning *et al.,* 1968; Wrischer, 1967, 1973). Crystalloids also have been observed in the plastids of plants in which there is no evidence of physiological stress (Gunning, 1965). Nevertheless, several lines of evidence indicate that the formation of the crystalline arrays, in most instances, is due to increased

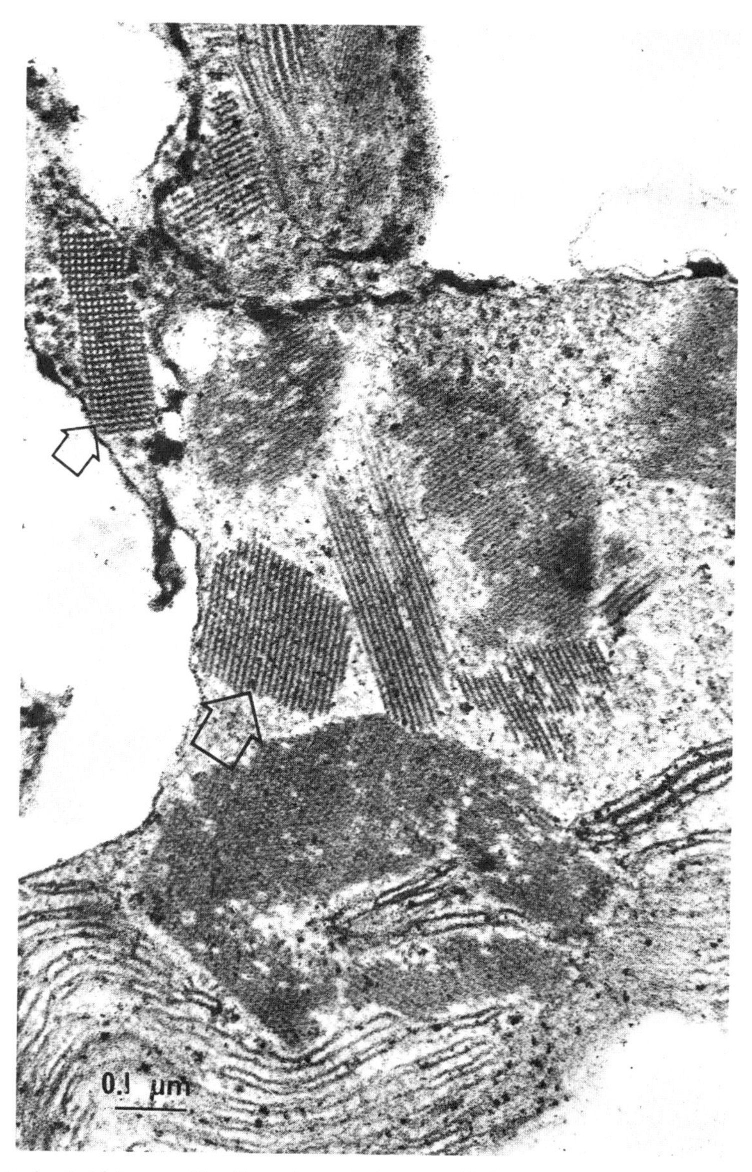

FIG. 3. A high magnification view of the crystalloids (arrows) in the stroma of a chloroplast in an ozone fumigated *Phaseolus* leaf. × 109,000.

water stress. Perner (1962, 1963) and later Shumway *et al.* (1967) found that the crystalline arrays formed in isolated chloroplasts subjected to a hypertonic medium. Crystalline arrays also form in chloroplasts of leaves that have wilted and in leaf cells that have been plasmolyzed (Gunning *et al.,* 1968; Wrischer, 1973) or subject to centrifugation (Gunning *et al.,* 1968). If the wilting is not too severe, the crystalline arrays disappeared on the addition of water (Wrischer, 1973). Thus it would appear that the formation of the crystalline arrays in the stroma is related to increased dehydration of the stroma due to the loss of water. This also seems possible in relation to the oxidant induced formation of these crystalline arrays, since increased permeability of the cells, at least in ozone treated material (Evans and Ting, 1973; Dugger and Palmer, 1967) and loss of water from the mesophyll cells is one of the first signs of damage (see Middleton, 1961, for review).

Although it would appear that the crystalline arrays form in chloroplasts of the oxidant treated material as a response to the loss of water from the organelle, this must be a secondary effect which follows a change and alteration of the plasmalemma or chloroplast envelope, or both. In other words, with a change in the permeability of these membranes, water is lost from the cell resulting in cell dehydration and the formation of the crystalline arrays. With an alteration and change in the permeability of the membranes, loss of ions would also be expected. This could result in an ionic imbalance and the loss of ions may be as important in the formation of the crystalline arrays as the loss of water.

The cell membranes have long been suggested as a primary site of action of the oxidants, particularly the olefinic groups of the membrane lipids (Giese and Christensen, 1954; Rich, 1964). This seems sensible, at least for the plasmalemma, since it represents the boundary of the protoplasm. However, after fumigation with the oxidants the plasmalemma has an increased clarity and definition (Thomson *et al.,* 1965, 1966; Swanson *et al.,* 1973). Although membrane breakdown occurred in the injured cells, it happened quite late in the damage sequence, and as far as membranes being generally sensitive it is significant that the granal membranes are still reasonably intact in the most severely damaged cells (Thomson *et al.,* 1965; Swanson *et al.,* 1973). Swanson *et al.* (1973) in their studies found that ozone did not affect the relative distribution of the fatty acids extractable from ozone treated material as compared to untreated controls. From these studies and the electron microscopic observations that the membranes are clearly resolved after ozone treatment, they concluded that the primary site of action of the oxidants was unlikely to be the olefinic groups of the membrane fatty acids. However, the technique is such that they could not resolve small or localized effects on the membrane structure and components which may be the primary critical factors involved in the injury response.

Swanson *et al.* (1973) also observed a swelling of the mitochondria and an increased density of the cytoplasm of the cells of ozone treated tobacco leaves. In the most severely injured cells, the cell contents were aggregated into a dense mass with few recognizable cell components except for the grana of the chloroplasts. The swelling of the mitochondria and increased density of the cytoplasm were attributed to permeability changes in the membranes. Lee (1968) has also observed that ozone induced swelling of mitochondria isolated from roots, leaves, and callus tissue. Lee (1968) attributed the swelling response of the mitochondria with short exposures or low concentrations of ozone to metabolic changes, but the rapid swelling observed following treatment with higher concentrations of ozone he suggested were due to changes in the permeability of the mitochondrial membranes.

Aspects of the injury response, particularly at the ultrastructural level, which have not been examined are what changes are reversible and at what point, and which alterations indicate irreversible damage. An ancillary question is, are there ultrastructural changes in the cells when exposed to low levels of the toxicants under conditions where the growth and physiology of the tissue is reduced but no visual symptoms of damage appear? Since the crystalline arrays appear prior to visible damage, it would be interesting to determine if these represent reversible aspects of the damage response and if they occur in the chloroplasts of plants with reduced growth rates but show no signs of visible damage. This would be of interest because of the implications relative to repair mechanisms concerning membranes. The observations that the crystalline arrays, induced by water stress, disappear when the leaf is returned to a more normal situation (Wrischer, 1973) suggests that this question is approachable.

III. Nitrogen Dioxide

Although NO_2 is a major component in the photochemical reactions involved in the atmospheric production and accumulation of peroxyacetyl nitrate and ozone (see reviews by Dugger and Ting, 1970a,b), it can cause plant injury by itself (Taylor and MacLean, 1970). Dolzmann and Ullrich (1966) studied the effects of NO_2 on the ultrastructure of bean leaves. After 1–2 hours of fumigation with 1% NO_2, they observed an alteration in chloroplast conformation consisting primarily of elongated protrusions. These protrusions were associated with mitochondria and appeared partially or entirely to enclose around the mitochondria. Similar protrusions have been observed in studies on chloroplast structure (Laetsch, 1969a,b), in chloroplasts of viral infected leaves (Shalla, 1964), and in the chloroplasts of leaves stressed by nutrient deficiency (Possingham *et al.*, 1964) and starvation (Ragetli *et al.*, 1970). It is not clear what factors are involved in the formation of these protrusions; however, their frequent occurrence with stress conditions, such as the treatment with NO_2, suggests that such conditions induce changes in the chloroplast envelope including probably surface, plastic, and elastic properties.

Dolzmann and Ullrich (1966) also observed crystalline arrays in the stroma of the chloroplasts after 1 or 2 hours of fumigation. We have also observed these in NO_2 treated material (Figs. 4 and 5). Their close correspondence to the crystalline arrays observed in the chloroplasts of ozone and PAN treated leaves would indicate they are composed of the same components and form in response to the same conditions (see above discussion). After 15 hours of treatment with NO_2, the cell organization was almost totally disrupted although the grana and the crystalline arrays of the chloroplasts were clearly recognizable. In experiments with NO_2, Wellburn *et al.* (1972) observed a swelling of the fret membranes of the internal membrane system of the chloroplasts, and with a higher concentration or prolonged exposure to the pollutants there was a swelling of the compartments of the grana and particularly the terminal compartments at the ends of the grana. Interestingly they found this swelling was reversible when the leaves were subsequently exposed to unpolluted air. They observed no alterations in other cell structures. Since the internal membrane system is the site of primary photosynthetic processes (Park and Pon, 1963), they suggested that these alterations may be related to the observed depression of photosynthesis by NO_2 that have been reported (Hill and Bennett, 1970).

IV. Fluoride

Wei and Miller (1972) observed a progressive series of alterations in the ultrastructure of the mesophyll cells of soybean leaves on fumigation with hydrogen fluoride. The first changes in ultrastructure occurred prior to the appearance of visible damage and consisted of an increase and aggregation in the endoplasmic reticulum. Subsequently there was a reduction in endoplasmic reticulum and after 2 days of treatment small vacuoles appeared in the cytoplasm and phytoferritin accumulations appeared in some chloroplasts. As the injury progressed, the tonoplast became disrupted and vesiculated, there was a swelling and deterioration of the mitochondria, a vesiculation of the endoplasmic reticulum, and a detachment of ribosomes and a decrease in free ribosomes; the chloroplasts became more round and some had a reduced grana-fretwork system and an increase in plastoglobuli. The final stages of injury were marked by a general deterioration of the cell structure and organization and a clumping of the remains in the center of the cells.

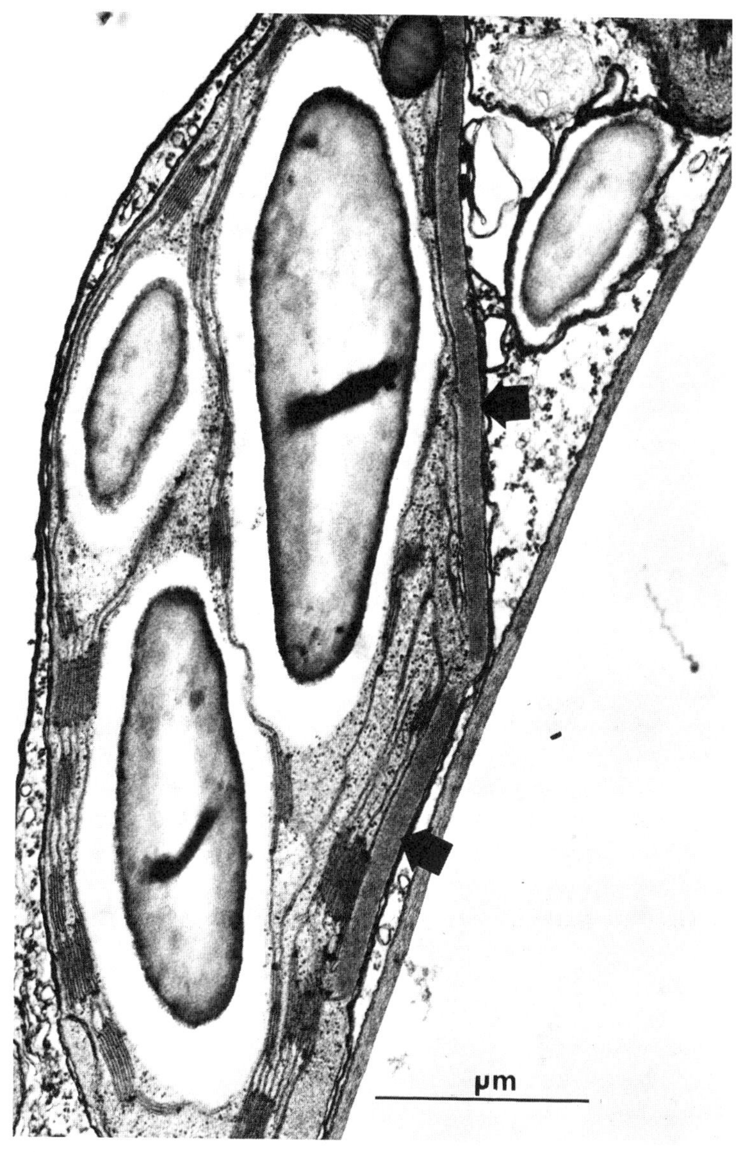

FIG. 4. A chloroplast of an NO_2 fumigated *Nicotiana* leaf. Note the presence of stromal crystalloids (arrows). × 34,000.

Wei and Miller (1972) pointed out that the initial change observed with fluoride fumigation was an alteration of the vacuolar membrane. They suggested that this disruption of the tonoplast membrane with the release of the vacuolar contents to the cytoplasm may have caused the continued

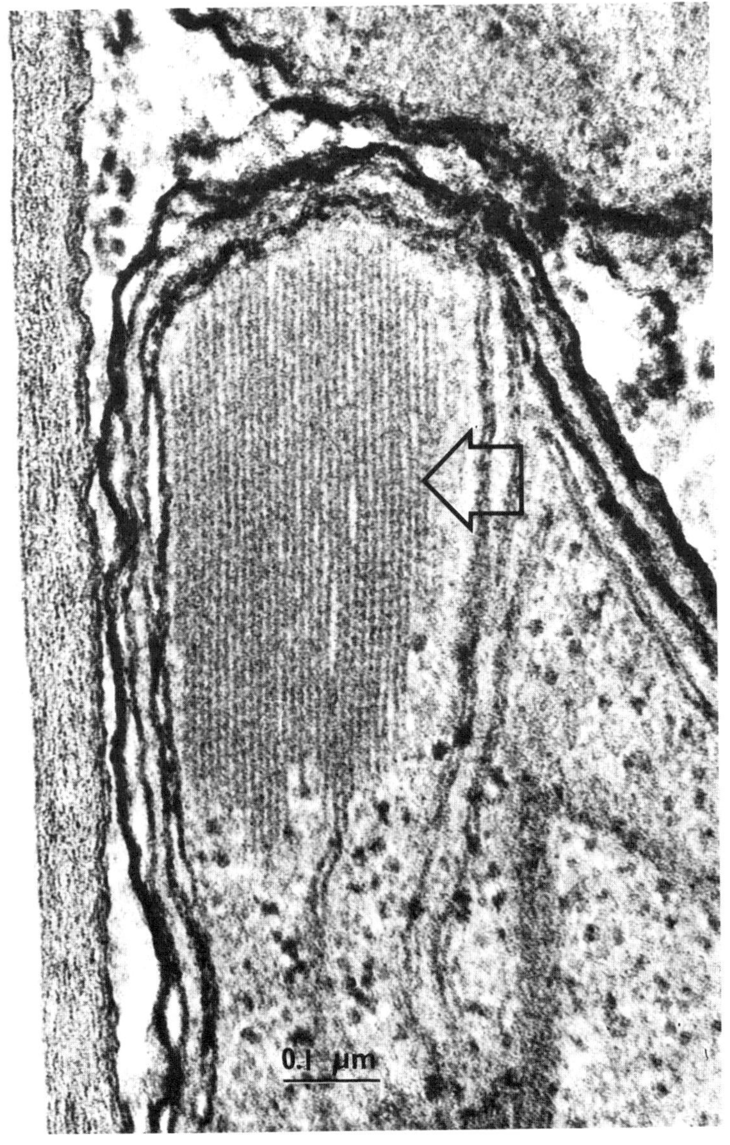

FIG. 5. A high magnification view of a crystalloid in the stroma (arrow) of a NO_2 fumigated *Nicotiana* leaf. × 153,000.

disintegration of the cytoplasmic organelles and cell organization. This suggestion would seem to have merit and should be further investigated, particularly since there is considerable evidence that vacuoles may represent a lysosomal compartment (e.g., Berjak, 1972; Buvat, 1971; Matile and Winkenbach, 1971; Matile and Moor, 1968; Villiers, 1967, 1972a,b;

Marty, 1970, 1971, 1972a,b; Coulomb, 1971a,b). Further there is a growing body of evidence and thought that alterations and disruptions of the tonoplast membrane play a key role in senescence leading to the deterioration and death of the cell (Butler and Simon, 1971; Dodge, 1971; Buvat, 1971; Matile, 1969; Matile and Winkenbach, 1971).

More recently Bligny *et al.* (1973) using scanning electron microscopy, found that fluoride delayed the formation of epicuticular waxes on the abaxial surfaces of young fir needles. In studies of thin sections of samples from the same material, they found a reduction in the size of the chloroplasts, a swelling of the granal-fretwork membranes, and a lack of close adherence of the granal compartments. The suggestion from these observations is that fluoride probably affects the general metabolism of the cell, including the synthesis and formation of the cuticle and the development and function of the chloroplasts.

V. Sulfur Dioxide

Sulfur dioxide is one of the very well studied air pollutants which is known to damage plants. Nevertheless, only a few studies have been directed at examining the effect of this compound at the fine structural level.

In leaves of *Vicia faba,* exposed to SO_2, Fisher *et al.* (1973) found that granulation of the stroma and swelling of the fret membranes of the chloroplasts were the first changes induced by SO_2 at the ultrastructural level. Wellburn *et al.* (1972), using *Vicia faba,* also observed a swelling of the fret membranes and granal compartments of the chloroplasts. They found that this swelling was more extensive with exposure to higher concentrations or longer duration of SO_2. It is interesting that Wellburn *et al.* (1972) found this swelling reversible after exposure to the pollutant for 1 hour, followed by exposure to unpolluted air. These investigators (Wellburn *et al.,* 1972) suggest that these changes, induced by SO_2, in the internal membrane system of the chloroplasts would imply a possible effect on photosynthesis.

Fisher *et al.* (1973) found a significant decrease in the apparent CO_2 fixation of SO_2 treated leaves. The inhibition was severe in leaves which showed large areas of visible damage, and they suggested that the damage and the probable reduced photosynthetic capacity of these regions were probably irreversible and that the observed partial recovery of photosynthetic capacity probably occurred in the regions where no damage was visible. Contrary to the observations of Wellburn, *et al.* (1972), Fisher and his associates reported they observed no alterations in the chloroplasts in undamaged cells immediately adjacent to the damage areas and a clarification of this discrepancy would be helpful. However, Fisher, *et al.* corroborated the observations of Wellburn *et al.* that the first observed alterations in the chloroplasts after SO_2 treatment of the leaves was a swelling of the granal compartments and fret channels between the grana. Subsequently the chloroplast became swollen, followed by a degradation of the chloroplast envelope. In the final stages of damage the density of the cell contents increased, and the micrograph they published indicated an aggregation of the cellular material although the granal membranes persisted for a long period of time.

VI. Ethylene

Although ethylene is a known and important air pollutant which is phytotoxic (Heck *et al.,* 1972), it is also produced endogenously by plants and functions as a growth hormone (Burg and Burg, 1965). No reports have appeared on the toxic effects of ethylene on the ultrastructure of plant cells. However, two reports have been published on the effects of ethylene on the fine structure of the abscission zone of leaves (Valdovinos *et al.,* 1972: Webster, 1973). Since ethylene is known to accelerate or induce many senescence phenomena (Sacher, 1973) and because abscission appears to be a senescence phenomenon, these observations are probably more directly related to senescence than to direct toxic effects of the pollutant. Nevertheless, since observed toxic effects may in themselves be senescence processes induced by ethylene (for discussions on senescence, see Butler and Simon, 1971; Osborne, 1968), the observation on the abscission zone will be briefly discussed here. Valdovinos *et al.* (1972) observed an accumulation of rough endoplasmic reticulum in the cells of the abscission zone of tobacco after 2 hours of treatment with ethylene, and there was a further increase in rough endoplasmic reticulum (RER) with 3–5 hours of treatment. They point out the increase in RER correlates with evidence that indicates an increase in RNA and protein in the abscission zone after treatment with ethylene (Osborne, 1968). Similar evidence exist for the increase in RNA and protein in other senescing systems (see Sacher, 1973, for review and discussion), and the suggestion that this increase is related to the production of degradative enzymes involved in senescence is worth considering. After 2 and 3 hours of treatment, Valdovinos and his associates observed a degradation of the walls, and as this continued vesicular structures appeared in the wall. Fibrous material and electron-dense bodies were also observed in the degenerating walls. During the first part of the treatment (3 hours) little or no changes were observed in the fine structure of the cells; however, after 5 hours of treatment in cells where advanced cell wall breakdown was observed, disorganization of the cytoplasm was also apparent. After 5 hours treatment the degradation of the wall of some cells was advanced, although little change in the cell organization was observed except for a decrease in density of the matrix of the microbodies.

In ethylene treated petiolar explants of *Phaseolus,* Webster observed that before cell wall rupture invaginations of the plasmalemma were apparent. However, she suggested that this may be related to plasmolysis and pointed out that they are present in intact and other plant tissues (Mahlberg, 1972). The mitochondria in the ethylene treated material had a dense matrix and an enlarged intracristal space.

VII. Conclusions

The studies to date, although limited in number, indicate that air pollutants induce characteristic alterations in the cell ultrastructure. In some of these studies many of these changes occur prior to the appearance of visible symptoms. It is not known whether this occurs for all the phytotoxicants studied, since many of these investigations involved examination of material several hours after treatment with the toxicant and when visible signs of injury were apparent. To obtain information which would be valuable in more clearly ascertaining probable primary sites of action of the pollutants, careful examination of material taken early and progressively throughout the treatment period and thereafter is needed. Another related problem, which has not been studied, is the question of whether there are changes in the ultrastructure of the cells of plants treated with low levels of pollutants which show no overt injury but have reduced growth rates and altered physiology. Further, if there are changes, are they similar to the early changes observed in studies on the sequential progression of changes, and are these fine structural alterations reversed with time on removal of the phytotoxic material? Such studies would be valuable in assessing the nature of less detrimental but probably far more economically important effects of the air pollutants, as well as in providing some insight into the type of changes which are reversible. The determination of early and possible reversible alterations could provide clues as to the primary sites of action of the pollutants as well as effects on physiological processes.

Such techniques as freeze-etching and cytohistochemical localization of enzymes could provide even more insight into action of the pollutants on cell ultrastructure, particularly in relation to their early effects on membrane structure and organization and compartmentation within the cells.

Acknowledgments

The author thanks Miss Kathryn Platt for excellent technical assistance.

References

Berjak, T. (1972). Lysosomal compartmentation: Ultrastructural aspects of the origin, development and function of vacuoles in root cells of *Lepidium sativum*. *Ann. Bot.* (*London*) [N.S.] **36,** 73–81.

Bligny, R., Bisch, A. M., Garrec, J. P., and Fourcy, A. (1973). Observations morphologiques et structurales des effets du flour sur les cires épicuticularies et sur les chloroplastes des aiguilles de sapin (*Abies alba* M.11). *J. Microsc.* (*Paris*) **17,** 207–214.

Burg, S. P., and Burg, E. A. (1965). Ethylene action and the ripening of fruits. *Science* **148,** 1190–1196.

Butler, R. D., and Simon, E. W. (1971). Ultrastructural aspects of plant senescence. *Advan. Gerontol. Res.* **3,** 73–129.

Buvat, R. (1971). Origin and continuity of cell vacuoles. *In* "Origin and Continuity of Cell Organelles" (J. Reinert and H. Ursprung, eds.), pp. 127–157. Springer-Verlag, Berlin and New York.

Coulomb, P. (1971a). Phytolysosomes dan le méristème radicularie de la courge (*Cucurbita pepo* L., Cucurbitacee). Activité phosphatasique acide et activité péroxydasique. *C. R. Acad. Sci.* **272,** 48–51.

Coulomb, P. (1971b). Sur la présence de phytolysosomes dans les cellules de tumeurs de la plantule de Pois (*Pisum sativum* L.) induites par l'*Agrobacterium tumefaciens. C. R. Acad. Sci.* **272,** 1229–1231.

De Greef, J. A., and Verbelen, J. P. (1973). Physiological stress and crystallites in leaf plastids of *Phaseolus vulgaris* L. *Ann. Bot.* (*London*) [N.S.] 37, 593–596.

Dodge, A. D. (1971). The mode of action of the bipyridylium herbicides, paraquat and diquat. *Endeavour* **30,** 130–135.

Dolzmann, P., and Ullrich, H. (1966). Einige Beobachtungen über Beziehungen zwischen Chloroplasten and Mitochondrien in Palisadenparenchym von *Phaseolus vulgaris. Z. Pflanzenphysiol.* **55,** 165–180.

Dugger, W. M., and Ting, I. P. (1970a). Air pollution oxidants—Their effects on metabolic processes in plants. *Annu. Rev. Plant Physiol.* **21,** 215–234.

Dugger, W. M., and Ting, I. P. (1970b). Physiological and biochemical effects of air pollution oxidants on plants. *Recent Advan. Phytochem.* **3,** 31–58.

Dugger, W. M., Jr., and Palmer, R. L. (1967). Carbohydrate metabolism in leaves of rough lemon as influenced by ozone. *Int. Proc. Citrus Symp.* **2,** 711–715.

Dugger, W. M., Jr., Koukol, J., Reed, W. D., and Palmer, R. L. (1963). Effect of peroxyacetyl nitrate on $C^{14}O_2$ fixation by spinach chloroplasts and pinto bean plants. *Plant Physiol.* **38,** 468–472.

Evans, L. S., and Ting, I. P. (1973). Ozone-induced membrane permeability changes. *Amer. J. Bot.* **60,** 155–162.

Fisher, F., Kramer, D., and Ziegler, H. (1973). Elektronenmikroskopische Untersuchungen SO_2-begaster Blatter von *Vicia faba*. I. Beobachtungun am Chloroplasten mit akuter Schädigung. *Protoplasma* **76,** 83–96.

Giese, A. C., and Christensen, E. (1954). Effects of ozone on organisms. *Physiol. Zool.* **27,** 101–115.

Gunning, B. E. S. (1965). The fine structure of chloroplast stroma following aldehyde osmium-tetroxide fixation. *J. Cell Biol.* **24,** 79–93.

Gunning, B. E. S., Steer, M. W., and Cochrane, M. P. (1968). Occurrence, molecular, structure and induced formation of the stromacentre in plastids. *J. Cell Biol.* **24,** 79–93.

Heck, W. W., Daines, R. H., and Hindawi, I. J. (1970). Other phytotoxic pollutants. *In* "Recognition of Air Pollution Injury to Vegetation: A Pictorial Atlas" (J. S. Jacobson and A. C. Hill, eds.), F-1-24. Air Pollut. Contr. Ass., Pittsburgh, Pennsylvania.

Hill, A. C., and Bennett, J. H. (1970). Inhibition of apparent photosynthesis by nitrogen oxides. *Atmos. Environ.* **4,** 341–348.

Hill, A. C., Heggestad, H. E., and Linzon, S. N. (1970). Ozone. *In* "Recognition of Air Pollution Injury to Vegetation: A Pictorial Atlas" eds., (J. S. Jacobson and A. C. Hill, eds.), B-1-22. Air Pollut. Contr. Ass., Pittsburgh, Pennsylvania.

Laetsch, W. M. (1969a). Specialized chloroplast structure of plants exhibiting the dicarboxylic acid pathway of photosynthetic CO_2 fixation. *Progr. Photosyn. Res.* **1,** 36–46.

Laetsch, W. M. (1969b). Relationship between chloroplast structure and photosynthetic carbon fixation pathways. *Sci. Progr. (London)* **57,** 323–351.

Lee, T. T. (1968). Effect of ozone on swelling of tobacco mitochondria. *Plant Physiol.* **43,** 133–139.

Malhberg, P. (1972). Further observations on the phenomenon of secondary vacuolation in living cells. *Amer. J. Bot.* **59,** 172–179.

Marty, M. F. (1970). Rôle du système membranaire vacuolaire dans la differénciation des lacticifères d'*Euphorbia characias* L. *C. R. Acad. Sci.* **271,** 2301–2304.

Marty, M. F. (1971). Vesicules autophagiques des laticifères différencies d'*Euphorbia characias* L. *C. R. Acad. Sci.* **272,** 399–402.

Marty, M. F. (1972a). Distribution des activités phosphatasiques acides au cour du processes d'autophagie cellulaire dans les cellules du méristème radiculaire d'*Euphorbia characias* L. *C. R. Acad. Sci.* **274,** 206–209.

Marty, M. F. (1972b). Localization ultrastructurale d'activités pyrophosphatasiques acides (=phosphohydrolasique acides) dans les cellules du méristème radiculaire d'*Euphorbia characias* L. *C. R. Acad. Sci.* **275,** 365–367.

Matile, P. (1969). Plant lysosomes. *In* "Lysosomes in Biology and Pathology" (J. T. Dingle and H. B. Fell, eds.), pp. 406–430. North-Holland Publ., Amsterdam.

Matile, P., and Moor, H. (1968). Vacuolation: Origin and development of the lysosomål apparatus in root-tip cells. *Planta* **80,** 159–175.

Matile, P., and Winkenbach, F. (1971). Function of lysosomes and lysosomal enzymes in the senescing corolla of the morning glory (*Ipomoea purpurea*). *J. Exp. Bot.* **22,** 759–771.

Middleton, J. T. (1961). Photochemical air pollution damage to plants. *Annu. Rev. Plant Physiol.* **12,** 431–448.

Osborne, D. J. (1968). Hormonal mechanisms regulating senescence and abscission. *In* "The Biochemistry and Physiology of Plant Growth Substances" (F. Wightman and G. Setterfield, eds.), pp. 815–840. Runge Press, Ottawa.

Park, R. B., and Pon, M. G. (1961). Correlation of structure with function in *Spinacea oleracea* chloroplasts. *J. Mol. Biol.* **3,** 1–10.

Park, R. B., and Pon, M. G. (1963). Chemical composition and substructure of lamellae isolated from *Spinacea oleracea* chloroplasts. *J. Mol. Biol.* **6,** 105–114.

Perner, E. (1962). Elekronenmikronskopische befunde uber kristallgitter strukturen in Stroma isolieter Spinatchloroplasten. *Port. Acta Biol., Ser. A* **6,** 359–372.

Perner, E. (1963). Kristallisationserscheinungen in Stroma isolierter Spinatchloroplasten guter Erhaltung. *Naturwissenschaften* **50,** 134–135.

Possingham, J. V., Vesk, M., and Mercer, F. V. (1964). The fine structure of leaf cells of manganese-deficient spinach. *J. Ultrastruct. Res.* **11,** 68–83.

Ragetli, N. W. J., Weintraub, M., and Lo, E. (1970). Degeneration of leaf cells resulting from starvation and excision. I. Electron microscopic observations. *Can. J. Bot.* **48,** 1913–1922.

Rich, S. (1964). Ozone damage to plants. *Annu. Rev. Phytopathol.* **2,** 253–266.

Sacher, J. A. (1973). Senescence and postharvest physiology. *Annu. Rev. Plant Physiol.* **24,** 197–224.

Shalla, T. A. (1964). Assembly and aggregation of tobacco mosiac virus in tomato leaflets. *J. Cell Biol.* **21,** 253–264.

Shumway, L. K., Weier, T. E., and Stocking, R. C. (1967). Crystalline structures in *Vicia faba* chloroplasts. *Planta* **76,** 182–189.

Swanson, E. S., Thomson, W. W., and Mudd, J. B. (1973). The effect of ozone on leaf cell membranes. *Can. J. Bot.* **51,** 1213–1219.

Taylor, O. C., and MacLean, D. C. (1970). Recognition of air pollution injury to vegetation: Nitrogen dioxide and peroxyactyl nitrates. *In* "Recognition of Air Pollution Injury to Vegetation: A Pictorial Atlas" (J. S. Jacobson and A. C. Hill, eds.), E-1-14. Air Pollut. Contr. Ass., Pittsburgh, Pennsylvania.

Thomson, W. W., and Swanson, E. S. (1972). Some effects of oxidant air pollutants (ozone and peroxyacetyl nitrate) on the ultrastructure of leaf tissues. *Proc. Electron. Microsc. Soc. Amer.* **30,** 360–361.

Thomson, W. W., Dugger, W. M., Jr., and Palmer, R. L. (1965). Effects of peroxyacetyl nitrate on ultrastructure of chloroplasts. *Bot. Gaz.* (*Chicago*) **126,** 66–72.

Thomson, W. W., Dugger, W. M., Jr., and Palmer, R. L. (1966). Effects of ozone on the fine structure of the palisade parenchyma cells of bean leaves. *Can. J. Bot.* **44,** 1677–1682.

Trebst, A. V., Tsujimoto, H. Y., and Arnon, D. I. (1958). Separation of light and dark phases in the photosynthesis of isolated chloroplasts. *Nature* (*London*) **187,** 351–355.

Valdovinos, J. G., Jensen, T. E., and Sicko, L. M. (1972). Fine structure of abscission zones. IV. Effects of ethylene on the ultrastructure of abscission cells of tobacco flower pedicels. *Planta* **102,** 324–333.

Villiers, T. A., (1967). Cytolysosomes in long-dormant plant embryo cells. *Nature* (*London*) **214,** 1356–1357.

Villiers, T. A. (1972a). Cytological studies in dormancy. II. Pathological aging changes during prolonged dormancy and recovery upon dormancy release. *New Phytol.* **71,** 145–152.

Villiers, T. A. (1972b). Cytological studies in dormancy. III. Changes during low temperature dormancy release. *New Phytol.* **71,** 153–160.

Webster, B. (1973). Ultrastructural studies of abscission in *Phaseolus:* Ethylene effects on cell walls. *Amer. J. Bot.* **60,** 436–447.

Wei, L.-L., and Miller, G. W. (1972). Effects of HF on the fine structure of mesophyll cells from *Glycine max* Merr. *Fluoride* **5,** 67–73.

Wellburn, A. R., Majernik, O., and Wellburn, F. A. M. (1972). Effects of SO_2 and NO_2 polluted air upon the ultrastructure of chloroplasts. *Environ. Pollut.* **3,** 37–49.

Wrischer, M. (1967). Kristalloids in Plastidenstroma. I. Electronmikro-skopisch-Cytochemische Untersuchungen. *Planta* **75,** 309–318.

Wrischer, M. (1973). Protein crystalloids in the stroma of bean plastids. *Protoplasma* **77,** 141–150.

10

EFFECTS OF AIR POLLUTANTS ON FORESTS

Paul R. Miller and Joe R. McBride

I. Introduction

The problem of adverse human influences on forest lands has its roots far in the past. The larger settlements in the temperate regions of the world were often established in or near hardwood or conifer forests which provided easy access to basic necessities, such as fuel, building materials, game, and a regulated supply of pure water. In the beginning, these forests conformed to the standard defined as "a plant association predominantly of trees or other woody vegetation occupying an extensive area of land" (Committee on Forest Terminology, 1944). Tree density was sufficient over large areas, so that distinct climatological and ecological conditions developed, which are easily distinguished from those of less densely vegetated areas. The forest ecosystem, with its distinct associations of plants, animals, and many other organisms, both macroscopic and microscopic, organized to utilize and transfer energy and raw materials, functioned efficiently under the influence of its controlling physical environment (Billings, 1970).

The typical forest in populated regions of the world today is not usually a broad expanse of green woodland stretching to the horizon. More often, we see remnants of a forest that has been modified in many ways to suit man's purposes. These forest remnants may no longer operate with the natural efficiency of the undisturbed ecosystem from which they were derived, but are defined and managed according to human needs. Changes in agricultural practices may allow the development of a new forest stand through a gradual process called secondary succession. For example, in some parts of Europe and the United States, recently abandoned agricultural land, carved from forest in pioneer times, is now becoming forested again through succession or by deliberate planting of trees for special purposes.

The forest resources of the United States are under increasing pressure, however. The demand for traditional forest products to serve as building materials, pulp, stable watershed, forage, and fuel has been intensified. Forests must also satisfy the growing deep psychological need of urban man to seek refuge from the increasing stresses of his everyday environment. More and more the commodity producing space of the forest is being transformed into recreation space for camping, vacation homes, and condominiums. Intensified recreational and residential use concentrates populations and resultant physical insults to forest stands. The successful management of both the developed and undeveloped surrounding forest is made more complicated because certain insect pests and diseases, which become more prevalent on trees stressed by man's activities and by air pollutants, are not easily controlled where there are mixed land ownerships and multiple uses.

Very frequently heavy industry has located in forested areas because of the coincident availability of rich ore deposits or sometimes cheaper electrical energy. The most damaging situations have been localized around nonferrous metal smelters, aluminum ore reduction plants, and coal burning power plants. The transport of oxidant air pollutants from urban centers in California to forests up to 80 miles downwind illustrates an exception where a long distance between the source and the forest offers no protection from severe ozone damage.

This chapter will consider the forest ecosystem as a receptor of air pollutants and will describe the forest types in the United States which are liable to injury. The history of important episodes of damage to forests will be described in some detail so that the effects of specific pollutants on individual species, forest communities, and ecosystems can be evaluated. Examples from Canada and Europe are included to broaden the coverage of common principles. The outlook for maintaining healthy forests in the continuing presence of pollutants is examined. Finally, forests are evaluated in their role as a source of certain air pollutants.

II. Forest Ecosystems and Forest Types in the United States

A. Classification of Forest Ecosystems

The forests and woodlands of the United States (Fig. 1) can be subdivided into 10 major ecosystems (Weaver and Clements, 1938). These are the boreal forest, subalpine forest, western montane forest, Pacific Coast forest, eastern deciduous forest, southeastern pine forest, lake states forest, tropical forest, California woodland, and southwestern woodland. Each ecosystem is distinguished by its regional climate and the life form of its dominant species. The Pacific Coast forest, for example, has a mild, moist climate influenced by the Japanese current, and its major tree species have a coniferous life form.

Variation within the regional environment defines several forest types within each major forest ecosystem; these are identified by their dominant tree species. The Society of American Foresters (Anonymous, 1964) recognizes a total of 156 forest cover types in North America. The most widely distributed types in the Pacific Coast forest ecosystem, for example, are

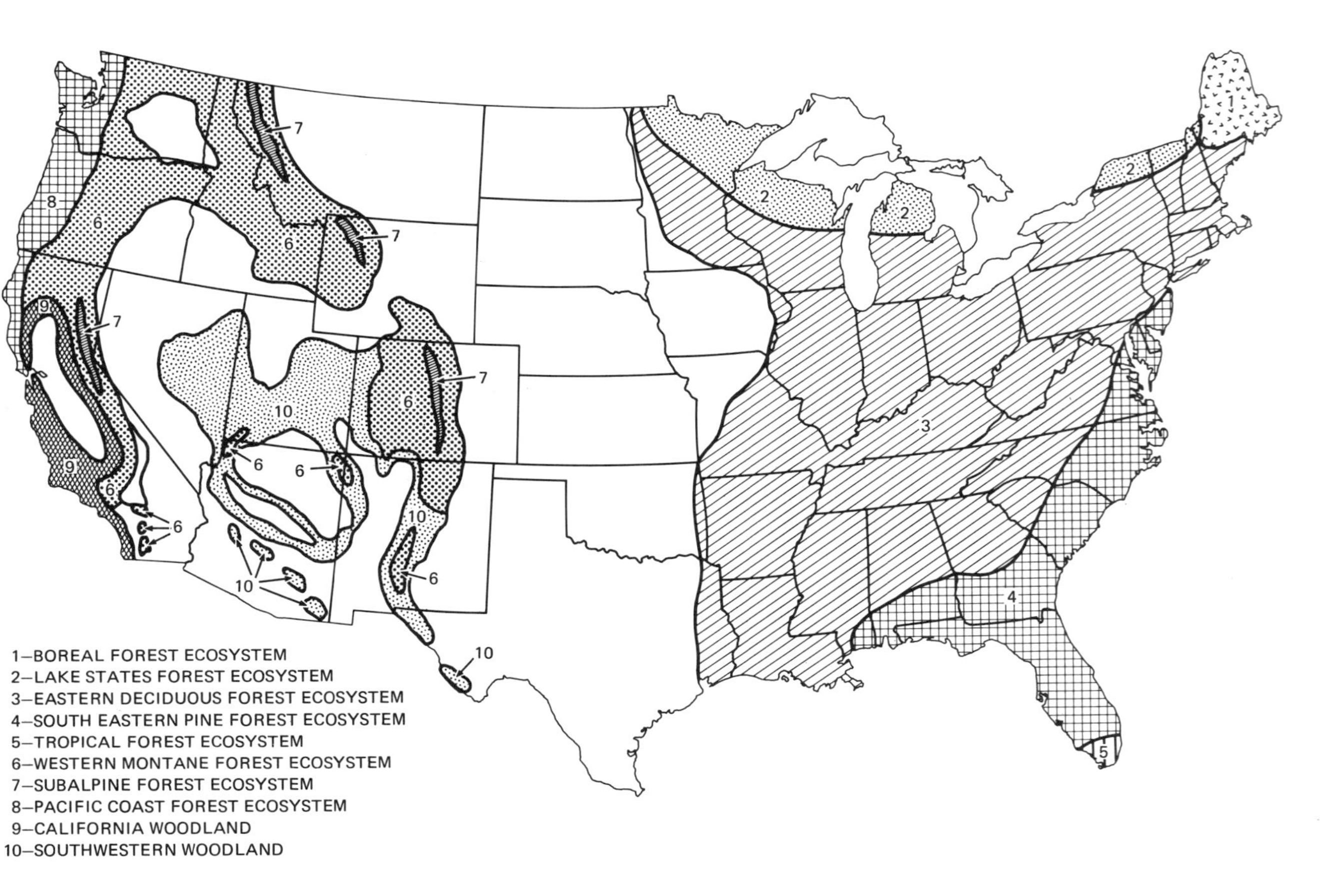

FIG. 1. Forest and woodland ecosystems of the United States.

the Sitka spruce forest type, western hemlock forest type, Pacific Douglas fir forest type, and the redwood forest type.

Each forest type is represented by many individual stands distributed over the landscape in which species composition and age distribution are relatively homogeneous. Each stand may vary in size from less than 1 acre to over 1000 acres, depending upon the uniformity of similar soils and climate and/or the ecological events, e.g., fire, which gave rise to the stand. The stand is a basic unit in forest management, and classification of stands for management purposes (Baker, 1950) is based on species composition (pure or mixed) and age distribution (even-aged or uneven-aged). Within the redwood forest type, for example, one finds pure stands of redwood (*Sequoia sempervirens*) as well as mixed stands of redwood and Douglas fir (*Pseudotsuga menziesii*). The type includes both even-aged and uneven-aged stands.

B. Dynamics of Forest Ecosystems

The forest stand is a dynamic unit which changes over time as individual trees die from a variety of competing biotic and abiotic causes thus allowing spaces for the continuing addition of new trees. Species composition may change completely, or species composition and age distribution of the stand may be essentially maintained by plant succession. A successional sequence is made up of a series of stands which replace one another until a final stand develops which is capable of replacing itself. This final stand or type is referred to as the climax (Daubenmire, 1968). The climax type is in equilibrium with its environment and continues to replace itself over time. When plant succession begins on a bare substrate, such as rock, sand, lava, or water, it is referred to as primary succession. Secondary succession refers to succession taking place following the destruction of existing vegetation by fire, hurricane, land clearing, or other destructive events. An understanding of plant succession is basic to forest management and the interpretation of air pollutant damage to forest types.

In pure, even-aged stands, planted and intensively managed, the concept of succession has little utility. The forest manager is mainly concerned with rotations or the interval required to bring the stand from planting to harvest age.

The wide variety in forest stands implies a similar variety of responses following injury by air pollutants (Figs. 2 and 3). The destructive influence of an air pollutant on the stability of a stand comprised of a complex variety of species compared to a single species plantation may have widely differing ecologic and economic importance. The relative susceptibility of each species and its importance in a successional sequence, plus intended

FIG. 2. A forest stand in the San Bernardino National Forest dominated by *Pinus ponderosa*. The clearing in the mid- and background with dead and dying trees is typical in stands severely damaged by oxidant air pollutants.

uses of the stand, are a few of the variables which must be considered when determining the significance of injury.

The examples of air pollution injury described in the following sections illustrate a variety of responses by forest stands to air pollutants. In the majority of cases, only the injury to the dominant or most economically important species is described in detail. In fewer cases the effects on other plant or animal members of the forest ecosystem are briefly considered. The ecosystem context is the preferred format in which future studies will hopefully be made because of the recent developments in systems ecology (Watt, 1966).

FIG. 3. A single large *Pinus ponderosa* tree severely injured by oxidant. It is flanked by other *Pinus ponderosa* trees having less injury.

III. Incidents of Sulfur Dioxide Injury

Before the turn of the century, sulfur dioxide was recognized as a serious agent of damage in German forests (Haselhoff and Lindau, 1903). Traditionally, the source of sulfur dioxide in forested areas has been the smelting of iron, copper, nickel, and zinc ores which are high in sulfur content. Other important sources include the combustion of coal and petroleum products containing sulfur, while manufacture of sulfuric acid, Kraft process pulp mills, and natural gas purification are sources of less frequent importance. Where there are groupings of such industries the resultant damage can involve whole regions with a common air basin (Calvert, 1967).

A detailed discussion of some major incidents of forest damage will be presented here, in order to develop a better perspective of the extent and severity of problems around point sources, e.g., smelters, refineries, and power generating plants.

A. Western United States and Canada

1. Copper Smelters at Redding, California and Anaconda, Montana

In 1903 and 1904 an investigation was made at a copper smelter situated northwest of Redding, in Shasta County, California to determine the exact cause and the full extent of damage to both native vegetation and nearby orchards (Haywood, 1905). In 1906 and 1907 Haywood (1910) investigated damage to both plant and animal life around the Washoe smelter at Anaconda, Montana, first established in 1884.

a. Extent of Injury. At Redding the area of greatest damage and the only damaged area acknowledged by the smelter operator formed an ellipse about 5½ miles long and 4 miles wide. The specific purpose of the United States Department of Agriculture investigation was to determine the validity of damage claims outside the acknowledged area. Careful field observations, controlled SO_2 fumigation of selected species, and analysis of the sulfur content of foliage at varying distances from the smelter all served as evidence to extend the limits of damage. It was confirmed that the severely damaged area extended for 12½ miles from north to south, 2½ miles east, and 5–6 miles west. Less severe injury extended beyond the above limits for a "considerable distance."

The Washoe smelter at Anaconda is located on a hill at 6600 ft (2012 m) elevation near the southern end of a narrow valley about 35 miles long. The surrounding upper slopes and ridge tops were the only forested areas immediately adjacent to the smelter. Smelter fumes drifted into tributary valleys or across comparatively high ridges to reach the main body of timber. The stack height was at about 7200 ft. Some dead timber near the smelter was due to past forest fires; however, there were sufficient numbers of trees within 10–15 miles of the smelter so that sulfur dioxide injury could be diagnosed. A number of apparently resistant trees of generally susceptible species were observed close to the smelter. After additional field surveys at Anaconda in 1908, it was found that Douglas fir was injured 15–19 miles north, 11½–14 miles east, 10 miles south, and 18–19 miles west of the smelter. In comparison, injury to less sensitive lodgepole pines extended 10–11 miles north, 9–10 miles south, and 10 miles west of the smelter. These distances are based on visible symptoms only and were not confirmed by analysis of foliage for sulfur content.

b. SPECIES AFFECTED. The investigator (Haywood, 1905) was a chemist; he did not identify the exact species injured at Redding other than in broad common terms, such as pine, scrub pine, and oak. It is possible that ponderosa pine (*Pinus ponderosa*) or Jeffrey pine (*Pinus jeffreyi*) may have been the pine. The scrub pine was probably knobcone pine (*Pinus attentuata*) or possibly Digger pine (*Pinus sabiniana*). It is more difficult to guess the species of oak which may have been damaged. On the positive side, a number of good photographs were printed which allows some speculation about the identity of species and the vegetation types affected.

This example of an incomplete investigation is a very instructive lesson even for modern investigators. At Anaconda, Haywood was accompanied by a forester.

The most sensitive species at Anaconda were Rocky Mountain Douglas fir (*Pseudotsuga menziesii* var. *glauca*) followed by lodgepole pine (*Pinus contorta* var. *latifolia*), but Rocky Mountain juniper (*Juniperus scopulorum*) was relatively undamaged even close to the smelter. Photographs accompanying Haywood's (1910) report show almost total destruction of Douglas fir and damage to lodgepole pine as far as 11 miles west of the smelter. The analysis of foliage for sulfur content confirmed that the injury was caused by sulfur dioxide.

More complete observations of timber damage at Anaconda (Scheffer and Hedgcock, 1955) were made in 1910 and 1911 by plant pathologists who investigated more species than Haywood. According to Scheffer and Hedgcock (1955), the order of sulfur dioxide susceptibility based on amounts and frequencies of foliage thinning at varying distances from the Washoe smelter, beginning with the most susceptible, was as follows: subalpine fir (*Abies lasiocarpa*), Douglas fir, lodgepole pine, Engelmann spruce (*Picea engelmannii*), ponderosa pine (*Pinus ponderosa* var. *scopulorum*), limber pine (*Pinus flexilis*), Rocky Mountain juniper, and common juniper (*Juniperus communis*). The principal species in this forest cover were lodgepole pine, Douglas fir, Engelmann spruce, subalpine fir, and limber pine.

Because the symptoms of winter injury to conifers could be confused with those caused by sulfur dioxide, a comparison was made. The order of sensitivity to winter injury starting with the most susceptible was as follows: ponderosa pine, Douglas fir, lodgepole pine, limber pine, Engelmann spruce, subalpine fir, Rocky Mountain juniper, and common juniper. The wide differences in the two lists is notable, particularly in the ranking of subalpine fir.

c. EFFECTS ON WATER AND SOILS. The pollution of streams by drainage from smelter tailings was a problem at both Redding and Anaconda, but at Anaconda there was also considerable fallout of arsenic which contaminated soils and forage crops. It was determined that some injury and mortality to cattle was attributable to arsenic poisoning.

The smelter at Redding is no longer in operation but the Anaconda smelter has continued operation at varying levels these many years.

d. ANACONDA REVISITED—1972. In spite of the installation of several control devices since the early 1900's nearly 1170 tons of SO_2 were emitted daily in 1972, although addition of a new sulfuric acid plant in 1973 reduced emissions to about 736 tons/day (Carlson, 1974b). In 1969, eight observation plots were established at distances ranging from $4\frac{1}{2}$–$11\frac{1}{2}$ miles away from the Anaconda smelter. An evaluation of these plots in 1972 (Carlson, 1974a) showed terminal branch dieback and needle tip necrosis on Douglas fir and limber pine at three plots within 5 miles of the smelter. The accumulative effects of more than 60 years of SO_2 exposure at some plots included a depression of soil pH, substantial reductions in numbers of plants in the understory and herb layers, and active soil erosion leading to windthrow of some trees. A more thorough investigation using old photographs (Haywood, 1910) to establish the former vegetation cover at particular sites, coupled with up to date aerial photography to show the present condition, may be an aid in reconstructing the full impact of the chronic SO_2 exposure suffered by these forest cover types.

2. COPPER SMELTER AT TRAIL, BRITISH COLUMBIA

The Trail, British Columbia smelter is located about 11 miles north of the United States–Canadian border in the gorge of the Columbia River. This smelter began operation in 1896 and was one of the largest nonferrous ore processing plants in the world. Emissions of sulfur dioxide increased sharply to nearly 10,000 tons per month in 1930. The installation of sulfur recovery equipment in 1931, coupled later with agreed upon production cutbacks during periods of unfavorable meteorological conditions, reduced subsequent sulfur dioxide emission to about 5000 tons or less per month (Scheffer and Hedgcock, 1955).

Pollution abatement measures followed complaints of injury to forests by residents of the state of Washington. An International Joint Commission with members from the United States and Canada assessed damages and awarded compensation payments. Voluminous reports were printed regarding this damage and dispute; Katz (1949) reviewed those aspects relating to forest damage.

a. EXTENT OF INJURY. During the time of highest emissions, about 1931, zone 3, the area where 1–30% of the trees were dead or in poor condition, extended 52 miles southward along the course of the Columbia River. The width of the damaged area varied from 1½–10 miles, depending upon the topography of the valley and tributary drainages. Zone 1, or 60–100% severe damage, extended about 33 miles south of the smelter. The outer limits of damage began to recede after 1931 in spite of abnormally high sulfur content in needle tissue (Scheffer and Hedgcock, 1955).

b. SPECIES AFFECTED. Ponderosa pine accompanied by inland Douglas fir, western larch (*Larix occidentalis*) and lodgepole pine, were the predominant species at lower elevations. At higher elevations only Douglas fir and western larch were present. Scheffer and Hedgcock (1955) prepared lists of relative susceptibility to SO_2 of these major conifers, other less abundant conifers, broad-leaf trees, shrubs, and herbs. Beginning with the most susceptible, the order of SO_2 susceptibility of the major conifers was Douglas fir, followed by ponderosa pine, lodgepole pine, and western larch. Because larch is the only deciduous conifer it had superior survival capability because mature foliage was tolerant and no needles were present in the winter to be injured by fumigations when temperatures were above 40°F.

Measurements of height and diameter growth decreases were used successfully to demonstrate the growth retardation of ponderosa pine. The near absence of cone production by damaged trees and the lack of seedlings and saplings in damaged stands were additional evidences of the full extent of damage (Scheffer and Hedgcock, 1955).

3. SULFUR GASES FROM A PULP AND PAPER MILL AT MISSOULA, MONTANA

AREA AND SPECIES AFFECTED. Throughout a 5200 acre area near Missoula, Montana a gradual decline in the health of ponderosa pine and Douglas fir has been observed since 1963 (Carlson, 1974a). Needle tip necrosis and accelerated foliage loss was observed especially on Douglas fir. Injury symptoms and foliage sulfur content increased with decreasing distance from the plant; injury was more prevalent west of the plant in one of the dominant downwind directions. Except for a sudden increase in visible symptoms in late 1972 or early 1973 coincident with the use of sulfur-containing fuel oil, the chronic decline in tree vigor appeared to be associated with exposure to H_2S or other sulfur-containing gases. The effects of chronic H_2S injury to forest tree species is not well known, and Carlson's (1974a) study does not supply sufficient information to identify precisely which sulfur-containing gas may be the cause of chronic injury.

4. Copper Smelter at Anyox, British Columbia

During 1914–1935, a copper smelter operated at Observatory Inlet, about 120 miles northeast of Prince Rupert on the British Columbia coast.

Area and Species Affected. The first record of damage by sulfur dioxide in 1916 indicated that western red cedar (*Thuja plicata*) had already been killed over a considerable area near the smelter. Subsequent observations showed that western hemlock (*Tsuga heterophylla*) was much more tolerant since it survived within 5 miles of the smelter, while western red cedar was killed as far away as 35 miles. The susceptibility of Pacific silver fir (*Abies amabilis*) and Sitka spruce (*Picea sitchensis*) was intermediate between these extremes. Very little is known of other plant species in the area (Errington and Thirgood, 1971). According to the description of the above authors, the damage intensified during the dry summers of 1922 and 1925 when fumes were able to travel farther, possibly because there was less "wash-out" by fog and rain. Forest fires occurred repeatedly in the vicinity of the smelter up to 1942. Dead timber was present at the water's edge and on up the steep slopes of the inlet; the soils were barren and badly eroded, particularly on the western slopes of the inlet. The investigators commented that "no area on the Pacific Coast (Canadian) has ever been subjected to such harsh treatment."

B. Eastern United States and Canada

1. Copper Basin, Tennessee

The devastation of a large area in Polk County, Tennessee began with mining and transportation of copper ore out of the basin in 1850. Later open-hearth furnaces were used for smelting ore at the mines; trees were cut for fuel. The entire basin was covered by a dense, mixed hardwood forest before mining began. The greatest smelting activity was during 1890–1895 (Hursh, 1948).

a. Area and Species Affected. Analysis of foliage and soil samples confirmed that sulfur dioxide was the damaging agent (Haywood, 1908). By 1910 the denuded zone was as large as it would get (Hursh, 1948). By 1913, when Hedgcock (1914) visited the area, the damage extended 12–15 miles to the north and 10 miles or more to the west of the smelter. Eastern white pine (*Pinus strobus*), which was particularly susceptible, was injured as far as 20 miles away (Hursh, 1948). Hedgcock (1914) also observed that hardwoods were more tolerant that conifers. From appearances in the field he listed the order of susceptibility of ten hardwoods along the Ocoee River and of nine additional oak species on the surrounding uplands.

b. POLLUTANT-INDUCED CHANGES IN MICROCLIMATE. During 1936–1939, the microclimate of three distinct zones around smelters was studied (Hursh, 1948): (1) the central barren zone, about $10\frac{1}{2}$ square miles in extent, which was expanding because of soil erosion; (2) a 17,000 acre belt of grassland, mostly broomsedge (*Andropogon scoparius*); (3) a 30,000 acre transition zone, partly grass and partly trees, formerly injured by sulfur dioxide, with indefinite boundaries where some recovery and reproduction was evident in both conifers and hardwoods. Hursh (1948) considered the central barren area to be "the largest completely bare area in any subhumid region of the United States." His measurements of differences in air and soil temperatures, wind, evaporation, air-moisture saturation deficit, and rainfall provided clear comparisons of three distinct microclimates brought about by changes in plant cover due to sulfur dioxide injury.

2. INDUSTRIAL AND URBAN POLLUTANT SOURCES IN THE NORTHEAST, CENTRAL, AND LAKE STATES: CHLOROTIC DWARF

CHLOROTIC DWARF. For the past 65 years, a stunting of random eastern white pines (*Pinus strobus*) growing in the northeast, central, and lake states has been a puzzling problem (Swingle, 1944). Chlorotic dwarf as found on young white pines in pure and mixed plantings and also in natural stands was shown to be related to air pollution (Dochinger, 1968). It is now believed that sulfur dioxide and ozone acting independently or possibly synergistically at relatively low concentrations for long periods of time are responsible for the dwarfing of genetically susceptible individuals (Dochinger and Heck, 1969). Similar symptoms of chronic air pollution injury have been identified on Scots pine (*Pinus sylvestris*) in Ohio (Dochinger, 1970).

Under somewhat similar circumstances, a reduction of growth rate of eastern white pine was observed without apparent injury to foliage near a coal burning power plant in Pennsylvania. The plant began operation in 1952 with four stacks whose heights did not exceed those of the surrounding ridgetops. The resulting pollution effects caused installation of a 600 ft stack by 1962. The reduction of sulfur dioxide concentrations near the plant resulted in significantly greater terminal and diameter growth after 1962 (Wood, 1967).

3. POLLUTANT SOURCES IN THE KINGSTON–OAK RIDGE–ROCKWOOD–HARRIMAN INDUSTRIAL AREA, TENNESSEE

AREA AND SPECIES AFFECTED. Beginning in about 1955, both the United States Forest Service and the Tennessee Valley Authority began an investigation of the decline of eastern white pines in scattered stands where particular individuals were exceptionally injured. The total area affected is several hundred square miles of the Cumberland Plateau (Berry and Hepting, 1964), and the pollutant sources include a pulp mill, a coal-burning power plant, a uranium refining mill, and a ferroalloy reduction plant (Ellertsen *et al.*, 1972). The types of symptoms observed here which were associated with the death of white pine are similar to postemergence chronic tipburn (PECT) (Berry and Hepting, 1964), white pine needle blight in Canada (Linzon, 1960) and chlorotic dwarf in Ohio (Dochinger, 1968). The suggestion was strong that sulfur dioxide and perhaps ozone were both involved as the primary cause of the decline.

Between 1956 and 1965, 10% of the dominant and co-dominant trees in 25 permanent plots scattered throughout the area had died; the economic impact was slight, however, because eastern white pine represented less than 5% of the total volume available for harvest (Ellertsen *et al.*, 1972). These results do not reveal whether the growth rates and reproductive capacity of associated tree species may also be diminished at a less perceptible rate because other species were not similarly studied.

4. Power Generation Facilities at Mount Storm, West Virginia–Gorman, Maryland

Damage to the Christmas Tree Industry. Complaints were first voiced in 1968 by Christmas tree growers that symptoms resembling pollution effects had begun to degrade the quality of their trees in the vicinity of a coal-burning power plant at Mount Storm, West Virginia. Growers had successfully produced high quality Christmas trees for the previous 10 years and suspected that the power plant which started operation in 1966 was responsible. The injured species included eastern white pine, Scots pine, Virginia pine (*Pinus virginiana*), Douglas fir, and Norway spruce (*Picea abies*). Investigations by many competent scientists culminated in hearings pursuant to an official Air Pollution Abatement Activity of the Environmental Protection Agency (U.S. Environmental Protection Agency, 1971).

A diverse collection of tree symptoms was described: needle tipburn or tip necrosis, chlorosis, early abscission, random short needles, twisting and elongation of needles, bud failure, adventitious budding, and basal spotting of needles. One symptom which was unique to this problem was the random short needle condition. It was possible to simulate this condition by topical application of weak acid to young elongating needles, although eriophyid mites (*Setoptus* sp.) were also suggested as the cause of basal spotting and, in turn, needle dwarfing. It was suggested that under field conditions acid aerosols and acids or alkali leached from fly ash particles could induce short needles. Most of the other symptoms were more easily ascribed to air pollution but the definite cause of the random short needle symptom does not definitely relate to air pollution. Damage was found regularly as far as 10 miles from the power plant among 30 commerical growers occupying more than 1000 sites in a two county area (U.S. Environmental Protection Agency, 1971).

5. Nickel and Copper Smelters, Sudbury District, Ontario

a. Area and Species Affected. Since 1888, the roasting and smelting of nickel and copper ore have liberated sulfur dioxide over the forested Sudbury area. In 1965, there were three large smelters discharging 6000 tons of sulfur dioxide daily (Linzon, 1965b). In this area, eastern white pine is the most susceptible species and was used as a bioindicator to delimit two zones of injury. The area of severe damage is an ellipse extending mainly north of the smelters for nearly 25 miles and covering 720 square miles. The second zone outside of the severely affected area receives occasional, less severe injury and it occupies 1600 square miles (Linzon, 1966).

b. Environmental Factors and Doses Required for Injury. Recent studies have helped to define the conditions leading to acute damage to forest vegetation in the Sudbury District. The necessary environmental factors include sunshine, high relative humidity, moderate temperatures, and available soil moisture; biological factors are sensitive genotypes and sensitive growth stages. The required dosages are 0.95 ppm for 1 hour, 0.55 ppm for 2 hours, 0.35 ppm for 4 hours, and 0.25 ppm for 8 hours (Dreisinger, 1965). Additional details are given in a review by Linzon (1973) which includes his own and other studies of sulfur dioxide effects on forest ecosystems.

6. Iron-Sintering Plant at Wawa, Ontario

Damage was confined to the familiar elliptical pattern downwind from the plant which released about 100,000 tons of "sulfur" per year (Gordon and Gorham, 1963). These investigators observed effects on the overstory, understory, and ground vegetation at varying distances from the source. Damage was estimated as "detectable" for 20 miles downwind, "severe" within 11 miles, and "very severe" within 5 miles. The only trees surviving as close as 11 miles to the smelter were a few white birch (*Betula papyrifera*) and white spruce (*Picea glauca*). Beyond 11 miles from the source, a number of species displayed foliage injury. They included white birch, white spruce, black spruce (*Picea mariana*), trembling aspen (*Populus tremuloides*), jack pine (*Pinus banksiana*), white cedar (*Thuja occidentalis*), larch (*Larix laricina*), and balsam fir (*Abies balsamea*). Associated eastern white pines were nearly all killed.

Ground flora variety declined from 20–40 species/40 m^2 plot, 10 miles from the source, to 0 or 1 species/plot within 2 miles from the source. Fringed bindweed, (*Polygonum cilinode*) and red-berried elder (*Sambucus pubens*) were highly resistant to sulfur dioxide. Seedlings of white pine were not observed within 30 miles, and those of white spruce, black spruce, and trembling aspen were not observed within 15 miles of the pollutant source. Lake and pond waters were abnormally high in sulfate as far as 11 miles away.

Sulfur dioxide from petroleum refineries caused injury to sensitive forest vegetation in the vicinity of Clarkson, Ontario and East St. Paul, Manitoba beginning about 1961 (Linzon, 1965a).

C. Europe

1. Recent Literature

Scurfield (1960), in his review, "Air Pollution and Tree Growth," cited examples of sulfur dioxide damage to forests in Scotland, Germany, Serbia, Sweden, Japan, Czechoslovakia, and Turkey. More recently, the proceedings of the sixth and seventh International Workshops of Forest Fume Damage Experts at Katowice, Poland, and Essen, West Germany in 1968 and 1970, respectively, provide much detail about recent damage from sulfur dioxide and other pollutants in central and eastern Europe. Russian literature has been translated by the American Institute of Crop Ecology: "Survey of U.S.S.R. Air Pollution Literature," particularly Volume 3. Specific segments of the above named volumes have been cited in this chapter.

2. Examples in West Germany and England

In the Ruhr Valley, a heavily industrialized region of West Germany which encompasses an area of approximately 3000 km^2, stands of Scots pine have survived only in locations remote from the industrial complex (Knabe, 1970). The severity of damage made it necessary to attempt development of resistant pine selections or to replant only hardwoods which are more resistant to sulfur dioxide.

The necessary removal in 1924 of the National Pinetum of Great Britain from Kew, near London, to Bedgebury in Kent is another example of the inability of conifers to persist near industrial centers (Scurfield, 1960).

IV. Incidents of Fluoride Injury

In the United States, the most notable incidents of fluoride damage to forests have occurred in the Pacific Northwest. The pollutant sources have been mainly aluminum ore reduction plants and occasionally phosphate fertilizer plants. From the former source, sodium fluoride and aluminum fluoride are released as particulate matter and deposited locally, whereas hydrogen fluoride and traces of carbon tetrafluoride are emitted as gases over a broader area. Examples of either past or present fluoride damage to forests are known at Spokane, Ferndale, and Longview, Washington; Troutdale and The Dalles, Oregon; Columbia Falls, Garrison, Hall, and Silver Bow, Montana; and Georgetown Canyon, Idaho. Two of the most devastating incidents in the United States will be discussed.

A. Western United States

1. An Aluminum Ore Reduction Plant in the Spokane–Mead, Washington Area

In about 1943, it was recognized that a "blight" condition of ponderosa pine in Franklin Park on the north side of Spokane and beyond was reaching serious proportions. In 1949, a citizen's group called the Inland Empire Pine Damage Committee was organized to encourage an investigation (Adams *et al.*, 1952).

a. Area and Species Affected. The damage extended about 10–12 miles north and south in the valley, which was about 3½ miles wide. By 1952, nearly all trees within a 3 square mile area at the center were dead, and significant damage was detectable within a 50 square mile area. The fluoride content of damaged pine needles was considerably higher than that of undamaged needles 10 or more miles from the center (Adams *et al.*, 1952). The foliar fluoride concentrations ranging up to 600 ppm on a dry weight basis and the elimination of other biotic tree pests as the possible cause firmly implicated the Kaiser aluminum ore reduction plant as the cause of the problem (Shaw *et al.*, 1951).

b. Methods used to Determine the Cause. The atmospheric concentrations of gaseous fluorine compounds and sulfur dioxide were measured at 12 locations and soluble fluoride was determined at 82 rainwater collecting stations (Adams *et al.,* 1952). Sulfur dioxide concentrations were too low to be damaging alone, but interaction between SO_2 and gaseous fluorine compounds was not ruled out by the investigators. In any case, the investigators were unwilling to draw any final conclusions about the cause of pine damage from aerometric data alone.

Confirming evidence was obtained by fumigation of ponderosa pine with anhydrous hydrogen fluoride, comparison of symptoms with those in the field, and analysis of fumigated needles for fluoride content (Adams *et al.,* 1956). The fumigations also confirmed field observations that the youngest needles were most sensitive, particularly during the period of elongation in the spring. These injured needles were later shed. An investigation of the meteorology of the area (Adams *et al.,* 1952) indicated that the frequency of temperature inversions and hence periods of high pollution were lowest in the spring and early summer. This serendipitous circumstance may have prevented even greater damage during the critical time of new needle growth and high sensitivity to fluorine.

c. OTHER SPECIFIC EFFECTS. Both Shaw *et al.* (1951), and Lynch (1951) reported extreme reductions in annual diameter growth of damaged ponderosa pine. There was no attempt by any of the investigators to describe fluorine effects on associated natural vegetation or foraging animals. The washout of fluorine in rainfall in considerable quantities over a large area around the source suggested that multiple effects could be occurring in the ecosystem (Adams *et al.,* 1952). In the field and also following fumigations, one or two uninjured, apparently resistant, ponderosa pines were usually observed among the many damaged trees.

2. AN ALUMINUM ORE REDUCTION PLANT AT COLUMBIA FALLS, MONTANA

Forested lands within the Flathead National Forest and Glacier National Park north and east of Columbia Falls, Montana, have been subjected to fluorine from an Anaconda Company aluminum reduction plant since 1955 (Carlson and Dewey, 1971).

a. AREA AND SPECIES AFFECTED. By the summer of 1957, ponderosa pines in the vicinity of the plant were dying. There was no evaluation of damage until 1970, when Carlson and Dewey (1971) determined that visible injury encompassed 69,120 acres, whereas elevated levels of fluoride were found on nearly 214,000 acres in trees, shrubs, and grasses. Insect samples taken at each vegetation plot indicated that pollinators contained as high as 406 ppm fluoride, while cambium feeders and predator insects had around 50 ppm.

The most serious visible injury to vegetation occurred where tissue fluoride concentration was at least 60 ppm (19,840 acres). Between 30 and 60 ppm, injury was confined to conifers, although lily of the valley was also sensitive to injury at levels close to 30 ppm. Among the conifers, the order of decreasing susceptibility was as follows: western white pine, ponderosa pine, lodgepole pine, Douglas fir, "spruces," western red cedar, and subalpine fir, according to field observations.

A resampling of the vegetation plots around Columbia Falls in 1971 (Carlson, 1972) was done to determine the effects of the reduction of emissions from 7500 lb. of fluorides/day in early 1970 to 2500 lb./day in the summer of 1971. Although a significant reduction of tissue levels of fluoride had occurred since the 1970 sampling, the accumulation of above normal amounts was still occurring up to 12 miles from the plant and visible injury was observed over a 15,000 acre area.

b. OTHER SPECIFIC EFFECTS. The accumulation of fluoride by pollinator and predator insects, the observed increase of scale populations on pine needles (Carlson and Dewey, 1971), and the accumulation of fluoride by some wild animals (Gordon, 1972) together suggest potential perturbations in the forest ecosystem which could be expected to increase. The possibility of long lasting fluoride effects is suggested by Davis and Barnes (1973) in controlled environment studies in which growth of both loblolly pine (*Pinus taeda*) and red maple (*Acer rubrum*) was reduced in two forest soils to which sodium fluoride was added at concentrations ranging from 2×10^{-4} to 2×10^{-2} *M*.

Investigations of two other incidents of fluoride damage in similar forest cover types have been described at The Dalles, Oregon (Compton *et al.*, 1961) and Georgetown Canyon, Idaho (Treshow *et al.*, 1967).

B. Europe

1. EXAMPLES OF AREAS AND SPECIES AFFECTED

Aluminum plants are frequently cited as the source of fluoride damage to European forests (Scurfield, 1960). In Norway, five plants were reported in 1968 to have caused considerable damage to two natural Scots pine forests and to have totally destroyed three other stands (Robak, 1969). Scots pine was very sensitive and Norway spruce (*Picea abies*) was almost as badly injured as the Scots pine. In experimental plantations, near an aluminum works, silver fir (*Abies alba*), Douglas fir, Norway spruce, Engelmann spruce, and Serbian spruce (*Picea omorika*) were all sensitive. Those species with resistance, in order of increasing resistance, were Nordmann fir (*Abies nordmanniana*), noble fir (*Abies procera*), white fir (*Abies concolor*), western hemlock, and Japanese larch (*Larix decidua*). A wide range in sensitivity, especially in populations of Norway spruce, may offer possibilities for selection of superior types.

In the Raushofen area, Austria, nearly 800 ha of "spruce and white pine with some fir and beech trees" were damaged or destroyed by fluorine from an aluminum plant. More tolerant, mainly deciduous species were planted as replacements for the original forest, namely, "sallow, birches and alders" close to the plant, while "larches and firs" were placed farther out (Jung, 1968).

2. Environmental Factors and Dosages Required for Injury

Robak (1969) emphasized that damage was enhanced when heavy emissions occurred during the May–June period of growth, when rainfall was low, and where aluminum plants are situated in deep, narrow valleys. Knabe (1968) studied fluorine accumulation in potted plants placed at 0, 4.3, 8.3, and 12.3 m above the ground. Plants placed at higher levels accumulated more fluorine, particularly pine. Knabe suggested that differences in wind velocities at various heights may account for differences in fluorine accumulated by test plants.

V. Incidents of Ozone and Oxidant Injury

A. Eastern United States and Canada

1. Injury to Eastern White Pine

A disease called white pine blight, white pine needle dieback, and white pine emergence tipburn occurs from time to time throughout the normal range of eastern white pine. The injury is characterized as a tip dieback of newly elongating needles during June and July (Berry, 1961). The first descriptions were made by Dana (1908) and Spaulding (1909), who were unable to identify a biological cause. Affected trees occurred at random in the stand, and the same trees showed symptoms in successive years. Faull (1922) discovered that many primary roots of affected trees were dead, as did Berry (1961).

a. Implication of Ozone as the Causal Agent. Linzon (1960) showed that the necrosis is initiated in semimature leaf tissue only and that it subsequently spreads through the older needle tissue towards the tip. He renamed the disease semimature tissue needle blight or SNB (Linzon, 1965b). SNB may occur several times in any one season, usually when several days of wet weather are followed immediately by a continuous sunny period (Linzon, 1960). During 8 years from 1957–1964, SNB occurred during 6 years, usually twice each year (Linzon, 1965b) at the Petawawa Forest Experiment Station, Chalk River, Ontario. The simultaneous onset of symptoms on susceptible trees suggested that some atmospheric constituent may be responsible, but during 3 years of ozone monitoring at Chalk River there were no incidences of SNB which coincided with a peak level of ozone occurring on the same day or the day before symptoms appeared (Linzon, 1965b). Berry and Ripperton (1963) observed that emergence tipburn (assumed to be the same as SNB) occurred in the field on known susceptible trees in Pocahontas County, West Virginia several days after ambient total oxidants (Mast ozone meter) had reached a daily peak of 6.5 parts per hundred million (pphm). Container-grown susceptible clones of eastern white pine were protected from emergence tipburn when placed in a chamber supplied with carbon-filtered air. Ozone fumigations at concentrations similar to those observed in the field (6.5 pphm for 4 hours) produced typical symptoms (Berry and Ripperton, 1963).

In laboratory experiments, Linzon (1967) induced a severe tip necrosis on maturing needles of eastern white pine with 40 pphm ozone for 2 hours. The apparent discrepancy between suspected causes in Canada and the United States remains unexplained; however, the thesis that low concentrations of ozone may be the cause was strengthened by Costonis and Sinclair (1969) who showed that silvery or chlorotic flecks, chlorotic mottling, and tip necrosis appeared after fumigations with 7 pphm for 4 hours or 3 pphm for 48 hours.

b. Extent of Injury. Even though the pine blight or SNB disease occurs throughout the natural range of eastern white pine (Berry, 1961), there is no clear indication of its significance to affected stands. Because the disease affects only sensitive genotypes, its occurrence is random in the forest, and its most obvious effect is the slow selection and gradual elimination of genotypes that may have otherwise superior silvicultural characteristics. This could be a serious loss to future tree improvement efforts.

c. Implications of Intraspecific Variability in Ozone Susceptibility. The significance of strong genetic control over susceptibility to ozone and sulfur dioxide together was mentioned above in relation to chlorotic dwarf (III,B), a well-known field problem. Differential susceptibility of individual clones of eastern white pine to ozone and sulfur dioxide has been shown by Berry and Heggestad (1968) and Costonis (1970). Costonis found that in fumigations in a controlled environment chamber new needles on different branchlets grafted to the same rootstock could be independently injured by either ozone or sulfur dioxide at about 25 pphm for 2 hours. Sulfur dioxide usually caused greater damage. Costonis (1970) suggested that SNB (Línzon, 1965b) may be caused by low concentrations of sulfur dioxide. Subsequent work by Costonis (1973) with sulfur dioxide–susceptible clones showed less injury to new needles from an ozone–sulfur dioxide mixture (5 pphm, 2 hours) than from the same dose of SO_2 alone. It was suggested that ozone and sulfur dioxide reacted in the fumigation chamber environments, reducing injury to this particular clone. When Dochinger and Heck (1969) determined that chlorotic dwarf was probably caused by an interaction of ozone and sulfur dioxide, they chose to fumigate a "chlorotic dwarf susceptible clone" which would eliminate the genotype variable from the experiment.

It is reasonable to suggest from the above evidence that a spectrum of genotypes exists representing various susceptibilities to pollutants, singly or in combination. Although environmental factors are important in conditioning pollutant susceptibility (Heck, 1968), the control of environment over injury to sensitive genotypes may be less pronounced if specific biochemical or physiological mechanisms are involved which have weak interactions with the environment.

Berry (1973) has described clones of eastern white pine which have been selected for both tolerance and susceptibility to sulfur dioxide, oxidants (ozone), and fluorides. These selections will serve as sensitive bioindicators of pollution and as foundation stock for a breeding program to produce progeny more tolerant to each pollutant.

2. Relative Oxidant Susceptibility of Eastern Forest Species

No clearly defined examples of injury to eastern forests from chronic exposure to ozone or peroxyacetyl nitrate (PAN) have been detected and described. Determinations of the relative ozone susceptibility of conifers in particular have been made, however, by Berry (1971), (3 species); and Davis and Wood (1972), (18 species). In view of the amount of interest focused on ozone as the cause of emergence tipburn (Berry, 1961) and possibly the SNB (Linzon, 1965b) disease of eastern white pine, it is notable that in both fumigation trials this species was found to be less susceptible to ozone than some others, particularly Virginia pine and jack pine. The geographic distributions of white pine and Virginia pine overlap to a large extent in several eastern states, while the ranges of jack pine and white pine overlap in the lake states, Ontario, and Quebec (Critchfield and Little, 1966). Although an emergence tipburn-SNB symptom syndrome has not been observed on these two species, they may serve as some of the best early indicators of chronic ozone injury in the future in their natural ranges.

Wood and Coppolino (1972) tested the ozone susceptibility of 21 broad-leaf, deciduous species in a controlled environment chamber. The five species most sensitive to injury after exposure to 10 pphm for 8 hours were thornless honeylocust (*Gleditsia triacanthos inermis*), American sycamore (*Platanus occidentalis*), white ash (*Fraxinus americana*) hybrid poplar (*Populus maximowiczii* X *trichocarpa*) and (*P. maximowiczii* X cv. *berolinensis*), and tulip (yellow) poplar (*Liriodendron tulipifera*). These results and those of Hibben (1969) and Wilhour (1971) suggested that many eastern deciduous species are sensitive to dosages of 20–30 pphm for 2–4 hours. Jensen (1973) fumigated nine eastern hardwood species for 5 months (0.30 ppm, 8 hours daily, 5 days/week); significant growth reductions were noted with sycamore, sugar maple (*Acer saccharum*) and silver maple (*A. saccharinum*). Because Jensen's (1973) growth response data did not agree with results from visible symptoms by Wood and Coppolino (1972), he suggested that growth responses were a more reliable criterion of ozone susceptibility.

Kohut (1972) found that simultaneous exposure to ozone and PAN resulted in antagonistic, additive, and synergistic responses by rooted cuttings of hybrid poplar receiving sequential exposures of ozone and PAN in early July.

3. Potential Oxidant Injury Problems

It is widely believed that vegetation growing under the humid conditions of the eastern United States could be very severely injured if oxidant concentrations ever reach the high levels (daily peaks from 0.30–0.60 ppm) commonly experienced in the less humid inland sections of California. An air pollution episode that occurred July 27–30, 1970 in the Washington, D.C. area serves as a warning of what may happen (National Park Service, 1970). During a 4-day period, the daily total oxidant peak concentrations ranged from 0.14–0.22 ppm accompanied by low levels of sulfur dioxide (0.04 ppm). Symptoms similar to those of oxidant injury were observed on 31 tree species as well as on 15 shrub and 18 herbaceous species in a 72 square mile area. Increased emissions of the precursors for photochemical oxidant formation could result in repeated episodes of acute injury or even chronic injury to eastern forest species.

B. Western United States

1. Chronic Occurrence of High Oxidant Concentrations in Forested Areas Downwind from Urban Centers in California

Photochemical oxidants are transported from urban centers of coastal California to inland valleys and across forested mountains to the warmer deserts on most summer days and even during some warm winter days. In the South Coast air basin of California, McCutchan and Schroeder (1973) have described five classes of meteorological patterns common between May and September; two of these result in high levels of oxidant air pollution along the inland mountain slopes. The mechanism of pollutant transport up slope and across the mixed conifer forest has been further described in two successive studies representing different kinds of days (Miller *et al.,* 1972a; Edinger *et al.,* 1972). Also, seasonal averages of oxidant based on daily maxima from May through September have been reported for one station at Rim Forest in the San Bernardino Mountains since 1968, indicating means ranging from 0.18–0.22 ppm, including 1972; the corresponding number of hours when concentrations exceeded 0.10 ppm daily, ranged from 7.9–10.6 (Taylor, 1973b). It is not uncommon to observe momentary oxidant peaks as high as 0.60 ppm. Along an 18 mile long transect stretching east from Rim Forest to Big Bear Lake (Ranger Station), the number of hours daily when oxidant exceeded 0.08 ppm ranged from 12.6–6.5 from June through September 1972 (Taylor, 1973b). Severe conifer injury is present at Rim Forest but no visible conifer injury was noted at Big Bear Ranger Station at the eastern end of the transect.

Similar gradients of oxidant concentration have been described between the San Joaquin Valley and the Sierra Nevada Mountains in central California (Miller *et al.,* 1972b). This general pattern of pollutant transport has caused various levels of conifer injury at a number of locations in southern and central California (Miller and Millecan, 1971).

2. Oxidant Damage to the Conifer Forest Ecosystem of the Southern California Mountains

The oxidant damage to ponderosa pine and associated species in the San Bernardino Mountains has been summarized in the broad context of the known history, beginning about 1800, of the forest vegetation of the area (Taylor, 1973a). Past wildfires, early logging, recent sanitation-salvage logging, the distribution of human settlements and recreational pressure, insect pests, and diseases are all considered. The investigation suggested that the present distribution and composition of the forest has been strongly influenced by early logging and repeated fires, which generally favored the reproduction and regrowth of ponderosa pine in the western midelevation portions of the forest, from 4500–6000 ft, and Jeffrey pine in the eastern portions above 6000 ft. For example, near Rim Forest and Lake Arrowhead, the number of ponderosa pines in the overstory is presently more than twice that of incense cedar (*Libocedrus decurrens*), white fir, and sugar pine (*Pinus lambertiana*) considered separately (Miller, 1973a).

a. STUDIES WHICH CONFIRMED OZONE AS THE CAUSE. Conspicuous damage to needles of ponderosa pine in the conifer zone was noticed in 1953 (Asher, 1956). This new damage of unknown cause or origin was at first believed to be related to the black pine leaf scale (*Nuculaspis californica*), but a survey and subsequent map by Stevens and Hall (1956) showed no coincidence between the occurrence of heavy scale infestation and the new "needle dieback" of ponderosa pine. When an air pollutant was suspected (Asher, 1956) an investigation began in 1957 to identify the pollutant. The Kaiser Steel Corporation investigated the possibility that fluoride from a nearby steel mill was responsible for the damage. Analysis of needle tissue did not indicate sufficient fluoride accumulation to cause injury, but the data gathered from 1957 until about 1961 suggested that damage was related to oxidant air pollution.

Parallel observations described the "chlorotic decline" of ponderosa as characterized by a progressive reduction in terminal and diameter growth, loss of all but the current season's needles, reduction in number and size of the remaining needles, yellow mottling of the needles, deterioration of the fibrous root system, and eventual death of the tree (Parmeter *et al.*, 1962). Examination of roots, stems, and needles failed to disclose the presence of pathogenic organisms. An extended period of drought from 1946–1960 coincided with the first observations of the condition. Examination of annual terminal growth indicated that healthy trees responded to increases in precipitation, but trees in decline did not. The inception of the decline corresponded with reports of air pollution injury to grapes in the San Bernardino Valley not more than 15 air miles away (Richards *et al.*, 1958). In 1960, reciprocal bud and twig grafts were established between healthy and chlorotic decline trees growing side by side (Parmeter and Miller, 1968). Four years of observation of the successful twig grafts led to the conclusion that a graft transmissible agent, namely a virus, was not present and could be discounted as a possible cause of the decline disease.

The needles of ponderosa pine treated with 0.5 ppm ozone in plastic enclosures for 9–18 days under field conditions developed a chlorotic mottle, terminal dieback, and accelerated abscission similar to the needle symptoms of chlorotic decline (Miller *et al.*, 1963), and the chlorophyll content of needles treated with ozone for 18 days was generally less than that of ambient air controls. A series of studies extending from 1957–1966 (Richards *et al.*, 1968) also identified ozone as the cause of the symptom syndrome referred to by Parmeter as "X-disease" and "chlorotic decline," and an additional name, "ozone needle mottle of pine," was coined (Richards *et al.*, 1968).

b. Extent and Severity of Injury. Early studies concentrated mainly in the Crestline-Lake Arrowhead area (Asher, 1956; Stevens and Hall, 1956). These indicated that damage, which extended for 12 miles from west of Crestline to just east of Lake Arrowhead along the undulating ridge crest, was most severe at Crestline. In the summer of 1969, damage to the ponderosa-Jeffrey pine types within the forest boundaries, indicated that 46,230 acres had heavy smog damage; 53,920, moderate damage; and 60,800, light or no damage. An estimated 1,298,000 individual trees were affected: 82% moderately, 15% severely, and 3% completely dead (Wert *et al.*, 1970).

Recent estimates of mortality, usually caused by attacks of pine bark beetles (*Dendroctonus* sp.) on ponderosa pines weakened by ozone in the "heavy" damage areas (Stark *et al.*, 1968) have been 8 and 10% during a 4-year period (Taylor, 1973b), 8% during a 3-year period (Miller, 1973a), and 24% during a 3-year period (Cobb and Stark, 1970), in separate locations. Ponderosa pines on the ridgetop at the forest border overlooking the polluted basin were more injured than trees on adjacent north facing slopes in a heavily damaged area (Miller, 1973a).

The severe damage to ponderosa and Jeffrey pines in the Angeles National Forest, which is contiguous with the San Bernardino National Forest on the west, can be considered an even more serious problem. Even though tree damage is no worse in the Angeles than in the San Bernardino, there are naturally smaller stands of timber because the extremely steep terrain limits the number of satisfactory sites for timber growth. Such sites are highly prized for recreational use.

c. Relative Susceptibility of Native Conifer Species. The relationship between oxidant dose (from May to September 1972) and injury scores across the San Bernardino mountain area (Taylor, 1973b) suggested that ponderosa pines in an area receiving concentrations exceeding 0.08 ppm for 12.6 hours daily were moderately to severely damaged, whereas Jeffrey pines receiving more than 0.08 ppm for durations between 5.7 and 6.5 hours daily showed no visible symptoms. In related studies, ponderosa and Jeffrey pines have shown insignificant differences in ozone susceptibility where they overlap ranges. White fir exhibits intermediate susceptibility with mortality an exceptional event. Incense cedar and sugar pine are equally oxidant tolerant, although reduced needle complement and chlorotic foliage are visible in areas where companion ponderosa pines are moderately to severely damaged. According to ozone fumigation tests at 0.45 ppm, using container-grown seedlings, the order of decreasing susceptibility of several western conifers was as follows, with larger injury scores representing greater susceptibility (Miller, 1973b):

15 down to 10. Western white pine (*Pinus monticola*), Jeffrey × Coulter pine hybrid (*P. jeffreyi* × *P. coulteri*), red fir (*Abies magnifica*), Monterey × knobcone hybrid (*P. radiata* × *P. attenuata*), and ponderosa pine.

9.99 down to 5. Coulter pine, Douglas fir, Jeffrey pine, ponderosa pine (Rocky Mountain), white fir, big-cone Douglas fir (*Pseudotsuga macrocarpa*), and knobcone pine.

4.99 down to 1.5. Incense cedar, sugar pine, and giant sequoia (*Sequoia gigantea*).

It seems unlikely that oxidant air pollutant concentrations will decline enough in the near future to prevent further damage to conifer forest ecosystems in southern California. The population growth in the central valley of California and increasing emission of air pollutants already is responsible for injury to ponderosa pine at several points in the extensive belt of mixed conifer forest along the western slope of the Sierras (Miller and Millecan, 1971).

VI. Techniques for Analysis of Air Pollutant Effects on Forest Ecosystems

A. Prepollution Background Studies

The many examples of pollutant injury discussed in the preceding sections were essentially postmortem examinations. Only on rare occasions has a suitable unaffected control area been available to provide a baseline description of the preexisting ecosystem and permit a more critical evaluation of the total injury. Generally, the total effect of long-term pollutant exposure on the preexisting ecosystem will never be known.

Recent environmental legislation has dictated that the effects of proposed pollutant sources be assessed prior to construction of the source. The environmental impact statement, a collection of the known facts and an evaluation of alternate proposals which may minimize pollutant impacts, may sometimes provide an adequate appraisal of the baseline situation and potential hazards. More often, however, the constraints of inadequate time and funding prevent a thorough analysis of the problem.

The need for a more thorough examination of potential injury to forest ecosystems comes into sharp focus where huge coal burning power plants are being planned in several western states. Gordon (1973) fears widespread effects where huge stacks up to 1200 ft tall may spread both sulfur dioxide and acid precipitation over wide areas. The proper amount of emission control must be engineered into these proposed power plants to protect downwind areas. In southeastern Montana, the National Ecological Research Laboratory of the Environmental Protection Agency is undertaking the kind of baseline studies including simulated exposures of natural vegetation under field conditions which are needed to set the proper standards for effecting adequate controls. Hill *et al.* (1974) have completed a short-term field study of the possible effects of SO_2 and $SO_2 + NO_2$ from a coal burning power plant on cold desert plants in Utah and New Mexico. Treshow and Stewart (1973) have described the relative ozone sensitivity of important species in grassland, oak, aspen, and conifer communities in Utah. Linzon (1972) described two prepollution background studies at Nanticoke and Timmins, Ontario.

The above are examples of the kinds of studies that must be widely encouraged and supported so that new incidents of pollutant injury to forest ecosystems can be greatly minimized and possibly prevented. Adequate lead time must be permitted so that data may be gathered for a longer period; for example, 1 or 2 full years of on-site meteorological data should be gathered because these factors strongly influence pollutant dispersion and plant sensitivity.

B. Postpollution Injury Evaluations

The evaluation of pollutant impact traditionally involves two phases—first, diagnosis to establish that an air pollutant is the cause of the observed injury, and, second, surveys to determine the extent and severity of injury sometimes including evaluations to determine damage or economic loss. Because a number of the important multiple use values of forest ecosystems cannot be clearly expressed in economic units, an additional method of documenting total impact, namely, systems analysis, is currently being adapted to this problem to evaluate the full spectrum of direct and indirect effects on the forest ecosystem (Taylor and Miller, 1973).

1. Methods for Diagnosis of Air Pollutant Injury

No effort will be made here to describe the symptoms of air pollutant injury because excellent color-illustrated references are available for this purpose (Jacobson and Hill, 1970; Van Haut and Stratmann, 1970). In diagnosis, a thorough description of symptoms on all species present and on leaf tissue of different ages and at different seasons is essential.

The analysis of foliage from the polluted area and a nearby unpolluted area provide an excellent proof of sulfur dioxide, fluoride, and chloride injury. However, tissue content of these pollutants cannot be used to quantify loss of tree growth (Stephan, 1971). The histological examination of leaf tissue injured by ozone and other pollutants or environmental factors (Stewart *et al.,* 1973; Miller and Evans, 1974), is a helpful procedure but it should always be supplemented by other procedures.

At the site of suspected pollutant injury, all, or a portion of an injured plant, may be placed in an enclosure provided with filters of activated charcoal or other appropriate materials; the clean air may decrease or eliminate visible symptoms after a few months (Costonis and Sinclair, 1969) or chlorophyll concentrations may increase immediately (Miller *et al.,* 1963). The sensitivity responses of plants can be enhanced by placing them in side by side filtered and unfiltered chambers (Miller and Yoshiyama, 1973). Injured plants may also be transplanted to unpolluted areas to see if they recover (U.S. Environmental Protection Agency, 1971). Indicator plants, namely, species or varieties particularly susceptible to specific pollutants may be placed in the area under investigation (Jacobson and Hill, 1970). Clonal selections of white pine differentially susceptible to sulfur dioxide, fluorides, and oxidants have been developed (Berry, 1973).

Measurement of height and diameter growth may also be useful in detecting injury to trees. Because this technique is useful in determining damage or economic loss it will be discussed more fully below.

Continuous air monitoring for one or more pollutants at the site of injury is an absolute necessity for completing a diagnosis of suspected air pollutant injury. More importantly, a network of monitoring stations throughout the damaged area will help to describe the dosages required for injury to important species. Knabe (1970) has established the relationship between annual dosage of SO_2 and the survivability of Scots pine in the Ruhr area of West Germany. Accumulation of air monitoring data and associated plant injury responses provide the criteria needed for establishing air quality standards which can lead to emission controls that minimize injurious pollution episodes.

Continuous measurement of wind speed and direction, temperature, and relative humidity should be included in the monitoring network because of positive effects they have on pollutant dispersion and plant sensitivity.

2. Methods for Determining the Extent and Severity of Injury and the Economic Loss

In the past, the extent and severity of injury to forest stands have usually been described by concentric zones centered around the pollutant source. The shape and size of such zones are often determined by topographic features and wind patterns; general terms described each zone, e.g., slight, moderate, or severe injury (Scheffer and Hedgcock, 1955; Gordon and Gorham, 1963). Tollinger (1968) evaluated damage around an industrial region in northwestern Bohemia by first dividing the area into 690 squares (each 25 ha); each square was assigned a score value based on vegetation composition, condition of the remaining vegetation, terrain, distance from the source of pollution, and wind direction. The resulting intepretation strongly emphasized the degree of difficulty of reforestation but did not estimate timber values lost or costs of reforestation. Aerial photography is also a useful tool for documenting injury (Wert *et al.,* 1970).

In a commercial forest, the most important aspect of pollution injury is the economic loss due to the diminished annual increase of wood volume. Pollanschutz (1971) has refined the technique of measuring annual ring width and the interpretation of the numerous variables affecting ring width so that reliable estimates of volume loss due to air pollution can be made. This is presently the only kind of data which can be converted into monetary loss with good confidence.

3. Systems Analysis of Air Pollutant Effects on Forest Ecosystems

The good health and productivity of green vegetation are the key factors which maintain normal function and structure in the ecosystem. Woodwell (1970) has summarized the expected effects of air pollutants on ecosystems; the same effects have been common in the foregoing examples: (1) elimination of sensitive species and reduction of diversity in numbers of species, (2) selective removal of larger overstory plants and a favoring of plants small in stature, (3) reduction of the standing crop of organic matter leading to a reduction of nutrient elements held within the system, and (4) enhancement of the activity of insect pests and some diseases which hasten mortality. Examples of this last item not encountered in the foregoing examples are reviewed by Grzywacz (1971) who cites both beneficial and deleterious effects on forest fungi. For example, most rusts and leaf diseases tend to disappear near sources of sulfur dioxide and fluorine, but often the root pathogens *Armillaria mellea* and *Fomes annosus* increase in incidence. Beneficial mycorrhizal fungi decrease their frequency of association with tree rootlets.

Increased attack by insect pests has been reported in the case of the pine bark beetles, *Dendroctonus* sp., on ponderosa pine (Stark *et al.,* 1968) and *Pissodes, Ips,* and *Cryphalus* species in Austria (Donaubauer, 1968).

The above examples of both primary and secondary effects on green plants suggest a complicated web of interactions involving herbivores, carnivores, and decomposer populations. For example, herbivores depend on the energy fixed by green plants often in the form of seeds; seed production is frequently reduced following chronic pollutant exposure (Scheffer and Hedgecock, 1955).

The hydrological balance of a mixed conifer forest may be altered because shade producing evergreen conifers are selectively removed over deciduous species; thus, snow melt and runoff are hastened, and evaporative losses increase. The more rapid runoff may cause increased stream and lake sedimentation; decreases in shade increase water temperature, undoubtedly affecting some stream fisheries.

Because there are a multitude of primary and secondary pollutant effects to be evaluated in chronically exposed ecosystems it is useful to employ the technique of systems analysis (Watt, 1966). The ecosystem must be carefully observed and described in order to understand the system and predict its fate. The formal steps include: systems measurement, data analysis, systems modeling, systems simulation, and systems optimization. The last step implies the application of forest management techniques to ameliorate damage.

A model of the impact of oxidant air pollutants on a western mixed conifer forest ecosystem in the San Bernardino mountains of southern California has been initiated (Taylor and Miller, 1973). The investigation is comprised of eleven subprojects collecting data on important interactions at the organism and community levels with the final objective of predicting the ultimate effects on ecosystem structure and function. The project has progressed significantly into the systems measurement and data analysis phases (Taylor, 1974). This forest ecosystem has already suffered substantial injury with no real hope for immediate decreases in total oxidant exposure. However, it is hoped that the data from this study will bolster the criteria and standards needed to prevent similar injury in new locations. The study also will hopefully identify the most important bioindicators for future use in determining the status of similar ecosystems exposed to oxidant pollutants. Minimal deterioration of the quality of life for this and future generations will be achieved only if we can include in the planning process a sufficient amount of information that will enable us to understand the full consequences of our actions.

VII. Anticipation of New Air Pollutant Injury to Forest Ecosystems in the United States

A. Gaseous Pollutant Damage

The enlargement and subsequent coalescence of urban centers, the tendency to group power generating stations near mine mouths, and the growth of industrial complexes which involve whole regions, together represent a massive capacity for advecting injurious levels of pollutants over wide areas, particularly during stagnant periods. Many new interfaces between forests and polluted air can be expected.

Both regional and local weather patterns must be considered to evaluate the potential for damage by new or enlarging pollutant sources. Isopleths of the total number of forecast-days of high meteorological potential for air pollution in a 5-year period (Holzworth, 1972) may be superimposed on forest ecosystems of the United States (Fig. 4) to provide a useful means of prediction. In general, the meteorological potential is greater along the Pacific Coast than in the eastern states. The eastern deciduous forest ecosystem is less insular, however, than the western montane forest ecosystem, for example, and may present more opportunities for new pollutant–forest interfaces. The level of resolution in Fig. 4 is not small enough to recognize or display some of the regional landforms which create air basins such as those recognized in California (Miller and Millecan, 1971). None of the local features such as narrow valleys, which limit pollutant dispersion, can be shown. The map may be most practically used to determine where massive industrialization and urbanization may subject particular forest ecosystems to new or increased air pollutant stress.

B. Acid Precipitation

The release of sulfur dioxide and nitrogen dioxide into the atmosphere from fossil fuel combustion is followed by a chemical transformation (hydrolysis) into sulfuric and nitric acids. Even without this man-made input, rainwater would be slightly acid (pH 5.8) if it were in equilibrium with atmospheric carbon dioxide. This relatively weak carbonic acid solution is one of the major factors in the chemical weathering of rocks.

The acidity of rain has increased significantly in recent years particularly in England, Scandinavia, and the eastern United States; for example, at the Hubbard Brook Experimental Forest in New Hampshire, the annual weighted average pH between 1965 and 1971 ranged between 4.03 and 4.19 (Likens *et al.,* 1972). The lowest pH recorded at Hubbard Brook was 3.0. In a review of the literature, Likens *et al.* (1972) have suggested that large areas of the United States, which are meteorologically isolated from large industrial centers, notably the Pacific and eastern Gulf Coastal regions, still have relatively high rainfall pH values evidently controlled by carbonic acid.

Likens *et al.* (1972) have suggested several effects on forest ecosystems including changes in the rates of leaching of nutrients from plant foliage and soils. In Sweden, an annual reduction in forest growth during the last

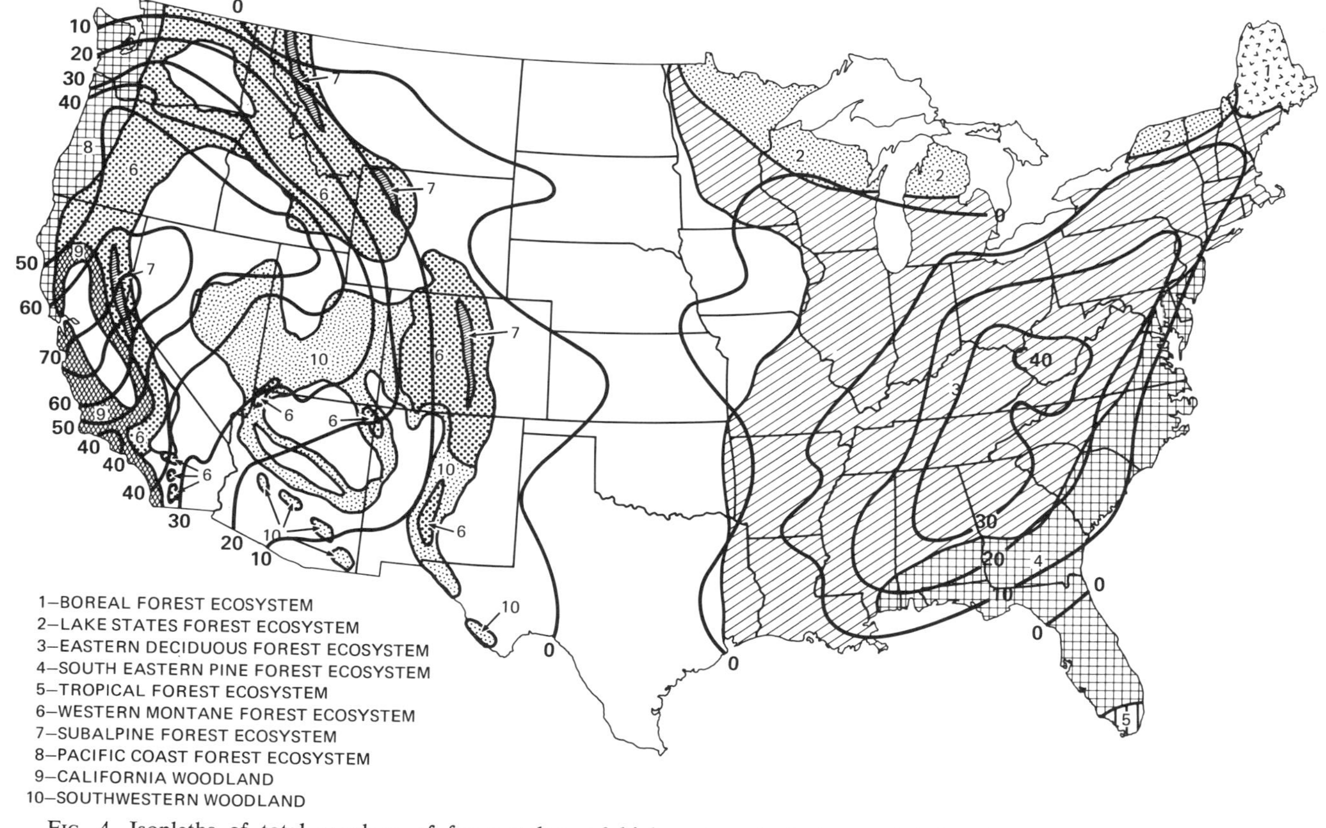

FIG. 4. Isopleths of total numbers of forecast-days of high meteorological potential for air-pollution in a five year period (Holzworth, 1972) compared with boundaries of forest ecosystems.

15 years has been possibly attributable to acid rainfall (Engstrom, 1971). An imbalance in the availability of calcium, magnesium, and potassium in shallow soils, already acid, because of a conifer forest cover, may best represent the conditions under which growth would be reduced.

The distribution of acid rainfall in the United States would not necessarily be similar to gaseous pollutants (Fig. 2) because it would not be a high frequency of stagnant days but rather the synoptic pattern typical of storm systems in relation to pollutant source areas that would control distribution. The full significance of acid precipitation to forest ecosystems remains to be defined.

VIII. Forests as Sources of Air Pollutants

A. Emission of Volatile Hydrocarbons

Water vapor, oxygen, and carbon dioxide are well known by-products of metabolism emitted by plants into the surrounding atmosphere. Very large quantities of photochemically reactive hydrocarbons, also metabolic by-products, are released simultaneously. The rate of release is light and temperature dependent. On a global basis, it has been estimated that tree foliage emits 175×10^6 tons of reactive hydrocarbons compared to 27×10^6 tons from the activities of man (Rasmussen, 1972).

Monoterpene and Hemiterpene Emissions and Their Fate

For simplification, the major forest tree species in the United States have been classified according to the most common terpene emitted (Rasmussen, 1972). According to this classification, both eastern and western conifers emit mainly α-pinene, but also appreciable quantities of other monoterpenes, namely, β-pinene, camphene, myrcene, limonene, and β-phellandrene. Eastern and western hardwoods emit the hemiterpene C_5, isoprene, while a very small number of hardwoods and one conifer (spruce) emit both isoprene and α-pinene. The principal source of terpenes is healthy foliage, although resin blisters and injured stems exuding xylem oleoresin also contribute.

The most celebrated hypothesis for the fate of emanated volatiles from plants is that they are precursors of the blue haze usually visible over forests and even deserts, especially in still, hot weather (Went, 1960a). The photooxidation and polymerization of terpenes results in formation of submicroscopic aerosol particulates which scatter light (Went, 1955).

The enormous quantity of terpenes emitted has suggested that they have some relation to origin of the world's petroleum formed in earlier geological periods (Went, 1960b).

The proper balance of ozone concentrations between the stratosphere and the troposphere may be maintained by the huge reservoir of reactive terpenes which consume ozone entering from the stratosphere (Ripperton *et al.,* 1967). On the other hand, mixtures of terpenes and nitric oxide react, as do olefins from gasoline combustion, to form ozone when irradiated in a closed system (Stephens and Scott, 1962).

B. Emission of Wax Particles

The surface wax deposits of conifer needles have been investigated (Hanover, 1972) with the scanning electron microprobe. There was variation between and within species in the amount and type of surface wax. In some cases, the wax was so thick that it partially occluded the stomata. Although the wax deposits are generally even and smooth at the tips of needles, wax fingers extending from the tips of *Pinus echinata* and other species have been described (Fish, 1972). These were present mainly on the needles exposed at the tops of trees or on the outside of lone trees. Wax fingers may be "the preserved record of a conduction path which became molten during the atmospheric phenomenon usually referred to as brush discharge" (Fish, 1972). The voltage gradients which develop during electrical storms cause an emission of wax particles. It was suggested that these particles may be part of the blue haze phenomenon described by Went (1960a).

C. Contribution of Wildfires and Forest Residue Combustion

Both wildfires and controlled or "prescribed" burning contribute large quantities of certain air pollutants to the atmosphere for short periods of time. If weather conditions are stagnant, the smoke from large fires may blanket large regions, causing reduction in visibility and other objectionable irritations to man. In the case of prescribed burning, the goals may be to improve regeneration in some forest types, control diseases such as mistletoe, hasten recycling of nutrients, or reduce the fuel load and lower the probability of catastrophic fires.

The relative amounts of five pollutants from the burning of forest fuels nationwide have been compiled by Hall (1972). In terms of millions of tons per year, the quantities are carbon monoxide 7.2, particulates 6.7, sulfur dioxide negligible, hydrocarbons 2.2, and nitrogen oxides 1.2. The percent of the national total emissions that each represents is carbon monoxide 7.2, particulates 23.7, sulfur dioxide none, hydrocarbons 6.9, and nitrogen oxides 5.8. These estimates may require adjustment as more information becomes available.

Hall (1972) suggests that there is little possibility that nitric oxide (NO) and hydrocarbons from smoke could react to form ozone photochemically because of the small amounts of NO formed and the rapid dispersion of smoke by atmospheric turbulence surrounding the fire. In Australia, Evans *et al.* (1974) measured ozone by instrumented aircraft in the upper layer of smoke plumes from forest fires. The ozone formed when smoke rose to inversion height and drifted downwind. The thickness of the ozone layer and concentrations increased with distance and greater irradiation time. The highest concentration observed downwind was 0.10 ppm. The rapid reaction of ozone with "combustion products" near the fire was considered responsible for the lack of ozone observed there. Ozone was observed to form when smoke samples in transparent plastic film bags were irradiated.

It is tempting to speculate that wood smoke from sawdust burners and other mill operations could cause slight ozone damage to downwind vegetation, particularly in confined valleys with persistent temperature inversions. More investigation is needed to establish the facts in the matter of smoke as a precursor of ozone.

References

Adams, D. F., Mayhew, D. J., Gnagy, R. M., Richey, E. P., Koppe, R. K., and Allan, I. W. (1952). Atmospheric pollution in the ponderosa pine blight area, Spokane County, Washington. *Ind. Eng. Chem.* **44,** 1356–1365.

Adams, D. F., Shaw, C. G., and Yerkes, W. D., Jr. (1956). Relationship of injury indexes and fumigation fluoride levels. *Phytopathology* **46,** 587–591.

Anonymous (1964). "Forest Cover Types of North America." Soc. Amer. Foresters, Washington, D.C.

Asher, J. E. (1956). "Observation and Theory on 'X' Disease or Needle Dieback," File Report. Arrowhead Dist., San Bernardino Nat. Forest, California.

Baker, F. S. (1950). "Principles of Silviculture." McGraw-Hill, New York.

Berry, C. R. (1961). White pine emergence tipburn, a physiogenic disturbance. *U.S., Forest Serv., Southeast. Forest Exp. Sta., Sta. Pap.* **130,** 1–8.

Berry, C. R. (1971). Relative sensitivity of red, jack and white pine seedlings to ozone and sulfur dioxide. *Phytopathology* **61,** 231–232.

Berry, C. R. (1973). The differential sensitivity of eastern white pine to three types of air pollution. *Can. J. Forest Res.* **3,** 543–547.

Berry, C. R., and Heggestad, H. E. (1968). Air pollution detectives. *Yearb. Agr.* (*U. S. Dep. Agr.*) pp. 142–146.

Berry, C. R., and Hepting, G. H. (1964). Injury to eastern white pine by unidentified atmospheric constituents. *Forest Sci.* **10,** 1–13.

Berry, C. R., and Ripperton, L. A. (1963). Ozone, a possible cause of white pine emergence tipburn. *Phytopathology* **53,** 552–557.

Billings, W. D. (1970). "Plants, Man and the Ecosystem." Wadsworth, Belmont, California.

Calvert, S. (1967). The ring around the air basin. *J. Air Pollut. Contr. Ass.* **17,** 366–367.

Carlson, C. E. (1972). Monitoring fluoride pollution in Flathead National Forest and Glacier National Park. *U.S., Forest Serv., Div. State, Priv. Forest., Missoula* pp. 1–25.

Carlson, C. E. (1974a). Sulfur damage to douglas-fir near a pulp and paper mill in western Montana. *U.S., Forest Serv., Div. State, Priv. Forest. Missoula* No. 74–13, pp. 1–41.

Carlson, C. E. (1974b). Evaluation of sulfur dioxide injury to vegetation on federal lands near the Anaconda copper smelter at Anaconda, Montana. *U.S., Forest Serv., Div. State, Priv. Forest., Missoula* No. 74–15, pp. 1–7.

Carlson, C. E., and Dewey, J. E. (1971). Environmental pollution by fluorides in Flathead National Forest and Glacier National Park, *U.S., Forest Serv., Div. State, Priv. Forest., Missoula.* pp. 1–57.

Cobb, F. W., Jr., and Stark, R. W. (1970). Decline and mortality of smog-injured ponderosa pine. *J. Forest.* **68,** 147–149.

Committee on Forest Terminology. (1944). "Forest Terminology." Soc. Amer. Foresters, Washington, D.C.

Compton, O. C., Remmert, L. F., and Rudinsky, J. A. (1961). Needle scorch and condition of ponderosa pine trees in The Dalles area. *Oreg., Agr. Exp. Sta., Misc. Pap.* **120.**

Costonis, A. C. (1970). Acute foliar injury of eastern white pine induced by sulfur dioxide and ozone. *Phytopathology* **60,** 994–999.

Costonis, A. C. (1973). Injury to eastern white pine by sulfur dioxide and ozone alone and in mixtures. *Eur. J. Forest. Pathol.* **3,** 50–55.

Costonis, A. C., and Sinclair, W. A. (1969). Relationship of atmospheric ozone to needle blight of eastern white pine. *Phytopathology* **59,** 1566–1574.

Critchfield, W. B., and Little, E. L., Jr. (1966). Geographic distribution of the pines of the world. *U.S., Forest Serv., Misc. Publ.* **991,** 1–97.

Dana, S. T. (1908). Extent and importance of the white pine blight. *U.S., Forest Serv., Leafl.* pp. 1–4.

Daubenmire, R. (1968). "Plant Communities." Harper, New York.

Davis, D. D., and Wood, F. A. (1972). The relative susceptibility of eighteen coniferous species to ozone. *Phytopathology* **62,** 14–19.

Davis, J. B., and Barnes, R. L. (1973). Effects of soil-applied fluoride and lead on growth of loblolly pine and red maple. *Environ. Pollut.* **5,** 35–44.

Dochinger, L. S. (1968). The impact of air pollution on eastern white pine: The chlorotic dwarf disease. *J. Air Pollut. Contr. Ass.* **18,** 814–816.

Dochinger, L. S. (1970). The impact of air pollution on Christmas tree plantings. *Amer. Christmas Tree J.* **14,** 5–8.

Dochinger, L. S., and Heck, W. W. (1969). An ozone-sulfur dioxide synergism produces symptoms of chlorotic dwarf disease of eastern white pine. *Phytopatholgy* **59,** 399.

Donaubauer, E. (1968). Secundarschaden in Osterreichischen Rauchschadensgebieten Schwierigkeiten der Diagnose und Berwertung. *Miedzynarodowej Konf. Wplyw. Zanleczyszczen Powietrza Na Lasy, 6th, Katowice, 1968* pp. 277–284.

Dreisinger, B. R. (1965). Sulfur dioxide levels and the effects of the gas on vegetation near Sudbury, Ontario. *58th Annu. Meet. Air Pollut. Contr. Ass.* Paper No. 65–121.

Edinger, J. G., McCutchan, M. H., Miller, P. R., Ryan, B. C., Schroder, M. J., and Behar, J. V. (1972). Penetration and duration of oxidant air pollution in the South Coast air basin of California. *J. Air Pollut. Contr. Ass.* **22,** 882–886.

Ellertsen, B. W., Powell, C. J., and Massey, C. L. (1972). Report on study of diseased white pine in Tennessee. *Mitt. Forstl. Bundesversuchanst., Wien* **97,** 195–206.

Engstrom, A. (1971). "Air Pollution Across National Boundaries: The Impact on the Environment of Sulfur in Air and Precipitation," Report. Swedish Preparatory Comm. for U.N. Conf. on Human Environment, Kungl. Boktryckeriet P. A. Norstedt et Soner, Stockholm.

Errington, J. C., and Thirgood, J. V. (1971). Search through old papers helps reconstruct recovery at Anyox from fume damage and forest fires. *Northern Miner, Annu. Rev.* pp. 72–75.

Evans, L. F., King, N. K., Packham, D. R., and Stephens, E. T. (1974). Ozone measurements in smoke from forest fires. *Environ. Sci. Technol.* **8,** 75–76.

Faull, J. H. (1922). Some problems of forest pathology in Ontario—Needle blight of white pine. *J. Forest.* **20,** 67–70.

Fish, B. R. (1972). Electrical generation of natural aerosols from vegetation. *Science* **175,** 1239–1240.

Gordon, A. G., and Gorham, E. (1963). Ecological aspects of air pollution from an iron-sintering plant at Wawa, Ontario. *Can. J. Bot.* **41,** 1063–1078.

Gordon, C. C. (1972). "Fluoride Pollution in Glacier National Park," Final report to EPA. Botany Dept., University of Montana, Missoula.

Gordon, C. C. (1973). Implications of air pollution in forest land planning. *Proc. Annu. Meet. Soc. Amer. Foresters* pp. 247–253.

Grzywacz, A. (1971). The influence of industrial air pollution on pathological fungi of forest trees. *Sylwan* **155,** 55–62.

Hall, J. A. (1972). Forest fuels, prescribed fire and air quality. *U.S. Forest Serv., Pac. Northwest* pp. 1–44.

Hanover, J. W. (1972). Factors affecting the release of volatile chemicals by forest trees. *Mitt. Forstl. Bundesversuchanst., Wien* **97,** 625–644.

Haselhoff, E., and Lindau, G. (1903). "Die Beschadigung der Vegetation durch Rauch." Borntraeger, Berlin.

Haywood, J. K. (1905). Injury to vegetation by smelter fumes. *U.S., Dep. Agr. Bur. Chem., Bull.* **89,** 1–23.

Haywood, J. K. (1910). Injury to vegetation and animal life by smelter wastes. *U.S., Dep. Agr., Bur. Chem., Bull.* **113,** 1–63.

Heck, W. W. (1968). Factors influencing expression of oxidant damage to plants. *Annu. Rev. Phytopathol.* **6,** 165–187.

Hedgcock, G. G. (1914). Injuries by smelter smoke in southeastern Tennessee. *J. Wash. Acad. Sci.* **4,** 70–71.

Hibben, C. R. (1969). Ozone toxicity to sugar maple. *Phytopathology* **59,** 1423–1428.

Hill, A. C., Hill, S., Lamb, C. and Barrett, T. W. (1974). Sensitivity of native desert vegetation to SO_2 and to SO_2 and NO_2 combined. *J. Air Pollut. Contr. Ass.* **24,** 153–157.

Holzworth, G. C. (1972). "Mixing, Wind Speeds and Potential for Urban Air Pollution Throughout the Contiguous United States," Publ. No. AP-101. Off. Air Program, U.S. Environmental Protection Agency, Washington, D.C.

Hursh, C. R. (1948). Local climate in the copper basin of Tennessee as modified by removal of vegetation. *U.S., Dep. Agr., Circ.* **774,** 1–38.

Jacobson, J. S., and Hill, A. C., eds. (1970). "Recognition of Air Pollution Injury to Vegetation: A Pictorial Atlas." Air Pollut. Contr. Ass., Pittsburgh, Pennsylvania.

Jensen, K. F. (1973). Response of nine forest tree species to chronic ozone fumigation. *Plant Dis. Rep.* **57,** 914–917.

Jung, E. (1968). Bestandesumwandlungen Im Rauchschadensgebiete Von Ranshofen. *Miedzynarodowej Konf. Wplyw. Zanieczyszczen Powietrza Na Lasy, 6th, Katowice, 1968* pp. 407–413.

Katz, M. (1949). Sulfur dioxide in the atmosphere and its relation to plant life. *Ind. Eng. Chem.* **41,** 2450–2465.

Knabe, W. (1968). Experimentelle Prufung der Fluoranreicherung in Nadeln und Blattern von Pflanzen in Abhangigkeit von deren Expositionshoh uber Grund. *Miedzynarodowej Konf., Wplyw. Zanieczyszczen Powietrza Na Lasy, 6th, Katowice, 1968* pp. 101–116.

Knabe, W. (1970). Distribution of Scots pine forest and sulfur dioxide emissions in the Ruhr area. *Staub-Reinhalt. Luft* **30,** 43–47.

Kohut, R. J. (1972). "Response of Hybrid Poplar to Simultaneous Exposure to Ozone and PAN," Publ. No. 288–72. Cent. Air Environ. Stud., Pennsylvania State University, University Park.

Kress, L. W. (1972). "Response of Hybrid Poplar to Sequential Exposures to Ozone and PAN," Publ. No. 259–72. Cent. Air Environ. Stud., Pennsylvania State University, University Park.

Likens, G. E., Bormann, F. H., and Johnson, N. M. (1972). Acid rain. *Environment* **14,** 33–40.

Linzon, S. N. (1960). The development of foliar symptoms and the possible cause and origin of white pine needle blight. *Can. J. Bot.* **38,** 153–161.

Linzon, S. N. (1965a). Sulfur dioxide injury to trees in the vicinity of petroleum refineries. *Forest. Chron.* **41,** 245–247.

Linzon, S. N. (1965b). Semimature-tissue needle blight of eastern white pine and local weather. *Ont. Dep. Forest., Res. Lab., Inform. Rep.* **0-X-1**.

Linzon, S. N. (1966). Damage to eastern white pine by sulfur dioxide, semimature-tissue needle blight and ozone. *J. Air Pollut. Contr. Ass.* **16,** 140–144.

Linzon, S. N. (1967). Histological studies of symptoms in semimature-tissue needle blight of eastern white pine. *Can. J. Bot.* **45,** 133–143.

Linzon, S. N. (1972). Pre-pollution background studies in Ontario. *Proc. 27th Annu. Meet. Soil Conserv. Soc. Amer.* pp. 1–7.

Linzon, S. N. (1973). The effects of air pollution on forests. *Pap., 4th Jt. Chem. Eng. Conf., 1973* pp. 1–18.

Lynch, D. F. (1951). Diameter growth of ponderosa pine in relation to the Spokane pine-blight problem. *Northwest Sci.* **25,** 157–163.

McCutchan, M. H., and Shroeder, M. J. (1973). Classification of meterological patterns in southern California by discriminant analysis. *J. Appl. Meteorol.* **12,** 571–577.

Miller, P. R. (1973a). Oxidant-induced community change in a mixed conifer forest. *Advan. Chem. Ser.* **122,** 101–117.

Miller, P. R. (1973b). Susceptibility to ozone of selected western conifers. *Abstr. Int. Congr. Plant Pathol., 2nd,* Abstract No. 0579.

Miller, P. R., and Evans, L. S. (1974). Histopathology of oxidant injury and winter-fleck injury on needles of western pines. *Phytopathology* **64,** 801–806.

Miller, P. R., and Millecan, A. A. (1971). Extent of oxidant injury to some pines and other conifers in California. *Plant Dis. Rep.* **55,** 555–559.

Miller, P. R., and Yoshiyama, R. M. (1973). Self-ventilated chambers for identification of oxidant damage to vegetation at remote sites. *Environ. Sci. Technol.* **7,** 66–68.
Miller, P. R., Parmeter, J. R., Jr., Taylor, O. C., and Cardiff, E. A. (1963). Ozone injury to the foliage of ponderosa pine. *Phytopathology* **53,** 1072–1076.
Miller, P. R., McCutchan, M. H., and Ryan, B. C. (1972a). Influence of climate and topography on oxidant air pollution concentrations that damage conifer forests in southern California. *Mitt. Forstl. Bundesversuchanst., Wien* **97,** 585–607.
Miller, P. R., McCutchan, M. H., and Milligan, H. P. (1972b). Oxidant air pollution concentrations in the central valley, Sierra Nevada foothills and Mineral King Valley of California. *Atmos. Environ.* **6,** 623–633.
National Park Service. (1970). "1970 Episode of Air Pollution Damage to Vegetation in Washington, D.C. Area," *Annual Report.* Chief Scientist, NPS, Washington, D.C.
Parmeter, J. R., Jr., and Miller, P. R. (1968). Studies relating to the cause of decline and death of ponderosa pine in southern California. *Plant Dis. Rep.* **52,** 707–711.
Parmeter, J. R., Jr., Bega, R. V., and Neff, T. (1962). A chlorotic decline of ponderosa pine in southern California. *Plant Dis. Rep.* **46,** 269–273.
Pollanschutz, J. (1971). Die Ertragskundlichen Messmethoden zur Erkennung und Beurteilungvon Von Forslichen Rauchshaden. *Mitt. Forstl. Bundesversuchsanst., Wien* **92,** 155–206.
Rasmussen, R. A. (1972). What do hydrocarbons from trees contribute to air pollution *J. Air Pollut. Contr. Ass.* **22,** 537–543.
Richards, B. L., Middleton, J. T., and Hewitt, W. B. (1958). Air pollution with relation to agronomic crops. V. Oxidant stipple of grape. *Agron. J.* **50,** 559–561.
Richards, B. L., Sr., Taylor, O. C., and Edmunds, G. F., Jr. (1968). Ozone needle mottle of pines in southern California. *J. Air Pollut. Contr. Ass.* **18,** 73–77.
Ripperton, L. A., White, O., and Jeffries, H. E. (1967). Gas phase ozone-pinene reactions. *Proc. 147th Meet., Amer. Chem. Soc., 1967* pp. 54–56.
Robak, H. (1969). Aluminum plants and conifers in Norway. *Proc. Eur. Congr. Influ. Air Pollut. Plants, Anim., 1st, 1968* pp. 27–31.
Scheffer, T. C., and Hedgcock, G. G. (1955). Injury to northwestern forest trees by sulfur dioxide from smelters. *U.S. Forest Serv. Tech. Bull.* **1117,** 1–49.
Scurfield, G. (1960). Air pollution and tree growth. *Forest. Abstr.* **21,** 339–347 and 517–528.
Shaw, C. G., Fischer, G. W., Adams, D. F., Adams, M. F., and Lynch, D. W. (1951). Fluorine injury to ponderosa pine: A summary. *Northwest Sci.* **15,** 156.
Spaulding, P. (1909). The present status of the white pine blights. *U.S., Dep. Agr., Circ.* **35,** 1–12.
Stark, R. W., Miller, P. R., Cobb, F. W., Jr., Wood, D. L., and Parmeter, J. R., Jr. (1968). Photochemical oxidant injury and bark beetle (Coleoptera: Scolytidae) infestation of ponderosa pine. I. Incidence of bark beetle infestation in injured trees. *Hilgardia* **39,** 121–126.
Stephan, K. (1971). Chemische Nadelanalyse Schadstoffbestimmung. *Mitt. Forstl. Bundesversuchsanst., Wien* **92,** 84–102.
Stephens, E. R., and Scott, W. E. (1962). Relative reactivity of various hydrocarbons in polluted atmospheres. *Proc. Amer. Petrol. Inst., Sect. 3* **42,** 665.
Stevens, R. E., and Hall, R. C. (1956). "Black Pine-leaf Scale and Needle Dieback, Arrowhead-Crestline Area San Bernardino National Forest, May 1956, Appraisal

Survey," File Report. Pacific Southwest Forest Range Exp. Sta., Berkeley, California.

Stewart, D., Treshow, M., and Harner, F. M. (1973). Pathological anatomy of conifer needle necrosis. *Can. J. Bot.* **51,** 983–988.

Swingle, R. U. (1944). Chlorotic dwarf of eastern white pine. *Plant Dis. Rep.* **28,** 824–825.

Taylor, O. C. (1973a). "Oxidant Air Pollutant Effects on a Western Coniferous Forest Ecosystem," Task B Report. Univ. Calif. Statewide Air Pollution Research Center, Riverside.

Taylor, O. C. (1973b). "Oxidant Air Pollutant Effects on a Western Coniferous Forest Ecosystem," Task C Report. Univ. Calif. Statewide Air Pollution Research Center, Riverside.

Taylor, O. C. (1974). "Oxidant Air Pollutant Effects on a Western Coniferous Forest Ecosystem," Task D Report. Univ. Calif. Statewide Air Pollution Research Center, Riverside.

Taylor, O. C., and Miller, P. R. (1973). Modeling the oxidant air pollutant impact on a forest ecosystem. *Calif. Air Environ.* **4,** 1) 1–3.

Tollinger, V. (1968). Beitrag zur Grossraumigen Diagnose der Rauchschaden. *Miedzynarodowej Konf., Wplyw. Zanieczyszczen Powietrza Na Lasy, 6th, Katowice, 1968* pp. 261–275.

Treshow, M., and Stewart, D. (1973). Ozone sensitivity of plants in natural communities. *Biol. Conserv.* **5,** 209–214.

Treshow, M., Anderson, F. K., and Harner, F. (1967). Responses of Douglas-fir to elevated atmospheric fluorides. *Forest Sci.* **13,** 114–120.

U.S. Environmental Protection Agency. (1971). Preconference report: Official air pollution abatement activity, Mt. Storm, West Virginia-Gorman, Maryland. *Air Pollut. Contr. Off. Publ.* **1,** No. APTD-0656, 21–48.

Van Haut, H., and Stratmann, H. (1970). "Farbtafelatlas uber Schwefeldioxid-Wirkungen an Pflanzen." Verlag W. Girardet, Essen, West Germany.

Watt, K. E. F. (1966). "Systems Analysis in Ecology," Academic Press, New York.

Weaver, J. R., and Clements, F. E. (1938). "Plant Ecology," 2nd ed. McGraw-Hill, New York.

Went, F. W. (1955). Air pollution. *Sci. Amer.* **192,** 63–72.

Went, F. W. (1960a). Blue hazes in the atmosphere. *Nature* (*London*) **187,** 641–643.

Went, F. W. (1960b). Organic matter in the atmosphere and its possible relation to petroleum formation. *Proc. Nat. Acad. Sci. U.S.* **46,** 212–221.

Wert, S. L., Miller, P. R., and Larsh, R. N. (1970). Color photos detect smog injury to forest trees. *J. Forest.* **68,** 536–539.

Wilhour, R. G. (1971). "The Influence of Ozone on White Ash," Publ. No. 188–71. Cent. Air Environ. Stud., Pennsylvania State University, University Park.

Wood, F. A. (1967). Air pollution and shade trees. *Proc. Int. Shade Tree Conf., 43rd,* pp. 66–82.

Wood, F. A., and Coppolino, J. B. (1972). The influence of ozone on deciduous tree species. *Mitt. Forstl. Bundesversuchanst., Wien* **97,** 233–253.

Woodwell, G. M. (1970). Effects of pollution on the structure and physiology of ecosystems. *Science* **168,** 429–433.

11

EFFECTS OF AIR POLLUTANTS ON LICHENS AND BRYOPHYTES

Fabius LeBlanc and Dhruva N. Rao

I. Introduction

By the mid-nineteenth century botanists had become aware of the sensitivity of lichens to a city environment. Nylander (1866) was perhaps the first person to make a definite statement on the subject. After seeing the impoverished lichen vegetation in the outskirts of the Jardins du Luxembourg in Paris, he wrote that lichens, in their own way, give a measure of the quality of the air and constitute a kind of sensitive "hygienometer." In 1879, Johnson (Gilbert, 1970a) visiting Gibside Woods, 8 km west of Newcastle, and observing no foliose nor fruticose lichens wrote: "The lichens which flourished here in the fine condition spoken of by Winch (in 1807 and 1831) have perished, and this evidently from the pollution of atmosphere by the smoke and fumes of Tyne-side and the collieries of the surrounding district." Arnold (1891–1901) attached epiphytic lichens to tree trunks in Munich and stated that their death occurred shortly afterward. Tobler (1925) demonstrated that, in towns, cultivation of some lichens was possible only in filtered air.

Over the years pollution and urbanization can cause much damage to plants. For example, according to Barkman (1969), within the last century the Dutch flora lost 3.8% of its species of flowering plants, 15% of its terrestrial bryophytes, 13% of its epiphytic bryophytes, and 27% of its epiphytic lichens. In Amsterdam alone, 23 species of bryophytes which occurred in the year 1900 are now extinct. Delvosalle *et al.* (1969) observed that among some 600 species of bryophytes which were indigenous to Belgium in 1850, nearly 114 have disappeared and 34 are now very rare. They attribute this to direct and indirect effects of human activities, especially to pollution factors. LeBlanc and De Sloover (1970) reported that some bryophyte species which were common on Mount Royal in the city of Montréal about 50 years ago are now extremely rare, and some species have disappeared. Similarly, many of the most common bryophytes in Northumberland, England are totally absent from a large part of the lower Tyne Valley, where scattered colliery towns, burning pit heaps, and the huge conurbation of Newcastle-upon-Tyne are sources of air pollution (Gilbert, 1968). In a recent study on the distribution of corticolous, saxicolous, lignicolous, and terricolous lichen communities near a zinc factory in Lehigh Water Gap, Pennsylvania, Nash (1972) discovered a marked reduction in species diversity up to 6 km to the west and 15 km to the east of the factory.

Epiphytic lichens and bryophytes are especially sensitive to pollution. The following epiphytes, *Caloplaca cerina, Lobaria pulmonaria, Parmelia vittae, Caloplaca ferruginea, Physcia ciliata, Ramalina pollinaria, Ulota ludwigii, Pyrenula nitida,* and eight species of *Usnea,* disappeared from the Netherlands during the last hundred years (Barkman, 1969). In the past, these epiphytes showed a marked decline in their abundance and vitality, a reduction in size (*Usnea* 50 cm in 1870, reduced to 10 cm in 1969), a loss of fertility (*Evernia prunestri* 6 cm long and fertile in 1869, only 3 cm long and sterile in 1930), and a shrinkage of the diameter of their apothecia (*Ramalina fraxinea* with apothecia 6 mm diameter in 1869, only 1 mm in 1930).

Laundon (1967) offers a striking example of affected lichen colonies on limestone memorials erected in a London churchyard between 1751 and 1950. According to him *Lecanora dispersa* consistently occurs on over 80% of the stones, whereas the presence of *Caloplaca heppiana,* a more sensitive species, has dropped from a presence of about 90% on stones erected before 1751 to 0% on stones dated after 1901. Perhaps the rising air pollution has made the colonization of new surfaces by *C. heppiana* increasingly difficult, but it has not yet eliminated already established colonies.

There is an interesting field observation which has confirmed the regeneration of lichens after normal, unpolluted conditions were restored in an area formerly polluted with SO_2. Skye and Hallberg (1969) observed that the lichens near the Kvarntorp shale-oil works disappeared as a result of increased levels of atmospheric SO_2 until production at the factory was terminated in 1966. By the summer of 1967 they already saw signs of recovery in certain lichen species which showed renewed growth in the lobes of their thalli.

Studies by Sernander (1926), Jones (1952), Skye (1958), Fenton (1960), LeBlanc (1961), Gilbert (1965), Rao and LeBlanc (1966), Coker (1967), Daly (1970), Taoda (1972), and others clearly demonstrate that the absence of lichens and bryophytes in cities and industrialized areas is mainly the result of the presence of pollugenic agents. Rydzak (1953, 1959), Klement (1956, 1958), and others have counter-suggested that the absence of lichens in cities is entirely due to a desiccation phenomenon ("drought hypothesis"). However, there is a growing body of evidence to indicate that air pollution ("pollution hypothesis") is the most important environmental factor in determining whether these organisms will survive or perish in urban environments (LeBlanc and Rao, 1973b).

Since lichens are slow-growing and long-lived organisms with a special ability to accumulate substances from their environment (Smith, 1962), they are susceptible to many pollutants present in the atmosphere or brought down in the rain. This sensitivity is heightened by the fact that, unlike tropophytes, they never shed their toxin laden parts. Similarly bryophytes, especially the mosses, with their delicate and uncuticularized plant body, seem to have a great capacity for absorbing and accumulating pollugenic substances from the environment (Shacklette, 1965).

II. Methods of Investigation and Some Results

It is now recognized that despite wide differences in their botanical affinity, lichens and bryophytes are outstandingly similar in their response to air pollution and that certain species, especially the epiphytic ones, can be reliable pollution indicators. In fact, they could provide the simplest and the most economical tool for assessing an air pollution problem in time and space.

The methods for studying the air pollution effects on lichens and bryophytes have been mainly phytosociological and ecophysiological. By these methods it is possible to relate the presence or absence of species, their number, frequency, coverage, and external and internal symptoms of injury to the degree of pollution *vis-à-vis* the purity of the atmosphere.

A. Ecological and Phytosociological Methods

One of the methods of phytosociology is to analyze the vegetation of an area which is then related to as many ecological factors as possible. Species may simply be listed, individuals may be counted, areas covered by plants may be estimated, or vegetation maps may be drawn. To understand the structure and variation of the vegetation of a landscape in terms of an environmental gradient, such as levels of pollution, the gradient analysis approach is commonly used (Whittaker, 1967). The characteristics of species inhabiting a specific substrate, for example, the bark of trees, are correlated with the pollution concentration along a line transect laid out downwind from the polluted to the normal area.

For quantifying phytosociological variables, each species is given a numerical value that characterizes it in certain ways. For example, *frequency* of a species in a community is the percentage of its occurrence, *cover* is the percentage area it occupies, and *density* is the average number of individuals per unit area. Also, several other qualitative factors, such as periodicity, sociability, vitality, or fidelity of species, can be estimated (Braun-Blanquet, 1951).

A number of workers have used ecological and phytosociological data to map lichen and bryophyte distribution in polluted areas, while some have used them to derive poleotolerance indices and biological scales for purposes of assessing approximate levels of environmental pollution.

1. Vegetation Mapping

Mapping is perhaps the most refined method to summarize vegetation field data. The various maps that have been published on lichen and/or bryophyte vegetation with respect to pollution may be grouped into distribution or vegetation maps. The distinctive features of these types of maps are described below.

a. Distribution Maps. This is the simplest type of map. By means of symbols, the presence or absence of species growing in and around a large conurbation, or in the neighborhood of an industrial complex, or even over a larger territory, such as a state or a province, is simply plotted on an outline map. Sometimes vitality, periodicity, coverage, frequency, and other qualitative and quantitative factors are also illustrated. On the basis of this information the territory investigated is delineated into zones which in turn are correlated with pollution data from the area.

Skye (1958) correlated the distribution of the epiphytic lichen vegetation around the shale-oil works near Kvarntorp, Sweden, with the air pollution data of the area and found that the lichens near the factory showed a weaker growth than those in a control area, 16 km distant. He observed a close conformity between the pattern of atmospheric pollution and the distribution of lichens. Using a number of phytosociological variables from the epiphytic lichen and bryophyte flora of Limburg Province, Belgium, Barkman (1963) prepared a detailed map covering about 1600 km^2 of that area. The map clearly shows the poorest epiphytic flora to be around mines and factories, and the richest flora to be in woods, parks, and valleys. Beschel (1958) in the Austrian cities of Innsbruck and Salzburg, and LeBlanc and De Sloover (1970) in the city of Montréal, used distribution maps to express the exact limits reached by numerous species of epiphytes growing there.

Rose (1973) correlated the distribution pattern of epiphytic lichens such as *Anaptychia ciliaris, Lobaria* spp., *Parmelia caperata, P. perlata,* and *Usnea* spp. with the mean sulfur dioxide levels in southeast England. He found that the distribution maps, where the inner limits of species were considered, provided an indication of the mean SO_2 levels in the air. Similarly, Morgan-Huws and Haynes (1973), using the empirical expression of "presence or absence," determined the distribution limits of 11 species of lichens on oak trees in an area surrounding an Esso oil refinery at Fawley, England and obtained a direct relationship between the annual average SO_2 levels and the zonation of the lichen flora in the area.

To render the mapping method relatively simple and rapid, it was suggested that instead of collecting data for a large number of species one could collect basic phytosociological data of only a single, widespread, and easily recognizable species (Skye, 1958; Granger, 1972). Where single species is studied, percentage cover is of considerable practical value as a relative indicator of the degree of air pollution. Suggestions have been made that in areas where *Lecanora conizaeoides* is the only epiphytic species present, the percentage cover of the lichen could be related to the degree of SO_2 pollution of the air (Gilbert, 1969; Hawksworth and Rose, 1970; Hawksworth, 1973). In principle, this is perhaps correct, but species, such as *L. conizaeoides,* which presumably take advantage of reduced competition and perhaps derive some nutritional benefit in polluted environments, must be handled with due prudence.

b. VEGETATION MAPS. Many authors have made detailed phytosociological investigations of epiphytic vegetation of cities or of larger areas around factories and have used the data to delineate the areas into several

TABLE I
LICHEN VEGETATION AND AIR POLLUTION ZONES IN THE STOCKHOLM REGION[a]

Vegetation and pollution zone	Characteristics	Species present
Central	More or less lichen free	*Lecanora conizaeoides* (occasionally)
		Hypogymnia physodes (rarely)
Inner transitional	A few species occurring occasionally	*Parmelia sulcata*
Central transitional	A few species in considerable numbers	*Cetraria glauca*
Outer transitional	Many species but a few still missing	*Xanthoria polycarpa*
		Parmelia exasperatula
Normal	Normal lichen vegetation	

[a] Modified after Skye, 1968.

epiphytic zones which were eventually correlated with pollution zones. Such a vegetation-pollution map was first attempted by Sernander (1926) who studied the lichen vegetation of Stockholm, Sweden, and then related it to the degree of pollution of that city. He described three lichen zones in Stockholm: *lavöken* (lichen desert, corresponding to the area of maximum pollution), *kampzon* (transitional zone, corresponding to the area of moderate pollution), and *normalzon* (normal lichen distribution zone, corresponding to the area of minimum or relatively no pollution). However, with growing experience, workers have realized that the influence of air pollution extends considerably beyond the source of pollution, and that the affected area itself could be subdivided into many narrower zones from the poorly developed to the nearly normal vegetation zones. For example, Skye (1968) mapped five lichen zones in the Stockholm region, each zone having a specific level of pollution and a characteristic lichen vegetation (Table I). It has been recognized that these zones normally follow the pattern of the prevailing winds and assume an elliptic shape, the axis of the ellipse being in the southwest to northeast direction and the source of pollution being located near the southwest end (Fig. 1).

Small-scale vegetation maps usually cover large areas. Barkman (1958) produced a generalized map of the Netherlands showing conspicuous epiphyte deserts around large conurbations and important industrial centers. Rao and LeBlanc (1967) published a map of the area around an iron sintering plant at Wawa, Ontario on the basis of the number of epiphytes and the sulfate concentration of soil and showed that the sulfur dioxide affected the vegetation up to a distance of 55 km northeast of the plant (Fig. 1). Similarly, Hawksworth and Rose (1970) prepared a map

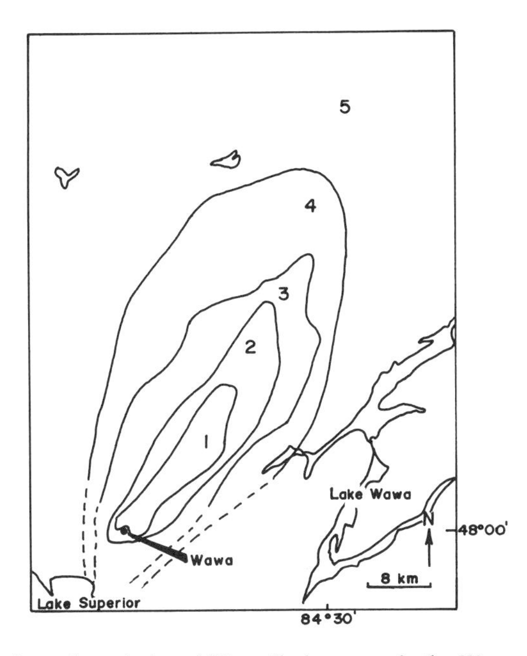

FIG. 1. Approximate boundaries of SO_2-pollution zones in the Wawa, Ontario, Canada area delineated on the basis of the number of epiphytic lichens and bryophytes, and the concentration of soil sulfate (after Rao and LeBlanc, 1967). Note the elliptical shape of the zones influenced by the SW prevailing winds. Zone 1: Sulfate concentration of soil more than 1.4 mEq per 100 g. Epiphytes absent. Zone 2: Sulfate concentration between 0.9 and 1.4 mEq per 100 g. The number of epiphytic species at any site ranges from 1–5. Zone 3: Sulfate concentration between 0.7 and 0.9 mEq per 100 g. The number of epiphytic species at any site ranges from 5–15. Zone 4: Sulfate concentration between 0.4 and 0.7 mEq per 100 g. The number of epiphytic species at any site ranges from 15–30. Zone 5: Sulfate concentration less than 0.4 mEq per 100 g. The number of epiphytic species at any site exceeds 30.

based on a qualitative scale for estimating the extent of SO_2 air pollution in England and Wales by using lichens growing on trees.

Domrös (1966) prepared a medium-scale map of the middle part of the Rhine-Westphalian industrial region on the basis of the coverage of crustose lichen vegetation growing on the southwest side of the trunks of maple and elm trees. He showed that coverage of 50–100% characterized areas largely free from industrial emissions and with less agricultural activity and low housing density; a coverage of 10–15% characterized areas with a large number of heavy industries and very high housing density. Gilbert (1969) published a map of the lower Tyne Valley and adjacent areas in England showing the limits of the lichen desert delineated on the basis of the presence of lichens on ash trees, on sandstone, and on asbestos. He observed that increasing pollution caused progressive reduction in luxuriance and decline in diversity of species. By using lichens, such as *Grimmia pulvinata* and *Parmelia saxatilis,* he mapped a lichen desert of about 1295 km^2 around Newcastle-upon-Tyne.

Many large-scale vegetative maps of comparatively small areas have been prepared on the basis of lichen and bryophyte vegetation. Some of the important urban and industrial centers which have been mapped are Oslo (Haugsjå, 1930), Helsinki (Vaarna, 1934), Stockholm (Høeg, 1936; Skye, 1968), Zurich (Vareschi, 1936), Debrecen (Felföldy, 1942), Uppsala (Krusenstjerna, 1945), Vienna (Sauberer, 1951), Caracas (Vareschi, 1953), Bonn (Steiner and Schulze-Horn, 1955), Munich (Schmid, 1957), Montréal (LeBlanc, 1961), and Rudnany (Pišút, 1962).

A special type of vegetation map is the so-called indices of atmospheric purity (IAP) map. This map is based on a mathematically derived IAP which expresses indirectly the long range effect of air pollution on epiphytic lichens and bryophytes (De Sloover and LeBlanc, 1968; LeBlanc and De Sloover, 1970). For accurate mapping, it is necessary that the area investigated be ecologically homogeneous and the study sites numerous and widespread. Preferably the epiphytes of a single tree species are investigated. However, the epiphyte vegetations of two or more species of trees may be mapped together provided that the bark characteristics of these trees (i.e., buffer capacity, density, moisture content, hardness) are somewhat similar.

To obtain the basic phytosociological data to prepare an IAP map all epiphytes present at a site are listed and their frequency-coverage is estimated. The IAP for the site is then calculated by using the following formula:

$$\mathrm{IAP} = \sum_{n}^{1} (Qf)/10$$

where n is the total number of epiphyte species, f is the frequency-coverage score of each species expressed by a number on a numerical scale chosen by the investigator, and Q is the resistance factor or ecological index of each species. Q for a species is determined by adding the numbers of its companion species present at all the investigated sites and then dividing this total by the number of sites. The sum of Qf is divided by 10 to reduce it to a small manageable figure. High IAP values represent rich epiphytic vegetation in relatively unpolluted areas, and low IAP values indicate poor and depauperate epiphytic vegetation in polluted environments (Table II).

For preparing a map the IAP values for different sites are plotted on an outline map of the area and all those falling within a certain interval

TABLE II
IAP AND THE CORRESPONDING SO_2 CONCENTRATION RANGES IN THE REGION OF SUDBURY, CANADA[a]

IAP	SO_2 conc (ppm)
0–9	>0.03
10–24	0.02–0.03
25–39	0.01–0.02
40–54	0.005–0.01
>55	<0.005

[a] After LeBlanc *et al.*, 1972a.

range are linked together by isometric lines distinguishing various IAP zones (Fig. 2). Since the number and size of these zones depend on the minimum and maximum limits of IAP values that one sets to delineate them, it is important that these limits be set in consonance with the field observations. The map thus drawn (Fig. 2) should simulate more or less precisely another map of the same area (Fig. 3) but prepared from actual SO_2 pollution data (Dreisinger, 1965), as has been shown for Sudbury, Ontario and Arvida, Québec by LeBlanc *et al.* (1972a,b).

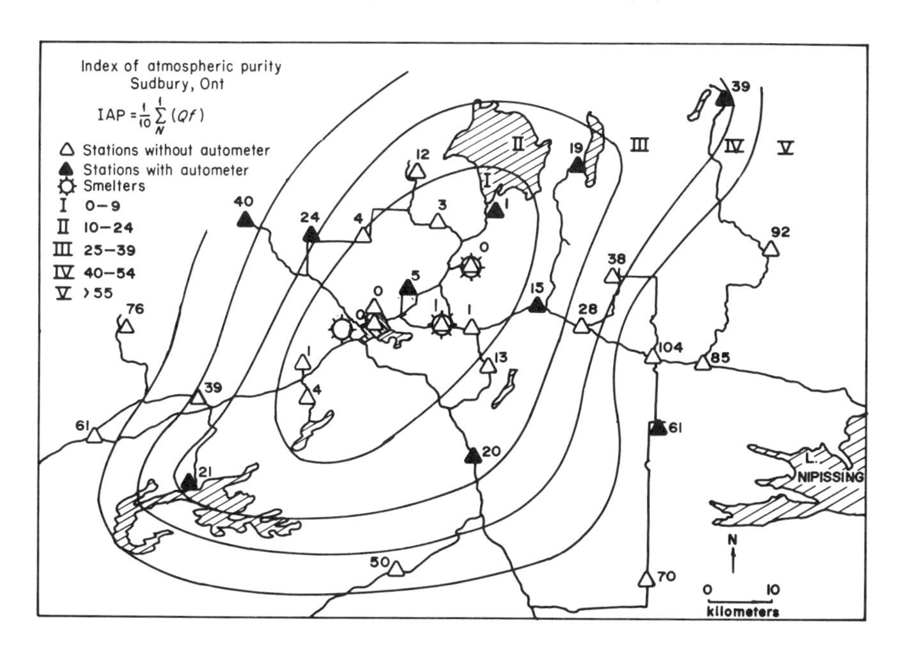

FIG. 2. The IAP zones in the Sudbury area (after LeBlanc *et al.*, 1972a).

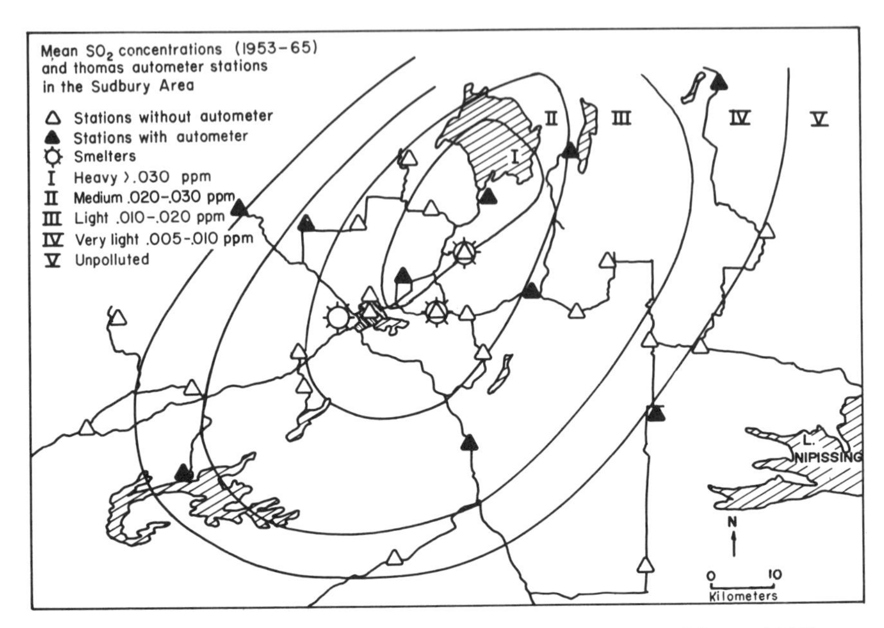

FIG. 3. The SO_2-pollution zones in the Sudbury area (after Dreisinger, 1965).

2. Poleotolerance Index and Biological Scales

Using lichen vegetation data, Trass (1968, 1973) devised the following formula to ascertain the extent of urban air pollution:

$$\text{I.P.} = \sum^{n} \frac{a_i - c_i}{C_i} \qquad i = 1$$

where I.P. is the poleotolerance index, a is the poleotolerance value of each species on a scale 1–10 determined by field experience, and C is the coverage value of each species on a scale 1–10 estimated by visual observation. The I.P. values are then brought to coincide with the annual averages of SO_2 prevailing in the area, as done by Trass for Estonia (Table III). To appreciate the expressiveness of poleotolerance indices one should know the long-range level of pollution in the area.

Because various species of lichens and bryophytes react differently to pollution, they can be ranked in a sensitivity scale to estimate changing levels of pollution. Gilbert (1970b) and Hawksworth and Rose (1970), on the basis of their observations on the distribution of lichens and bryophytes in certain areas of England, ranked groups of species from very sensitive to highly resistant. After correlating these species groups with

TABLE III
CORRELATION BETWEEN POLEOTOLERANCE INDICES (I.P.) AND SO_2 CONCENTRATIONS IN ESTONIA[a]

I.P. value	SO_2 concentration: mg/m^3 (given by Trass)	SO_2 concentration: ppm	Lichen zone
1–2	Absent		Normal zone
2–5	0.01–0.03	0.0035–0.0105	Mixed zone
5–7	0.03–0.08	0.0105–0.028	Mixed zone
7–10	0.08–0.10	0.028–0.035	Struggle zone
10	0.10–0.30	0.035–0.105	Struggle zone
0	>0.3	>0.105	"Lichen desert"

[a] After Trass, 1968.

annual average SO_2 concentrations tolerated, they produced biological scales which could be used to estimate the approximate sulfur dioxide air pollution prevailing in isoecological areas.

Gilbert (1970b) used lichen and bryophyte assemblages growing on sandstones, deciduous trees (ash or oak, Table IV) and old asbestos roofs, and he arranged them into six groups depending on their tolerance to SO_2

TABLE IV
MAXIMUM RANGES OF ANNUAL AVERAGE SO_2 CONCENTRATIONS TOLERABLE TO LICHENS GROWING ON THE NONEUTROPHIATED BARK OF *Quercus* OR *Fraxinus* SPP. IN THE TYNE VALLEY, ENGLAND[a]

SO_2-conc. range (ppm)	Tolerant lichen epiphytes
0.03–0.06	*Lecanora conizaeoides*
0.01–0.02	*Pertusaria* *Buellia punctata* *Dicranoweisia cirrata* *Hypnum cupressiforme*
<0.005 (relatively "pure" air)	*Frullania dilatata* *Metzgeria furcata* *Camptothecium sericeum* *Zygodon viridissimus* *Orthotrichum affine* *Tortula laevipila*

[a] Modified after Gilbert, 1970b.

TABLE V

MAXIMUM RANGES OF ANNUAL AVERAGE SO_2 CONCENTRATIONS TOLERABLE TO LICHENS GROWING ON THE NONEUTROPHIATED BARK OF *Quercus* OR *Fraxinus* SPP. AND EUTROPHIATED BARK OF *Ulmus* SP. IN ENGLAND AND WALES[a]

SO_2 conc. range (ppm)	Tolerant lichen epiphytes on	
	Quercus or *Fraxinus* spp.	*Ulmus* sp.
0.03–0.06	*Lecanora conizaeoides*	*Lecanora conizaeoides*
	Lepraria incana	*L. expallens*
		Buellia punctata
		B. canescens
0.02–0.03	*Hypogymnia physodes*	*Physcia adscendens*
	Parmelia saxatilis	*P. orbicularis*
	P. sulcata	*P. tenella*
	P. glabratula	*P. tribacia*
	P. subrudecta	*Physconia grisea*
	Lecidea scalaris	*P. farrea*
	Lecanora expallens	*Buellia alboatra*
	L. chlarotera	*Ramalina farinacea*
	Chaenotheca ferruginea	*Haematomma coccineum* var. *porphyrium*
	Parmeliopsis ambigua	*Schismatomma decolorans*
	Calicium viride	*Opegrapha varia*
	Lepraria candelaris	*O. vulgata*
	Pertusaria amara	*Parmelia acetabulum*
	Platismatia glauca	*Xanthoria candelaria*
	Ramalina farinacea	*X. parietina*
	Evernia prunastri	
0.01–0.02	*Parmelia caperata*	*Pertusaria albescens*
	P. revoluta	*Physconia pulverulenta*
	P. tiliacea	*Physciopsis adglutinata*
	P. exasperatula	*Arthopyrenia alba*
	P. perlata	*A. biformis*
	P. reticulata	*Caloplaca luteoalba*
	Pertusaria albescens	*Xanthoria polycarpa*
	P. hymenea	*Lecania cyrtella*
	P. hemisphaerica	*Bacidia rubella*
	Graphis elegans	*Ramalina fastigiata*
	Pseudevernia furfuracea	*R. obtusata*
	Alectoria fuscescens	*R. pollinaria*
	Rinodina roboris	*Candelaria concolor*
	Arthonia impolita	*Physcia aipolia*
	Normandina pulchella	*Anaptychia ciliaris*
	Usnea ceratina	*Parmelia perlata*
	U. subfloridana	*P. reticulata*
	U. rubiginea	*Gyalecta flotowii*
		Desmaziera evernioides

TABLE V *Continued*

SO_2 conc. range (ppm)	Tolerant lichen epiphytes on	
	Quercus or *Fraxinus* spp.	*Ulmus* sp.
0.005–0.01	*Lobaria pulmonaria* *L. amplissima* *Pachyphiale cornea* *Dimerella lutea* *Usnea florida*	*Caloplaca cerina* *C. aurantiaca* *Physcia Leptalea* *Ramalina calicaris* *R. fraxinea* *R. subfarinacea*
<0.005 (relatively "pure" air)	*Lobaria scrobiculata* *Sticta limbata* *Pannaria spp.* *Usnea articulata* *U. filipendula* *Teloschistes flavicans*	*Caloplaca cerina* *C. aurantiaca* *Physcia leptalea* *Ramalina calicaris* *R. fraxinea* *R. subfarinacea*

[a] Modified after Hawksworth and Rose, 1970.

concentrations. By using his scale, he concluded that large areas of the countryside in England experience SO_2 concentrations of 45μg/m^3 or 0.015 ppm yearly on the average.

Hawksworth and Rose (1970), on the other hand, produced a 10-point scale by ranking epiphytic lichen species with respect to their tolerance to SO_2 for concentrations ranging from >170 μg/m^3 (0.060 ppm) to concentrations that were very low. They proposed a series of epiphytic communities for trees, such as oak or ash, having noneutrophiated bark and another series for trees, such as elm, having eutrophiated bark (Table V). Similar epiphytic list of species have been presented by LeBlanc *et al.*, (1972a) for Sudbury, Ontario (Table VI) and by Taoda (1972) for Tokyo, Japan (Table VII).

Since on a given substrate in uniform habitats species usually behave in a predictable manner, biological scales should provide a cheap and easy method for estimating SO_2 in air. By knowing lichens and bryophytes of an area, one could read directly from the scale the approximate level of SO_2 at which these species survive. The scales so far proposed are of particular value in areas with mean SO_2 levels in the range of <30–170 μg/m^3 (0.01–0.06 ppm), and are most sensitive in the range of <30–70 μg/m^3 (0.01–0.024 ppm). However, such scales are of restricted value because they can be used only in areas ecologically similar to those for which they are devised.

TABLE VI

MAXIMUM RANGES OF ANNUAL AVERAGE SO_2 CONCENTRATIONS TOLERABLE TO LICHENS AND BRYOPHYTES ON THE NONEUTROPHIATED BARK OF *Populus balsamifera* IN THE SUDBURY AREA, CANADA[a]

SO_2 conc. range (ppm)	Tolerant epiphytic lichens and bryophytes
0.03–0.05	*Bacidia chlorococca*
	Lecanora saligna
	Lepraria aeruginosa
	Parmelia sulcata
0.02–0.03	*Candelariella vitellina*
	Cetraria pinastri
	C. saepincola
	Cladonia coniocraea
	Evernia mesomorpha
	Hypogymnia physodes
	Lecanora expallens
	L. subfusca variolosa
	L. subintricata
	Parmelia sulcata
	P. exasperatula
	Parmeliopsis ambigua
0.01–0.02	*Cladonia cristatella*
	Parmelia olivacea
	Pertusaria multipuncta
	Physcia adscendens
	P. grisea
	P. stellaris
	Rinodina hallei
0.005–0.01	*Caloplaca cerina*
	Lecanora symmicta
	Parmelia septentrionalis
	Physcia aipolia
	Alectoria americana
	A. nidulifera
	Usnea sp.
<0.005 (relatively "pure" air)	*Buellia stillingiana*
	Candelaria concolor
	Cetraria ciliaris
	Cladonia fimbriata
	Lecidea nylanderi
	Parmelia rudecta
	P. trabeculata
	Physcia farrea
	P. orbicularis
	Ramalina fastigiata
	Rinodina papillata
	Xanthoria fallax
	X. polycarpa

[a] Modified after LeBlanc *et al.*, 1972a.

TABLE VII

MAXIMUM RANGES OF ANNUAL AVERAGE SO_2 CONCENTRATIONS TOLERABLE TO EPIPHYTIC BRYOPHYTES ON THE NONEUTROPHIATED BARK OF *Acer, Quercus, Prunus, Ginkgo,* ETC., IN AND AROUND TOKYO, JAPAN[a]

SO_2 conc. range (ppm)	Tolerant epiphytic bryophytes
0.03–0.05	*Hypnum yokohamae* var. *kusatsuense*
	Glyphomitrium humillimum
	Entodon compressus
	Cololejeunea japonica
	Venturiella sinesis
0.02–0.03	*Sematophyllum subhumile* subs p. *japonicum*
	Hypnum plumaeforme var. *minus*
	Lejeunea punctiformis
	Bryum argenteum
	Frullania muscicola
0.01–0.02	*Eurohypnum leptothallum*
	Fabronia matsumurae
0.005–0.01	*Aulacopilum japonicum*
	Bryum capillare
	Haplohymenium sieboldii
	Herpetineuron toccoae
	Trocholejeunea sandvicensis
	Macromitrium japonicum
	Schwetschkea matsumurae
	Lophocolea minor
	Cheilolejeunea sp.

[a] Modified after Taoda, 1972.

B. Ecophysiological Methods

1. STUDIES UNDER FIELD CONDITIONS

To relate pollution to its source of emission and to corroborate the results of laboratory experiments, lichens and bryophytes have been transplanted along with their substrates from unpolluted sites to ecologically similar but polluted sites. Organisms growing on any substrate, for example, stone, soil, or bark, can be used in such experiments. The method of transplantation, however, varies with the type of substrate. In terricolous and saxicolous species, a mere physical transference of the substrate from normal to polluted environments would serve the purpose, but for corticolous species one has to punch out portions of bark without injuring the organisms and then fix these bark pieces carefully onto host trees in polluted areas. Most transplantation experiments have been performed with corticolous species. After a certain exposure period, say 4 or 12 months, the transplants from polluted areas are compared with those from nonpolluted areas with respect to injury and other changes induced under the influence of pollution.

Brodo (1961, 1966), using his newly developed technique, transplanted corticolous lichens onto trees in eastern Long Island, New York. He observed that transplants in the control areas remained healthy, sometimes even showing measurable growth in their thalli, while those in the vicinity of New York City died within 3–4 months of transplantation. He attributed their death to pollutants, especially SO_2, in the city environment. LeBlanc and Rao (1966) transplanted bark discs bearing lichens and bryophytes in the region of Sudbury, Ontario, where the atmosphere remained heavily polluted with SO_2. They observed that 1 year after transplantation most of the epiphytes were either seriously damaged or dead. Samples of *Parmelia caperata* and *P. sulcata* collected from the exposed discs showed abnormal microscopic features, such as reduction of the thallus thickness, deposition on the thallus surface of a whitish substance insoluble in water but soluble in acetone, and plasmolysis as well as chlorophyll degradation in the algal cells.

Schönbeck (1969) grafted bark cores bearing the lichen *Hypogymnia physodes* to vertical wooden frames supported by iron poles and exposed them, facing north, in sites in the Ruhr Valley. In areas polluted by SO_2, the thalli died within a few weeks, but in more or less unpolluted areas they remained normal even after 9 months. This way of proceeding is useful especially in severely polluted and built-up areas where transplantation studies could be restricted due to scarcity of suitable trees. Similarly Pyatt (1970) observed death of corticolous lichens within 2 years of their transplantation to the industrial area of Port Talbot, England. However, the reverse, that is transplantation of lichens from the Port Talbot vicinity to nonpolluted areas, showed that discs with crustose species, such as *Lecanora conizaeoides* and *Lepraria incana,* were generally overgrown by foliose species within 2 years, whereas discs already bearing foliose species were little affected and showed no apparent growth during this period. In his transplant studies with *Lobaria pulmonaria,* Hawksworth (1971) reported death of that foliose lichen after 19 months of its exposure to the SO_2-polluted air of Dovedale, England.

The external and internal injury symptoms of the transplanted lichens closely resemble those of lichens fumigated with SO_2 under laboratory conditions (Rao and LeBlanc, 1966). Since different levels of SO_2 produce different sets of external and internal injuries in lichens and mosses,

LeBlanc and Rao (1973a), on the basis of their transplant studies in the Sudbury area, suggest that the average concentrations of SO_2 above 0.154, between 0.087–0.154 or below 0.042 ppm, if calculated for the fumigation periods only, could produce acute, chronic, or no injury respectively (Table VIII, column a). However, in view of the fact that these organisms are being used as biological systems for monitoring the long-range effects of pollution, they emphasize that the long-range average concentrations, instead of the short-range ones, should be considered more important. Consequently, they think that the long-range average concentrations of SO_2 above 0.03, between 0.006 and 0.03, or below 0.002 ppm could cause respectively acute, chronic, or no injury to epiphytes in the Sudbury area (Table VIII, column c).

Gilbert (1968) observed that mosses, such as *Camptothecium sericium, Grimmia pulvinata, Hypnum cupressiforme,* and *Tortula muralis,* showed rapid deterioration after 3 months of their transfer to sites with high levels of SO_2 in Newcastle-upon-Tyne. He also noticed that in polluted areas *Hypnum cupressiforme* showed a rapid breakdown of chlorophyll, so much so that less than 10% of it remained after 10 weeks, and its respiration rate, though staying constant or increasing slightly in the beginning, fell rapidly later on. Daly (1970) made similar observations on the mosses *Brachythecium* and *Hypnum* transplanted in Christchurch, New Zealand.

TABLE VIII

LEVELS OF SO_2 CONCENTRATION FOR DIFFERENT TIME PERIODS AT VARIOUS SITES IN THE SUDBURY AREA, CANADA[a]

	Average SO_2 concentration, ppm[b]			
Site	a	b	c	Injury
Skead	0.194	0.042	0.030–0.042	Acute
Garson	0.154	0.030	0.030–0.042	Acute
Rayside	0.158	0.018	0.010–0.020	Chronic
Kukagami Lake	0.121	0.016	0.010–0.020	Chronic
Callum	0.103	0.012	0.010–0.020	Chronic
Burwash	0.081	0.008	0.005–0.010	Chronic
Lake Panache	0.121	0.007	0.005–0.010	Chronic
Grassy Lake	0.079	0.007	0.005–0.010	Chronic
Morgan	0.087	0.006	0.005–0.010	Chronic
St. Charles	0.042	0.002	<0.005	No injury

[a] Modified after LeBlanc and Rao, 1973a.

[b] a, over the periods when SO_2 was actually present; b, over the total period of 6 months; c, concentration range over a long period of time.

Usually the response of mosses to SO_2 pollution starts with a loss of color in the tips of the more exposed leaves, then gradually progresses down until all the leaves and the shoots have lost their chlorophyll. It has been seen that mosses which are completely bleached never recover even when replaced in their original habitats; however, in some acrocarpous mosses basal regenerative shoots may produce entirely new moss cushions. Field studies show that most bryophytes do not survive when the average winter concentrations of SO_2 exceed 50 $\mu g/m^3$ or 0.017 ppm (Gilbert, 1968; Daly, 1970).

The transplant method has been used also to study the effects of atmospheric fluorides on lichens and bryophytes. Nash (1971) transferred the terricolous lichens *Cladonia cristatella* and *C. polycarpoides* and the saxicolous species *Parmelia plittii* from their natural environment to a point 100 m from a fluoride-emitting factory in Pennsylvania and observed chlorosis and eventual disintegration of their thalli after 3 months, from July to September. The fluoride contents of these exposed plants ranged from 174–200 ppm, in contrast to 8–28 ppm of the control 6000 m distant from the factory. Similarly, LeBlanc *et al.* (1971) exposed epiphytic species, such as the lichen *Parmelia sulcata* and the mosses *Orthotrichum obtusifolium* and *Pylaisiella polyantha,* onto trees in the fluoride-polluted area of Arvida, Québec. They observed that after 12 months exposure the transplants near the factory showed complete destruction of chlorophyll (Fig. 4), plasmolysis and other cellular abnormalities, and high levels of fluoride accumulation. The F concentrations in *P. sulcata* and *O. obtusifolium* transplanted 1 km distant from the factory were 990 and 600 ppm respectively,

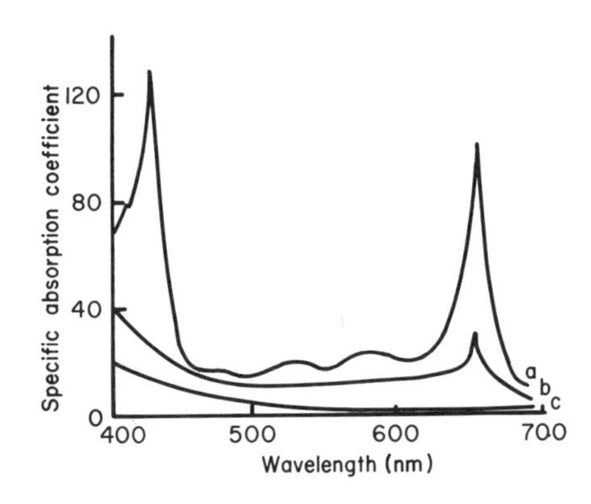

FIG. 4. Absorption spectra of chlorophyll extracts from *Parmelia sulcata* transplanted in the Arvida area: (a) thalli exposed for 12 months at a control site 40.0 km NE, (b) and (c) thalli exposed for 4 and 12 months respectively at a site 1.0 km NE of the aluminum factory (after LeBlanc *et al.,* 1971).

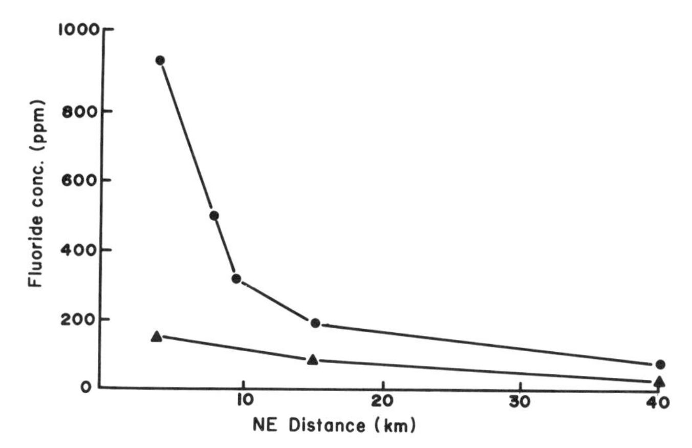

FIG. 5. Fluoride concentrations in the NE transplants of the lichen *Parmelia sulcata* ● and the moss *Orthotrichum obtusifolium* ▲ after 4 months exposure in the Arvida area (after LeBlanc *et al.*, 1971).

while those in the controls at 40 km were 70 and 20 ppm only. It appears that lichens are more efficient accumulators of fluoride than mosses (Fig. 5). Another noticeable effect was the failure of transplants to develop reproductive structures, vegetative as well as sexual, in the polluted areas.

Some workers have expressed doubts about transplant experiments. They feel that it may be the mere act of transplantation and/or possible alteration of moisture which could be affecting the organisms in polluted environments (Rydzak, 1968). Such doubts, however, seem to be groundless because the control transplants in unpolluted areas remain healthy and sometimes even show growth.

It should be remembered that the transplantation method helps in the estimation of pollution only roughly and that for precise correlation of injury symptoms with pollution levels, laboratory experiments are necessary.

2. STUDIES UNDER LABORATORY CONDITIONS

In an attempt to understand the reactions of lichens and bryophytes to pollugenic agents, many workers have initiated physiological and biochemical experiments by exposing plants to pollutants under controlled conditions. Three pollutants, namely sulfur dioxide, fluoride, and ozone, have so far been used in fumigation experiments.

a. EXPOSURE TO SULFUR DIOXIDE (SO_2). Three types of techniques have been used to study the effects of SO_2 on lichens and bryophytes. Samples have been exposed in closed vessels (Pearson and Skye, 1965;

Rao and LeBlanc, 1966; Coker, 1967), in continuous flow chambers (Syratt and Wanstall, 1969; Comeau and LeBlanc, 1971; Nash, 1971; Showman, 1972; Taoda, 1973a), or in aqueous solutions of SO_2 (Gilbert, 1968; Hill, 1971; Puckett *et al.*, 1973).

Pearson and Skye (1965) noted that thalli of *Parmelia sulcata* in SO_2-contaminated flasks showed morphological and photosynthetic abnormalities similar to those of lichens collected in polluted areas of a city. Consequently they suggested that the SO_2 present in city environments could destroy the chlorophyll of city lichens. Subsequently, Rao and LeBlanc (1966) exposed lichen thalli of *Xanthoria fallax, X. parietina, Parmelia caperata,* and *Physcia millegrana* to 5 ppm SO_2 for 24 hours at various humidities. They observed bleaching of chlorophyll, permanent plasmolysis, and formation of sporadic brown dots on the chloroplast of *Trebouxia,* the algal symbiont of the treated lichens. These injury symptoms gradually intensified at higher himidities (Fig. 6). The authors detected sulfurous acid and magnesium in the extracts of SO_2-exposed thalli, the chlorophyll extract of which showed maximum light absorption at 667 nm, the absorption peak characteristic of pheophytin a (Fig. 7). The following chemical reactions are suggested to be involved in the chlorophyll degradation process (Rao and LeBlanc, 1966):

$$SO_2 + H_2O \rightleftharpoons H_2SO_3$$
$$H_2SO_3 \rightleftharpoons HSO_3^- + H^+$$
$$\text{Chlorophyll a} + 2H^+ \rightarrow \text{Pheophytin a} + Mg^{2+}$$

It appears that the algal symbiont of the lichen thallus is the most vulnerable to SO_2 pollution. According to Showman and Rudolph (1972) this seems reasonable since most of the water present in a nonsaturated lichen thallus is held in the algal layer, thus providing an ideal situation for maximum SO_2 reaction.

Coker (1967) fumigated epiphytic bryophytes, *Orthotrichum lyellii, O. diaphanum,* and *Radula complanata,* to 120, 60, 30, 10, and 5 ppm SO_2, and concluded that the effects of SO_2 on bryophytes were similar to those described by Rao and LeBlanc (1966) for lichens. Coker noted that the elution of Mg^{2+} ions from the chloroplasts was proportional to the degree of SO_2 pollution. He suggested that atmospheric oxygen might be responsible for the decomposition of bisulfite ions:

$$2HSO_3^- + O_2 \rightleftharpoons 2SO_4^{2-} + 2H^+$$

Puckett *et al.* (1973) observed, from studies *in vitro* and *in vivo* with chlorophyll extracted from lichens, that sulfur dioxide at moderate concentrations induces the conversion of chlorophyll to pheophytin at pH 2.2 but not at pH 3.2 and above. In view of these observations on pH, the

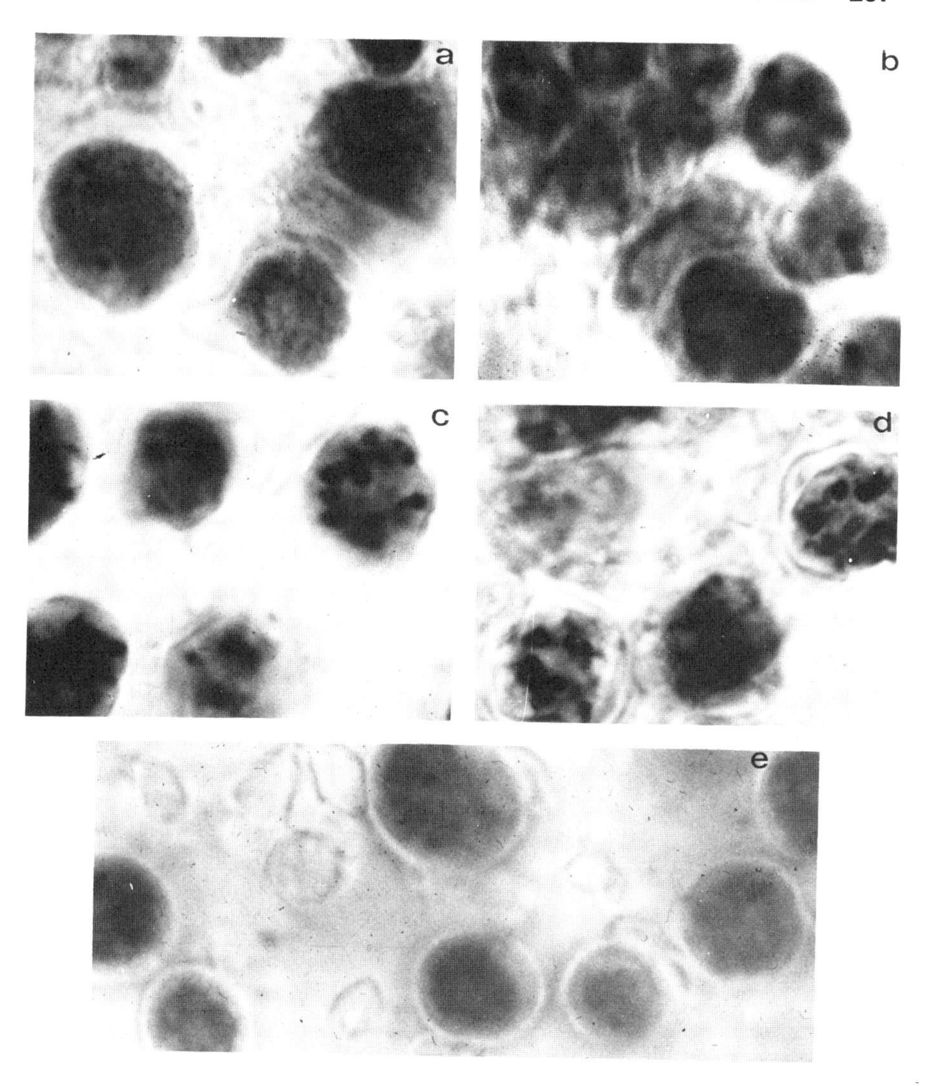

FIG. 6. *Trebouxia* cells from *Xanthoria fallax:* (a)–(d) from thalli exposed to 5 ppm SO_2 at relative humidities 20, 45, 92, and 100%, respectively; (e) from control thalli (after Rao and LeBlanc, 1966).

findings of Taoda (1973b) about the effects of sulfurous and sulfuric acids on epiphytic bryophytes, seem interesting. According to him, the toxicity of H_2SO_3 was stronger than H_2SO_4 at the same pH, bryophytes treated in the neutralized H_2SO_4 remained uninjured while those in the neutralized H_2SO_3 were severely injured, and, finally, the high pH did not reduce the toxic effect of H_2SO_3.

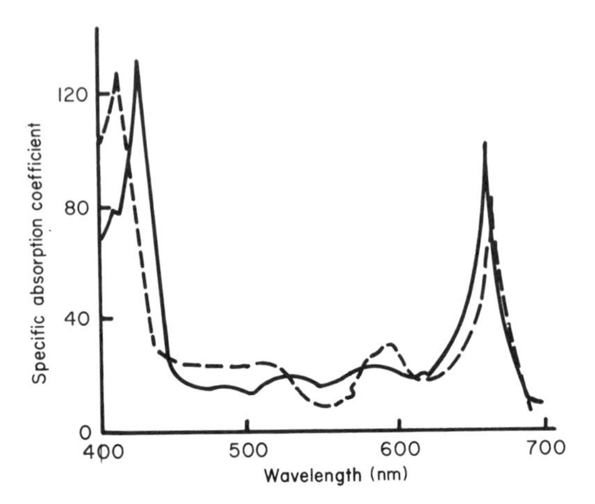

FIG. 7. Absorption spectra of chlorophyll from control (——) and SO_2 exposed (– – –) thalli of *Xanthoria fallax* (after Rao and LeBlanc, 1966).

Syratt and Wanstall (1969) studied the effects of 5 ppm SO_2 at 0, 20, 40, 60, 80, and 100% relative humidities on chlorophyll breakdown of bryophytes, such as *Dicranoweisia cirrata, Metzgeria furcata, Hypnum cupressiforme* var. *filifome, Frullania dilatata, Neckera pumila,* and *Ulota crispa.* They found that damage to chlorophyll showed dependence on the humidity at which the SO_2 was supplied, the higher the humidity the greater the damage. However, under similar conditions, *D. cirrata* was able to withstand higher levels of SO_2 than the other species could. They believe that the relatively high chlorophyll concentration (2.897 OD units/100 mg dry weight) and high SO_4-conversion efficiency (521.3 μg SO_4^{2-}/100 mg dry weight) of *D. cirrata* enable it to resist SO_2 pollution better than other species.

Taoda (1973b), while studying the SO_2 tolerance of epiphytic bryophytes in Japan, observed that most bryophytes were injured by 0.8 ppm SO_2 in 10–40 hours or by 0.4 ppm in 20–80 hours of total exposure; at 0.2 ppm, acute injury, such as discoloration of shoots, did not occur even after 100 hours exposure, but chronic injury, such as growth retardation, was noticed. He also observed that bryophytes in dry condition were much more tolerant to SO_2 than in wet condition.

Hill (1971) studied the effects of SO_3, an oxidation product of SO_2, on the photosynthetic carbon dioxide fixation of *Usnea subfloridana, Hypogymnia physodes,* and *Lecanora conizaeoides* lichen species which are considered to be respectively highly, moderately, and lowly sensitive to SO_2 pollution. He found that the photosynthetic responses of these lichens to sulfite solution matched their relative sensitivity to atmospheric pollution as observed in the field. Puckett *et al.* (1973) measured the net photosynthetic ^{14}C fixation of *Cladonia alpestris, C. deformis, Stereocaulon paschale,* and *Umbilicaria muhlenbergii* after incubating these lichens in 12 ml of aqueous sulfur dioxide at concentrations 0.75, 7.5, 75, or 750 ppm buffered to pH values of 3.2, 4.4, and 6.6. They found that the reduction in net ^{14}C fixation was greatest for the lichen pre-incubated in solutions of high SO_2 content and low pH. At a pH of 3.2 a substantial reduction was observed at 7.5 ppm SO_2, while at higher pH values this concentration had much less effect on net fixation. A comparison between lichen species showed that at all concentrations of SO_2 greater than 7.5 ppm *Cladonia alpestris, Umbilicaria muhlenbergii,* and *Cladonia deformis* were more affected than *Stereocaulon paschale.*

Both Hill and Puckett *et al.* studied the effects of SO_2 in aqueous solutions on net photosynthesis using radioactive sodium bicarbonate ($NaH^{14}CO_3$). The obvious drawback in the method is that the net fixation in saturated lichen thalli in solution may be lowered by more than 50% of the maximum which is usually obtained in the range 25–60% of thallus saturation (Kershaw, 1972).

Showman (1972) compared the net photosynthetic rate decreases of the whole thalli of *Cladonia cristatella, Caloplaca holocarpa,* and *Lecanora dispersa* with those of their isolated phycobionts after exposing them to various SO_2 concentrations. He found that the first response of the phycobionts was at 4 ppm, while the whole thalli appeared unaffected at that concentration. Farrar (1973) suggested the possibility that prevailing pollution conditions could interfere with the flow of nutrients, such as carbohydrates, between the algal and fungal symbionts resulting in the breakdown of symbiosis.

Evidently the action of SO_2 on photosynthesis in lichens largely depends on the intensity of fumigation. Based on the results to date, Richardson and Puckett (1973) summarize the following sequence of increasing damage that could occur in lichens at increasing SO_2 concentrations: (1) a temporary inhibition in photosynthesis with subsequent recovery (Puckett *et al.,* 1973); (2) a permanent reduction in photosynthesis but with no chlorophyll breakdown (Showman, 1972); (3) a permanent reduction in photosynthesis associated with chlorophyll breakdown (Rao and LeBlanc, 1966).

Unlike photosynthesis, there is often an initial increase in respiration in lichens and bryophytes under the stress of pollution. Syratt and Wanstall (1969) observed that the respiration rates of *Dicranoweisia cirrata* and *Metzgeria furcata* continued to rise for some time when they were exposed to 5 ppm SO_2 at 100% relative humidity. Similarly, Showman (1972) noted an increase in the dark respiration of *Cladonia cristatella* after exposure to 4 ppm SO_2. It is believed that the energy produced from enhanced respiration is used by plants for quick oxidation of SO_3^{2-} into SO_4^{2-}. This appears to be a metabolic device in plants to fight pollution. However, the findings of Baddeley *et al.* (1971, 1972) indicate that lichen respiration is generally inhibited by low concentrations of SO_2 in solution. They noticed that SO_2 in solution caused a 25% decrease in the respiration rates of *Cladonia impexa, Hypogymnia physodes,* and *Usnea fragilescens* at concentrations of 23–27 ppm and of *Parmelia saxatilis* and *Ramalina fastigiata* at concentrations of 40–45 ppm. In some instances, however, low concentrations of up to 10 ppm SO_2 in solution gave slightly enhanced rates of respiration.

Since the physiological response of bryophytes and lichens varies according to the physical state of a pollutant, one should be cautious when comparing the effects on respiration produced by SO_2 as a gas in air with those of SO_2 as a solution in water. For the sake of comparison it may be mentioned that 100 μg SO_2/m^3 (0.035 ppm) in air would be equivalent to an aqueous solution containing 35 ppm SO_2 (Saunders, 1966). Further, the pH of the solution determines the form in which SO_2 is distributed. According to Vas and Ingram (1949) sulfur dioxide in solution occurs as sulfite (SO_3^{2-}), bisulfite (HSO_3^-) or sulfurous acid (H_2SO_3) at pH 8, 3.5, or 0, respectively. At low pH values, therefore, sulfur dioxide in solution would become increasingly distributed into the relatively more toxic bisulfite and sulfurous acid forms. Also, comparisons of respiration rates should be made between species from similar habitat and occurring within a single geographical and climatic region.

Presently it is difficult to say whether respiratory sensitivity is as useful as photosynthetic sensitivity in relating the physiological responses of lichens exposed to laboratory conditions to responses as they really occur in nature. However, Baddeley *et al.* (1973) feel that, even though photosynthesis is generally three to five times more sensitive to SO_2 than respiration, the respiratory sensitivity can be a reliable measure of lichen pollution response.

According to Pearson (1973) there exists a significant correlation between the degree of SO_2 pollution and the change in the amino acids of lichens. He found that the amino acids composition (especially of cystine) of the proteins of *Lecanora melanophthalma* decreased in direct proportion to the degree of pollution.

b. Exposure to Hydrogen Fluoride (HF) and to Ozone (O_3). Comeau and LeBlanc (1972) exposed *Funaria hygrometrica* and *Hypogymnia physodes* to HF concentrations of 13, 65, and, 130 ppb for 4, 8, and 12 hours, and 13 ppb for 36, 72, and 108 hours duration, and investigated their reaction with respect to some macro- and microscopic changes. They observed that when the exposure factor (concentration × time) was low, the overall injury to the plants was minimum, the rate of fluoride accumulation in the thalli was low and, finally, the recuperation rate (loss of accumulated fluoride) was high. On the other hand, when the exposure factor was high, the foliar injury was severe, the fluoride concentration was high, and the recuperation rate was low. At an exposure factor of 780 (65 ppb × 12 hours), the plants showed chlorotic spots and a slight disintegration of their chloroplasts, as well as plasmolysis in their cells. After a recovery period of 3 weeks, the F concentration in *Funaria* had been reduced by 26–36% and in *Hypogymnia* by 36–47%.

Comeau and LeBlanc (1971) also exposed *Funaria hygrometrica* gametophytes to 0.25, 0.5, 1, and 2 ppm of ozone concentrations for 4, 6, and 8 hour periods and compared the regenerative capacity of fumigated and unfumigated leaves. They cultured the apical (upper third) and the basal (lower third) leaves of fumigated and control plants on solid agar with mineral salts in Petri dishes kept in a phytotron and determined in each case the percentage of leaves developing protonemal structures on their margins. The results indicate that the apical leaves are regeneratively more potent than the basal ones and that the regeneration percentage is more or less inversely proportional to the exposure factor. Another interesting observation reported by these authors is the fact that O_3 at very low exposure factors seems to stimulate rather than inhibit growth. This observation, however, needs to be checked with further experimental studies.

3. Reproduction and Fertility

Through mapping and transect studies it is abundantly clear that air pollution inhibits not only the growth but also the reproduction of lichens and bryophytes. Pyatt (1970) observed that in many species of lichens the ability to produce viable ascospores, as well as soredial and isidial structures, decreased with increasing atmospheric pollution. De Sloover and LeBlanc (1970), while relating their indices of atmospheric purity (IAP) with the frequency and fertility of species, observed that a decrease in the level of pollution is reflected in an increase in such species characters as the presence rate, vitality, and fertility, these changes ultimately increasing the IAP values in less polluted areas. The rates of specific changes, however, may vary in different plants (Table IX). LeBlanc *et al.* (1971), studying in the fluoride-affected areas of Arvida, Québec, and LeBlanc and Rao (1973a) in the SO_2-affected areas of Sudbury, Ontario, report that lichens generally tend to lack soredial and other vegetative structures

TABLE IX

Relative Frequency of Species (Presence) and of Capsules or Apothecia (Fertility) in the Six IAP Zones of Montreal

Species		IAP					
		0–20	21–40	41–60	61–80	81–100	>100
Brachythecium salebrosum	Presence, %	2.4	14.9	37.5	40.4	67.6	66.6
	Fertility, %	–	–	–	–	8.0	16.7
Candelaria concolor	Presence, %	3.9	44.8	57.8	76.6	91.9	83.3
	Fertility, %	–	–	3.1	2.8	11.8	40.0
Leskea polycarpa	Presence, %	12.5	67.6	89.1	93.6	97.3	83.3
	Fertility, %	–	19.6	61.4	75.0	83.3	80.0
Physcia adscendens	Presence, %	14.8	71.6	93.8	95.7	100	100
	Fertility, %	–	2.1	1.7	6.7	10.8	16.7
Xanthoria fallax	Presence, %	24.2	77.6	92.2	100	97.3	100
	Fertility, %	3.2	5.8	11.9	23.4	36.1	83.3
Physcia millegrana	Presence, %	23.4	95.5	95.5	100	100	100
	Fertility, %	3.3	7.8	39.7	63.0	73.0	100

[a] Modified after De Sloover and LeBlanc, 1970.

in polluted areas. In the latter study it was observed that *Parmelia sulcata* produced no soredia whatsoever in areas having more than 0.03 ppm SO_2, only a few soredia in areas with 0.03–0.005 ppm SO_2, but abundant soredia in areas having less than 0.005 ppm SO_2. According to LeBlanc and Rao (1973a), the extent of soredial development appears to be in part a function (mathematically speaking) of the pollugenic agents of the atmosphere: the purer the air the greater the development of soredia.

Recently Margot (1973) studied the multiplication rate of lichen algal cells by growing in nutrient cultures the soredia of *Hypogymnia physodes* exposed to 0.05, 0.1, and 1.0 ppm SO_2-contaminated atmospheres or sprayed with 0.15, 0.75, and 1.5 meq/100 ml solutions of ammonium, magnesium, or zinc sulfate. He observed that whether applied in the form of gaseous SO_2 or as a solution of sulfate, the pollutant lowered the rate of multiplication of the algal cells, and thus inhibited the germination of soredia; the effect was more severe at increasing pollutant concentrations and relative humidities.

In view of such inhibitory effects of pollugenic agents on the reproduction of species, it might be suggested that the degree of success of species in polluted areas would largely depend on their reproductive potential in such environments. It is likely that active metabolism, high reproductive capacity and fast growth rate of certain species enable them to resist the effects of pollution. According to Gilbert (1971a), *Bryum argenteum, Ceratodon purpureus, Dicranella heteromalla, Funaria hygrometrica, Leptobryum pyriforme, Lunularia cruciata, Marchantia polymorpha,* and *Pohlia annotina,* all SO_2-resistant bryophytes, share a common feature in their capability to transmute rapidly from the highly sensitive protonemal phase of their life cycle to the more resistant gametophytic phase and its subsequent fast growth.

4. Species Susceptibility

The different species of lichens and bryophytes show a considerable variations in their susceptibility to injury by air pollution, especially sulfur dioxide. Characteristically, few species can thrive in polluted areas; however, some manage to survive even under conditions of relatively high levels of pollution. Such species are either toxitolerant, taking advantage of the reduced competition, or toxiphilous, being actually stimulated metabolically by certain pollutants present in the urban and industrial environments. Among lichens, *Stereocaulon pileatum* (Kershaw, 1963), *Lecanora conizaeoides* (Fenton, 1964), *Buellia punctata, Candelariella aurella, Lecanora dispersa* (Gilbert, 1970a), *Bacidia chlorococca, Endocarpon pusillum* (LeBlanc and De Sloover, 1970), and *Micraria trisepta* (Nash, 1972) appear to be toxitolerant species. Though at the moment there is no clear-cut example of a toxiphilous lichen, there are indications that *Lecanora conizaeoides* has some nutritional requirements which are easily met within polluted environments (Fenton, 1964; Gilbert, 1970a; Pyatt, 1970). Among bryophytes, *Bryum argenteum, Lunularia cruciata,* and *Tortula muralis* are considered to be toxitolerant, and *Ceratodon purpureus, Funaria hygrometrica,* and *Leptobryum pyriforme* to be toxiphilous species (Gilbert, 1970a; Daly, 1970).

For certain localities, lichens and bryophytes have been arranged in the order of their field sensitivity to SO_2 pollution. As shown in Tables IV–VII those species which occur only in areas having SO_2 in the range of 0.005–0.01 or <0.005 ppm seem to possess little or no immunity to SO_2 toxicity. It may be inferred therefore that these species are sensitive enough to be useful as bioindicators of SO_2 pollution. Their sudden disappearance from an area could be an indication of environmental pollution if the area remains otherwise ecologically unaltered. Following the method of gradient analysis it is relatively easy to relate such changes in vegetation to the changes in levels of pollution of a landscape and to list species according to their sensitivity to pollution. Taoda (1973b) calculated the indices of relative tolerance (R.T.) of certain epiphytic bryophytes on the basis of their SO_2 tolerance under fumigation and then he correlated these indices with their frequency-occurrence in the struggle zone of Tokyo, i.e., the area around the city where the ambient SO_2 concentration usually remained between 0.01–0.03 ppm (Table VII). He found that *Glyphomitrium humillimum* and *Hypnum yokohamae* var. *kusatsuense,* most tolerant to SO_2 fumigation, were frequent in the struggle zone; *Haplohymenium sieboldii, Herpetineuron toccoae,* and *Trocholejeunea sandivicensis,* less tolerant to SO_2, were not very frequent in the struggle zone; and *Bryum argenteum, Schwetschkea matsumurae,* least tolerant to SO_2, were rare in the struggle zone (Tables VII and X).

While comparing pollution responses of species from different areas, like the ones given in Tables IV–VII, one should keep in view the bioclimatic differences of the areas. Further, it should be remembered that early developmental stages, such as the protonemata of mosses or the soredia and

TABLE X

RELATIVE TOLERANCE (R.T.) OF CERTAIN EPIPHYTIC BRYOPHYTES TO SO_2[a]

Species	Number of experiments	RT[b] in SO_2 fumigation		Frequency[c] (%)
		0.4 ppm	0.8 ppm	
Glyphomitrium humillimum	9	90	100	67
Hypnum yokohamae var. *kusatsuense*	10	81	45	87
Sematophyllum subhumile subsp. *japonicum*	6	75	67	20
Entodon compressus	8	60	69	37
Hypnum plumaeforme var. *minus*	4	50	88	20
Haplohymenium sieboldii	8	75	47	0
Fabronia matsumurae	5	75	10	3
Trocholejeunea sandvicensis	5	50	40	0
Lejeunea punctiformis	2	50	0	20
Bryum argenteum	6	38	0	17
Frullania muscicola	3	50	7	3
Venturiella sinensis	2	—	25	17
Herpetineuron toccoae	2	—	75	0
Schwetschkea matsumurae	4	12	0	0
Orthotrichum consobrinum	2	0	0	0
Lophocolea minor	3	—	17	0

[a] After Taoda, 1973b.

[b] RT = (100a + 50b)/(a + b + c) (a, number of experiments in which the species was very tolerant, b, number of experiments in which the species was intermediate in tolerance; c, number of experiments in which the species was intolerant).

[c] Frequency of occurrence in 30 stations in the struggle zone of Tokyo.

isidia of lichens, appear more sensitive to pollution than the mature plants and their absence could perhaps be a better indicator of pollution.

The sensitivity level of species usually differs in the case of different pollutants. For example, the normally SO_2-sensitive *Usnea floridana* and *Ulota crispa* are found to be resistant to fluorides (Gilbert, 1971b). Some species remain equally insensitive to different pollutants. For example, *Stereocaulon pileatum* can establish itself easily on spoil heaps of copper and lead mines, and in SO_2- or HF-polluted areas (Kershaw, 1963; Gilbert, 1971b). Obviously, such toxitolerant species would be pioneer colonizers of a variety of polluted habitats which may be inhospitable for other species.

5. Factors Affecting Species Pollution Response

Pollutants affect lichens and bryophytes either directly, by causing toxicity to already established plants, or indirectly, by rendering the substrates unfit for propagule establishment. Nevertheless, the effect intensity can be modified by certain substrates and growth forms of plants.

Field studies show that the pollution sensitivity of a species is closely related to the buffer capacity of its substrate. Species occupying acidic substrates are usually far more sensitive to SO_2 than those present on basic substrates. Substrate wise, the order of increasing sensitivity is from terricolous to saxicolous to corticolous species. In general, species growing on trees exhibit far greater sensitivity than those growing on other substrates, and that is why epiphytes are considered to be the most interesting group in pollution studies. In other words, species on tree trunks tolerate lower levels of pollution than species present on other substrates, stone walls (Table XI), for example. Between epiphytes, species growing on tree bases can cope with the conditions of pollution better than those on tree trunks (Rao and LeBlanc, 1967). Among the trunk species themselves, those occurring on the eutrophiated bark of elm resist higher levels of pollution than those on the noneutrophiated bark of oak or ash (Hawksworth and Rose, 1970). This is especially true for the *Xanthorion* community, which consists of nitrophilous lichens, such as *Caloplaca, Candelariella, Physcia,* and *Xanthoria* (Table V). The greater resistance capacity of eutrophiated-bark lichens to sulfur dioxide may be attributed to particulate contaminants, especially alkaline earth metals, trapped in the bark crevices. These cations increase the intrinsic buffering capacity of the bark which keeps up the rate of conversion of SO_2 to SO_4^{2-} (Coker, 1967; Skye, 1968).

The relative success, in polluted environments, of tree-base epiphytes, terricolous species, and saxicolous species may be attributed to the sheltering and physical protection afforded by the surrounding ground vegetation

TABLE XI

SUBSTRATE-INDUCED DIFFERENCES IN THE LEVEL OF SO_2 SENSITIVITY OF CERTAIN LICHENS AND BRYOPHYTES IN CHRISTCHURCH, NEW ZEALAND

Species	Tolerance to average winter SO_2 (ppm)	
	On stone walls	On tree trunks
Brachythecium rutabulum	0.02	0.004
Bryum rubrum	0.03	0.004
Hypnum cupressiforme	0.02	0.04
Leptogium tremelloides	0.02	0.002
Parmelia conspersa	0.02	0.004
Physcia planthiza	0.004	0.03
P. regalis	0.004	0.04
Pohlia cruda	0.05	0.04
Ramalina ecklonii	0.004	0.02
Rhynchostegiella muricata	0.04	0.02
Tortula princeps	0.04	0.004
Xanthoria parietina	0.04	0.03

[a] Modified after Daly, 1970.

or from snow covering during winter when the level of pollution is usually high (LeBlanc, 1961); to the reduced contact with pollution laden winds because of the location of plants closer to the ground where the fumigation rate remains minimum (Geiger, 1965); and to the buffering action of substrates which decreases the effectiveness of acidic pollutants, such as SO_2, by increasing the degree of ionization or the oxidation rate of pollutants (Barkman, 1958).

Evidently species can extend their normal limits in polluted areas if they are able to grow in sheltered niches. However, such locations, which are usually shady, are suitable for bryophytes, but not congenial for lichens, which tend to disintegrate under high humidity and low light conditions. Perhaps it would be fair to say that species occupying sheltered niches are only evaders and not tolerators of pollution.

Bases present in soil possess acid-buffering capacity. Once in contact with the soil, SO_2 is absorbed from the air and oxidized to sulfate within seconds; in comparison to this, SO_2 oxidation may require hours in air and water (Bohn and Cauthorn, 1972). Perhaps this explains why terricolous species, in general, appear to be little or not affected by air pollution.

According to Skye (1968) the high pH in the plant body also helps to reduce the toxicity by SO_2. Consequently, the species most sensitive to SO_2 are usually those which have the lowest buffer capacity for acid substances and are the first to disappear from an SO_2-polluted area. Besides pH, the moisture level in the plant body can modify pollution effects more or less in a linear fashion. It is not unlikely that the secret of the remarkable resistance of *Lecanora conizaeoides* to SO_2 lies in the nonwetting properties of its crustose thallus (Barkman, 1969; Baddeley *et al.*, 1972).

Field studies show that the tolerance of some lichen and bryophyte species to SO_2 is higher in drier more continental climates than in more oceanic ones (Barkman, 1958). Skye (1968) reports that SO_2 is highly correlated with temperature, increasing about 3 pphm or 84 $\mu g/m^3$ for every 10°C drop in temperature during the winter months, suggesting thereby that at the same SO_2-emission rate, the colder areas would experience a higher level of pollution than the warmer ones.

To some extent growth forms also can temper the effects of pollution. According to Gilbert (1970a) the general trend of increasing susceptibility to pollution among growth forms appears to be in the order of (1) leprose, (2) crustose, (3) foliose, and (4) fruticose for lichens; (a) short turf and thalloid, (b) smooth mat and small cushions, and (c) rough mat, tall turf, weft, large cushions, and leafy forms for bryophytes. These trends, however, are of little practical use because species tend to behave differently under different ecological conditions.

III. Conclusions

The decline or absence of lichens and bryophytes, especially of epiphytic species, in cities and industrial areas is a phenomenon mostly caused by air pollution. The presence or absence of certain common but sensitive species and their pattern of distribution in a given area can be related to the degree and extent of pollution of the area over a period of time. The inhibitions of photosynthesis, respiration, growth, reproduction, and other metabolic processes are pollution-induced responses of lichens and bryophytes.

Phytosociological and ecophysiological studies indicate that the pollution sensitivity of these cryptogamic plants is relatively higher than other plant groups. Thus, the pollution indicator potential of these plants is more than significant. The first European Congress on the Influence of Air Pollution on Plants and Animals held at Wageningen in April 1968 adopted the following resolution (Barkman *et al.*, 1969, p. 241): ". . . that cryptogamic epiphytes should be strongly recommended for general use as biological pollution indicators, because (1) they are so easy to handle, (2) they show a vast range of specific sensitivity to air pollutants greatly exceeding that of most higher plants."

Acknowledgment

This research was subsidized by a grant (A-1206) from the National Research Council of Canada with the senior author.

References

Arnold, F. (1891–1901). Zur lichenflora von München. *Ber. Bayer. Bot. Ges.* 1891: pp. 1–147; 1892: pp. 1–76; 1897: pp. 1–45; 1898: pp. 1–82; 1900: pp. 1–100; 1901: pp. 1–24.

Baddeley, M. S., Ferry, B. W., and Finegan, E. J. (1971). A new method of measuring lichen respiration: Response of selected species to temperature, pH and sulphur dioxide. *Lichenologist* **5,** 18–25.

Baddeley, M. S., Ferry, B. W., and Finegan, E. J. (1972). The effects of sulphur dioxide on lichen respiration. *Lichenologist* **5,** 283–291.

Baddeley, M. S., Ferry, B. W., and Finegan, E. J. (1973). Sulphur dioxide and respiration in lichens. *In* "Air Pollution and Lichens" (B. W. Ferry *et al.,* eds.), pp. 299–313. Oxford Univ. Press (Athlone), London and New York.

Barkman, J. J. (1958). "Phytosociology and Ecology of Cryptogamic Epiphytes." Assen, Netherlands.

Barkman, J. J. (1963). De epifyten-flora en-vegetatie van Midden-Limburg (België). *Verh. Kon. Ned. Akad. Wetensch., Afd. Natuurk, Tweede Sect.* **54,** 1–46.

Barkman, J. J. (1969). The influence of air pollution on bryophytes and lichens. *Air Pollut., Proc. Eur. Congr. Influence Air Pollut. Plants & Anim., 1st, 1968* pp. 197–209.

Barkman, J. J., Rose, F., and Westhoff, V. (1969). The effects of air pollution on non-vascular plants. *Air Pollut., Proc. Eur. Congr. Influence Air Pollut. Plants & Anim., 1st, 1968,* Discussion in Sect. 5, pp. 237–241.

Beschel, R. (1958). Flechtenvereine der Städte, Stadtflechten und ihr Wachstum. *Ber. Naturwiss.-Med. Ver. Innsbruck* **52,** 1–158.

Bohn, H. L., and Cauthorn, R. C. (1972). Pollution: The problem of misplaced waste. *Amer. Sci.* **60,** 561–565.

Braun-Blanquet, J. (1951). "Pflanzensoziologie," 2nd ed. Springer, Wien.

Brodo, I. M. (1961). Transplant experiments with corticolous lichens using a new technique. *Ecology* **42,** 838–841.

Brodo, I. M. (1966). Lichen growth and cities: A study on Long Island, New York. *Bryologist* **69,** 427–449.

Coker, P. D. (1967). The effects of sulphur dioxide pollution on bark epiphytes. *Trans. Brit. Bryol. Soc.* **5,** 341–347.

Comeau, G., and LeBlanc, F. (1971). Influence de l'ozone et de l'anhydride sulfureux sur la régénération des feuilles de *Funaria hygrometrica* Hedw. *Natur. Can.* **98,** 347–358.

Comeau, G., and LeBlanc, F. (1972). Influence du fluor sur le *Funaria hygrometrica* et l'*Hypogymnia physodes. Can. J. Bot.* **50,** 847–856.

Daly, G. T. (1970). Bryophyte and lichen indicators of air pollution in Christchurch, New Zealand. *Proc. N. Z. Ecol. Soc.* **17,** 70–79.

Delvosalle, L., Demaret, F., Lambinon, J., and Lawalrée, A. (1969). Plantes rares, disparues ou menacées de disparition en Belgique: L'appauvrissement de la flore indigène. Service des Réserves Naturelles domaniales et de la Conservation de la Nature. *Travaux* No. 4, pp. 1–128.

De Sloover, J., and LeBlanc, F. (1968). Mapping of atmospheric pollution on the basis of lichen sensitivity. *In* "Proceedings of the Symposium on Recent Advances in Tropical Ecology" (R. Misra and B. Gopal, eds.), pp. 42–56. Varanasi, India.

De Sloover, J., and LeBlanc, F. (1970). Pollutions atmosphériques et fertilité chez les mousses et chez les lichens epiphytiques. *Bull. Acad. Soc. Lorraines Sci.* **9,** 82–90.

Domrös, M. (1966). Luftverunreinigung und Stadtklima im Rheinisch-Westfälischen Industriegebeit und ihre Auswirkung auf den Flechtenbewuchs der Bäume. *Arb. Rheinschen Landeskunde* **23,** 1–132.

Dreisinger, B. R. (1965). "Sulfur Dioxide Levels and the Effects of the Gas on Vegetation Near Sudbury, Ontario" (mimeo.). Ont. Dept. Mines, Sudbury.

Farrar, J. F. (1973). Lichen physiology: Progress and pitfalls. *In* "Air Polution and Lichens" (B. W. Ferry *et al.,* eds.), pp. 238–282. Oxford Univ. Press (Athlone), London and New York.

Felföldy, L. (1942). A városi Levegö hatása az epiphyton-zuzmóvegetációra Debrecenben. *Acta Geobot. Hung.* **4,** 332–349.

Fenton, A. F. (1960). Lichens as indicators of atmospheric pollution. *Ir. Natur. J.* **13,** 153–158.

Fenton, A. F. (1964). Atmospheric pollution of Belfast and its relationship to the lichen flora. *Ir. Natur. J.* **14,** 237–245.

Geiger, R. (1965). "The Climate Near the Ground." Harvard Univ. Press, Cambridge, Massachusetts.

Gilbert, O. L. 1965. Lichens as indicators of air pollution in the Tyne Valley. *In* "Ecology and the Industrial Society" (G. T. Goodman *et al.,* eds.), pp. 35–47. Oxford Univ. Press, London and New York.

Gilbert, O. L. (1968). Bryophytes as indicators of air pollution in the Tyne Valley. *New Phytol.* **67,** 15–30.

Gilbert, O. L. (1969). The effect of SO_2 on lichens and bryophytes around Newcastle upon Tyne. *Air Pollut., Proc. Eur. Congr. Influence Air Pollut. Plants & Anim., 1st, 1968* pp. 223–235.

Gilbert, O. L. (1970a). Further studies on the effect of sulphur dioxide on lichens and bryophytes. *New Phytol.* **69,** 605–627.

Gilbert, O. L. (1970b). A biological scale for the estimation of sulphur dioxide pollution. *New Phytol.* **69,** 629–634.

Gilbert, O. L. (1971a). Urban bryophyte communities in north-east England. *Trans. Brit. Bryol. Soc.* **6,** 306–316.

Gilbert, O. L. (1971b). The effect of airborne fluorides on lichens. *Lichenologist* **5,** 26–32.

Granger, J.-M. (1972). Computer mapping as an aid in air pollution studies. *Sarracenia* **5,** 43–83.

Haugsjå, P. K. (1930). Über den Einfluss der Stadt Oslo auf die Flechtenvegetation der Bäume. *Nyt Mag. Naturvidensk.* **68,** 1–116.

Hawksworth, D. L. (1971). *Lobaria pulmonaria* (L.) Hoffm. transplanted into Dovedale, Derbyshire. *Naturalist* (*Hull*) pp. 127–128.

Hawksworth, D. L. (1973). Mapping studies. *In* "Air Pollution and Lichens" (B. W. Ferry *et al.,* eds.), pp. 38–76. Oxford Univ. Press (Athlone), London and New York.

Hawksworth, D. L., and Rose, F. (1970). Qualitative scale for estimating sulphur dioxide air pollution in England and Wales using epiphytic lichens. *Nature (London)* **227,** 145–148.

Hill, D. J. (1971). Experimental study of the effect of sulphite on lichens with reference to atmospheric pollution. *New Phytol.* **70,** 831–836.

Høeg, O. A. (1936). Zur Flechtenflora von Stockholm. *Nyt Mag. Naturvidensk.* **75,** 129–136.

Jones, E. W. (1952). Some observations on the lichen flora of tree boles, with special reference to the effect of smoke. *Rev. Bryol. Lichenol.* **21,** 96–115.

Kershaw, K. A. (1963). Lichens. *Endeavour* **22,** 65–69.

Kershaw, K. A. (1972). The relationship between moisture content and net assimilation rate of lichen thalli and its ecological significance. *Can. J. Bot.* **50,** 543–555.

Klement, O. (1956). Zur Flechtenflorula des Kölner Domes. *Decheniana* **109,** 87–90.

Klement, O. (1958). Die Flechtenvegetation der Stadt Hannover. *Beitr. Naturk. Niedersachs.* **11,** 56–60.

Krusenstjerna, E. (1945). Bladmossvegetation och bladmossflora i Uppsalatrakten. *Acta Phytogeogr. Suec.* **19,** 1–250.

Laundon, J. R. (1967). A study of the lichen flora of London. *Lichenologist* **3,** 277–327.

LeBlanc, F. (1961). Influence de l'atmosphère polluée des grandes agglomérations urbaines sur les épiphytes corticoles. *Rev. Can. Biol.* **20,** 823–827.

LeBlanc, F., and De Sloover, J. (1970). Relation between industrialization and the distribution and growth of epiphytic lichens and mosses in Montréal. *Can. J. Bot.* **48,** 1485–1496.

LeBlanc, F., and Rao, D. N. (1966). Réaction de quelques lichens et mousses épiphytiques à l'anhydride sulfureux dans la région de Sudbury, Ontario. *Bryologist* **69,** 338–346.

LeBlanc, F., and Rao, D. N. (1973a). Effects of sulfur dioxide on lichen and moss transplants. *Ecology* **54,** 612–617.

LeBlanc, F., and Rao, D. N. (1973b). Evaluation of the pollution and drought hypotheses in relation to lichens and bryophytes in urban environments. *Bryologist* **76,** 1–19.

LeBlanc, F., Comeau, G., and Rao, D. N. (1971). Fluoride injury symptoms in epiphytic lichens and mosses. *Can. J. Bot.* **49,** 1691–1698.

LeBlanc, F., Rao, D. N., and Comeau, G. (1972a). The epiphytic vegetation of *Populus balsamifera* and its significance as an air pollution indicator in Sudbury, Ontario. *Can. J. Bot.* **50,** 519–528.

LeBlanc, F., Rao, D. N., and Comeau, G. (1972b). Indices of atmospheric purity and fluoride pollution pattern in Arvida, Québec. *Can. J. Bot.* **50,** 991–998.

Margot, J. (1973). Experimental study of the effects of sulphur dioxide on the soredia of *Hypogymnia physodes. In* "Air Pollution and Lichens" (B. W. Ferry *et al.*, eds.), pp. 314–329. Oxford Univ. Press (Athlone), London and New York.

Morgan-Huws, D. I., and Haynes, F. N. (1973). Distribution of some epiphytic lichens around an oil refinery at Fawley, Hampshire. *In* "Air Pollution and Lichens" (B. W. Ferry *et al.*, eds.), pp. 87–108. Oxford Univ. Press, London and New York.

Nash, T. H., III. (1971). Lichen sensitivity to hydrogen fluoride. *Bull. Torrey Bot. Club* **98,** 103–106.

Nash, T. H., III. (1972). Simplification of the Blue Mountain lichen communities near a zinc factory. *Bryologist* **75,** 315–324.

Nylander, W. (1866). Les lichens du Jardin du Luxembourg. *Bull. Soc. Bot. Fr.* **13,** 364–372.

Pearson, L., and Skye, E. (1965). Air pollution affects pattern of photosynthesis in *Parmelia sulcata,* a corticolous lichen. *Science* **148,** 1600–1602.

Pearson, L. C. (1973). Air pollution and lichen physiology: Progress and problems. *In* "Air Pollution and Lichens" (B. W. Ferry *et al.,* eds.), pp. 224–237. Oxford Univ. Press (Athlone), London and New York.

Pišút, I. (1962). Bemerkungen zur Wirkung der Exhalationsprodukte auf die Flechtenvegetation in der Umgebung von Rudňany (Nordostslowakei). *Biologia (Bratislava)* **17,** 481–494.

Puckett, K. J., Nieboer, E., Flora, W. P., and Richardson, D. H. S. (1973). Sulphur dioxide: Its effects on photosynthetic ^{14}C fixation in lichens and suggested mechanisms of phytotoxicity. *New Phytol.* **72,** 141–154.

Pyatt, F. B. (1970). Lichens as indicators of air pollution in a steel producing town in South Wales. *Environ. Pollut.* **1,** 45–56.

Rao, D. N., and LeBlanc, F. (1966). Effects of sulfur dioxide on the lichen algae, with special reference to chlorophyll. *Bryologist* **69,** 69–75.

Rao, D. N., and LeBlanc, F. (1967). Influence of an iron-sintering plant on corticolous epiphytes in Wawa, Ontario. *Bryologist* **70,** 141–157.

Richardson, D. H. S., and Puckett, K. J. (1973). Sulphur dioxide and photosynthesis in lichens. *In* "Air Pollution and Lichens" (B. W. Ferry *et al.,* eds.), pp. 283–298. Oxford Univ. Press (Athlone), London and New York.

Rose, F. (1973). Detailed mapping in south-east England. *In* "Air Pollution and Lichens" (B. W. Ferry *et al.,* eds.), pp. 77–88. Oxford Univ. Press (Athlone), London and New York.

Rydzak, J. (1953). Rozmieszczenie i ekologia porostów miasta Lublina. *Ann. Univ. Mariae Curie-Sklodowska, Sect. C* **8,** 233–356.

Rydzak, J. (1959). Influence of small towns on the lichen vegetation. Part VII. Discussion and general conclusions. *Ann. Univ. Mariae Curie-Sklodowska, Sect. C* **13,** 275–323.

Rydzak, J. (1968). Lichens as indicators of the ecological conditions of the habitat. *Ann. Univ. Mariae Curie-Sklodowska, Sect. C* **23,** 131–164.

Sauberer, A. (1951). Die Verteilung rindenbewohnender Flechten in Wien, ein bioklimatisches Grosstadtproblem. *Wetter und Leben* **3,** 116–121.

Saunders, P. J. W. (1966). The toxicity of sulphur dioxide to *Diplocarpon rosae* Wolf causing blackspot of roses. *Ann. Appl. Biol.* **58,** 103–114.

Schmid, A. B. (1957). Die epixyle Flechtenvegetation von München (mimeo.). Dissertation, University of Munich.

Schönbeck, H. (1969). Eine Methode zur Erfassung der biologischen Wirkung von Luftverunreiningungen durch transplantierte Flechten. *Staub-Reinhalt. Luft* **29,** 14–18.

Sernander, R. (1926). "Stockholms natur." Almqvist and Wiksells, Uppsala.

Shacklette, H. T. (1965). Element content of bryophytes. *U.S., Geol. Surv., Bull.* **1198-D,** 1–21.

Showman, R. E. (1972). Residual effects of sulfur dioxide on the net photosynthetic and respiratory rates of lichen thalli and cultured lichen symbionts. *Bryologist* **75,** 335–341.

Showman, R. E., and Rudolph, E. D. (1972). Water relations of living, dead and cellulose models of the lichen *Umbilicaria papulosa. Bryologist* **74,** 444–450.

Skye, E. (1958). Luftföroreningars inverkan på bush-och bladlavfloran kring skifferoljeverket i Närkes Kvarntorp. *Sv. Bot. Tidskr.* **52,** 133–190.

Skye, E. (1968). Lichens and air pollution: A study of cryptogamic epiphytes and environment in the Stockholm region. *Acta Phytogeogr. Suec.* **52,** 1–123.

Skye, E., and Hallberg, I. (1969). Changes in the lichen flora following air pollution *Oikos* **20,** 547–552.

Smith, D. C. (1962). The biology of lichen thalli. *Biol. Rev. Cambridge Phil. Soc.* **37,** 537–570.

Steiner, M., and Schulze-Horn, D. (1955). Über die verbreitung und Expositionsabhangigkeit der Rindenepiphyten im Stadtgebiet von Bonn. *Decheniana* **108,** 1–16.

Syratt, W. J., and Wanstall, P. J. (1969). The effect of sulphur dioxide on epiphytic bryophytes. *Air Pollut., Proc. Eur. Congr. Influence Air Pollut. & Anim., 1st, 1968,* pp. 79–85.

Taoda, H. (1972). Mapping of atmospheric pollution in Tokyo based upon epiphytic bryophytes. *Jap. J. Ecol.* **22,** 125–133.

Taoda, H. (1973a). Bryo-meter, an instrument for measuring the phytotoxic air pollution, *Hikobia* **6,** 224–228.

Taoda, H. (1973b). Effect of air pollution on bryophytes. I. SO_2 tolerance of bryophytes. *Hikobia* **6,** 238–250.

Tobler, F. (1925). "Biologie der Flechten." Borntraeger, Berlin.

Trass, H. (1968). Indeks samblikurühmituste kasutamiseks õhu saastatuse määramisel. *Eesti Loodus* **11,** 628.

Trass, H. (1973). Lichen sensitivity to the air pollution and index of poleotolerance (I.P.). *Fol. Crypt. Est.* **3,** 17–24.

Vaarna, V. V. (1934). Helsingin kaupungin puiden ja pensaiden jäkäläkasvisto. *Ann. Bot. Soc. Zool. Bot. Fenn. "Vanamo"* **5,** 1–32.

Vareschi, V. (1936). Die Epiphytenvegetation von Zürich (Epixylenstudien II). *Ber. Schweiz. Bot. Ges.* **46,** 445–488.

Vareschi, V. (1953). La influencia de los bosques y parques sobre el aire de la ciudad de Caracas. *Acta Cient. Venez.* **4,** 89–95.

Vas, K., and Ingram, M. (1949). Preservation of fruit juices with less SO_2. *Food Mf.* **24,** 414–416.

Whittaker, R. H. (1967). Gradient analysis of vegetation. *Biol. Rev. Cambridge Phil. Soc.* **49,** 207–264.

12

INTERACTIONS OF AIR POLLUTANTS WITH CANOPIES OF VEGETATION

JESSE H. BENNETT AND A. CLYDE HILL

I. Introduction

The earth's vegetation is immersed in an environment of flowing energy and materials at the bottom of an atmospheric "ocean of air." The dynamic atmosphere continually absorbs and deposits a wide variety of natural and man-made substances. Many of these substances are converted to other forms en route. Interactions among the physical and biological components on the earth's surface and the atmosphere play important roles in the circulation, fate, and effects of the constituents in the biosphere, some of which are required for life and are considered to be beneficial, while others may cause adverse effects according to man's perspectives. The study of environmental pollution is largely a study of the cycling of unwanted energy and materials through the earth-atmosphere system in concentrations considered to be deleterious by man, and of their effects on the system.

A treatment of the impact of vegetation interactions with the total atmosphere is beyond the scope of this chapter. Its purpose is to describe interfacing interactions between plant canopies and major gaseous pollutants in the air in the immediate vicinity of the vegetation. A general introduction into the earth-atmosphere cycling of atmospheric components and into the complexity of vegetation biosystems is presented initially, however, to orient the reader and provide important supporting knowledge. This is followed by a discussion of gaseous pollutant interchange between simple plant canopies and adjacent canopy boundary air layers and of the potential plant sink sites and pollutant fate within the vegetation. Reference model exchange systems applied to simple, uniform canopies are employed in the treatment of the subject material to aid in comprehending the principles involved. The reader is referred to other standard texts and reference sources for discussions of the earth-atmosphere-pollution system in its broad context (Hodges, 1973; Magill *et al.,* 1956; Robinson and Robbins, 1969, 1971; Committee on Chemistry and Public Affairs, 1969).

II. Cycling of Materials Through the Earth-Atmosphere System

The earth's atmosphere is a fluid medium in intimate contact with the land, waters, living things, and other structures on the earth's surface. Atmospheric gases are mixed to a height of about 45 miles above the ground. This enveloping atmospheric layer, commonly referred to as the homosphere, contains more than 99.9% of the mass of the atmosphere (about 4.5×10^{15} metric tons). Weather occurs in the troposphere—a restless layer adjacent the earth, varying from 5 miles thick near the poles to 11 miles thick at the equator. Air heated at the earth's surface by the sun's energy expands, rises, and is replaced by colder air, thus causing the air to circulate. The general circulation in the troposphere is influenced by the earth's surface features and its motions in space, particularly its daily spin and annual revolution around the sun. These factors cause variations in the amount of solar energy received by an area and physically deflect the wind (which results from pressure fluctuations in the atmosphere). The motion of the air near the earth mixes the atmosphere. Local air masses and their contents must in the long run be treated as part of regional and global atmospheres but temporary deviations within a given airshed from the mean geochemical or geophysical constitution of the regional atmosphere may cause repercussions of immediate concern to local ecological systems (ecosystems). The importance of the biogeochemical and biogeophysical cycling of certain materials (for example, the oxygen, carbon, water, nitrogen, and sulfur cycles; dispersal of pollen and other particulates) has been recognized for years although the sources, sinks, interconversions, residence times, and impact of many constituents are poorly understood.

A. Composition of Lower Atmosphere and Cycling of Air Pollutants

Approximately 99% of the constituents of "clean dry air" within the atmosphere of the biosphere is composed of two gases: nitrogen (about 78%) and oxygen (about 21%). The balance, nearly 1%, consists of inert gases (mostly argon), small amounts of carbon dioxide (about 0.03%), and various trace gases* including natural background levels of a number of substances (e.g., CO, O_3, H_2S, sulfur oxides, nitrogen oxides, and various organic compounds) that are regarded as pollutants when they occur in sufficiently high concentrations. A real atmosphere may contain, however, up to 4% water vapor plus a wide array of other gaseous, liquid, and particulate matter, such as windblown dusts, seaspray, microorganisms, pollen, aerosols formed from chemical reactions and aggregation of materials in the atmosphere, and many additional substances originating from natural processes and human pursuits. The earth and its atmosphere form practically a closed system with respect to cycling of gaseous, liquid, and particulate materials. An environmental pollutant within this system is regarded as anything added to the environment that adversely affects something man values and is present in sufficiently high concentration to do so. For a substance to be formally classified as a pollutant its effects must be perceived. Pollutants may arise from natural processes, such as forest fires, volcanic activities, biological decay, or from man's activities, e.g., combustion and processing activities. Fortunately, mechanisms exist for the removal of pollutants from the atmosphere. Figure 1 illustrates the general cyclic nature of the system.

* Trace gases are defined as gases present in the atmosphere in concentrations less than 10 ppm (parts per million by volume). Natural background concentrations of trace gases listed occur in concentrations ranging from less than a part per billion (ppb) for SO_2 to a few pphm (parts per hundred million) for O_3 and CO. The carbon oxides, sulfur oxides, and nitrogen oxides arise largely from oxidation of precursors emitted from decay processes. Background O_3 probably originates predominantly in the upper atmosphere.

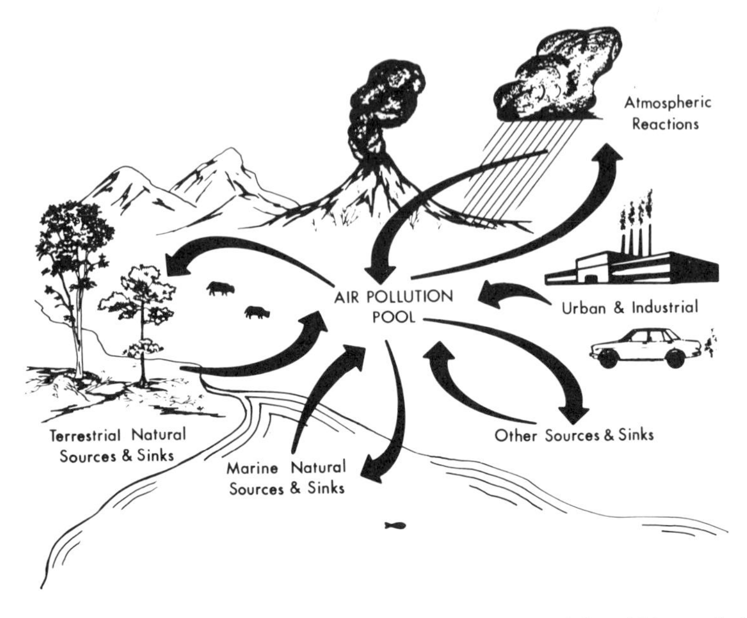

FIG. 1. Illustration depicting general air pollutant sources and sinks within an airshed.

More is known about the sources, movement, and effects of air pollutants than about their fate in the environment. Many people have expressed concern that toxic pollutants may accumulate in the atmosphere and threaten life when the air becomes sufficiently polluted. Available data however indicate that most air pollutants of major importance in the biosphere are removed rather rapidly (Hill, 1971; Meetham, 1950, 1954; Robinson and Robbins, 1969, 1971; Weinstock, 1969). Vegetation is known to effectively absorb a number of these pollutants and may interact to cleanse them from the air. Conversely, certain noxious plants may serve as local sources for odorous and malodorous compounds, allergenic pollen, and other substances which may undergo further changes in the atmosphere as a consequence of atmospheric chemical and photochemical reactions. In addition, plant and animal decay processes are known to be major contributors of certain "background" substances found in the atmosphere. These substances are mostly reduced or oxidized forms of carbon, sulfur, and nitrogen compounds, the precursor compounds being emitted during decomposition and respiratory functions. The volatilized compounds tend to be oxidized in the atmosphere. Of course, photosynthesizing plants recycle atmospheric CO_2 into organic compounds and regenerate O_2 which is vitally required for respiratory and other oxidative processes. Assimilative reactions occurring in photosythesizing plants can also "fix" certain gaseous air pollutants such as SO_2 into organic compounds (Bennett and Hill, 1973a; Daines, 1968; Thomas *et al.*, 1950). The aqueous cellular solution and multitudinous reactants, products, and transformations that occur in or on living plants and other living and nonliving things provide a wide variety of opportunities for the sorption or production of pollutants and pollutant precursors.

Although the cleansing action of vegetation may be important in the cycling of certain pollutants, it may not always be desirable. For example, the accumulation of fluoride in plants exposed to atmospheres containing HF may injure sensitive plants as well as result in fluorosis of grazing animals that consume the forage over a prolonged period of time. Certain airborne peticides and herbicides may also have adverse effects on plants and the food chain. Plant exposures to elevated levels of SO_2, NO_2, O_3, PAN, Cl_2, and other phytotoxic pollutants can also cause plant injury; however, many of these pollutants such as the ones just mentioned tend to be converted to other less toxic forms rather rapidly in plants and may not result in the direct accumulation of toxic pollutant residuals important in the biological food chain. Even though these effects may occur at times, the cleansing of the atmosphere by vegetation appears to be a highly desirable feature of the ecosystem.

B. Air Pollutant Uptake by Vegetation

Based on pollutant uptake experiments on alfalfa canopies (Hill, 1971), it has been estimated that a continuous cover of assimilating alfalfa growing under conditions equivalent to those of the studies could remove more than $\frac{1}{4}$ ton of NO_2 or SO_2 per square mile per day from air containing an average NO_2 concentration similar to that found in the atmosphere of the South Coastal Basin vicinity of Los Angeles (6 pphm) or to the mean SO_2 concentrations measured over a 2000 square mile area around Sudbury, Canada, polluted by smelter emissions (1–3 pphm). Greater amounts could be removed from more polluted atmospheres but plants generally become relatively less efficient in taking up pollutants at high concentrations. Land areas covered by dense forests would be expected to remove more pollutant from the air, whereas sparse vegetation cover may remove less. It has furthermore been reported that 82% of the SO_2 pollution in the atmosphere over the British landscape (excluding that portion which was carried out to sea) could be accounted for by reaction with plant and physiographic surfaces. The remainder was thought to be deposited on the land in rainwater and via the settling of particulates. Present data indicate that vegetation could be an important sink for at least the following air pollutants of major importance: HF, SO_2, NO_2, O_3, Cl_2, and to a lesser extent PAN. Undoubtedly, many others such as HCl could be added to the list. Two important pollutants which are known to not be taken up effectively by plants, however, are CO and NO. Soil microorganisms appear to be a major sink for CO. Nitric oxide, though absorbed slowly by plants, is converted in the atmosphere to other forms which may then be taken up more rapidly.

Since vegetation can be a source and scavenger of many airborne substances, interactions of large masses of vegetation with the environment can affect the composition of the atmosphere. Of the approximately 52 million square miles of land area in the world—exclusive of the 6 million square miles covered by polar icecaps—about one-third is estimated to be covered by forest vegetation, more than one-fourth of the earth is taken up by agriculture, and the remaining two-fifths consists of brushlands, deserts, mountaintops, tundralands, roads, towns, cities, and other terrestrial physiographic areas (Frey *et al.,* 1964). In addition to the land, plants also abound in both fresh and saline waters of the earth which occupy approximately 70% of the earth's surface. More than 300,000 diversified species of plants exist in the world.

III. Plant Biosystems: Complexity and General Interactions

Vegetation can be defined as the sum total of plants growing in an area. The vegetal biosystem, however, should be viewed as much more than a mere grouping of plants. Many factors interact to produce the plant life occupying a given habitat. The emerging state of the dynamic living system results from complex, interwoven, biofeedback processes in which environmental conditions intimately affect the developing flora. The plants which are able to tolerate and compete within the range of conditions to which they are subjected, in turn, modify their habitats and interact upon one another. Coexisting animal life and microorganisms further alter the character of the vegetation.

Each plant, being a product of the conditions under which it grows, is a measure of its local environment. A plant community, however, more reliably reflects the general environment of an area and the stage of development attained by the biosystem. While natural vegetation develops according to general structural patterns, striking differences can be readily observed within any vegetation type. Habitat differences resulting in local variability in microclimate and edaphic conditions are reflected in the type, quantity, and general status of the plants that grow and compete. Artificially manipulated vegetation, such as that resulting from agricultural practices or ornamental arrangements used in landscaping, also reflect the growing conditions, but many of the most important limiting factors—both biotic and abiotic—are manipulated by man in order to produce the kind and quantity of plant life desired. Most of these artificial biotic systems can exist only under rather intensive management. The cultural practices require an understanding of the physiology of the plants and the ecological implications inherent in attempting to optimize conditions for development. Such biosystems are precariously balanced by conscious means and are subject to rapid deterioration when management practices become inadequate to maintain control.

The relationships between structure and function in any living system are highly complex; yet, it is this very complexity that provides the uniqueness required for development and stability. The integration of biological structure and function is indissoluble. As biological structures function, the functional activities create new structures. The intricate state of a biosystem or subsystem is mediated by all processes which affect its structural and functional integrity. Living systems possess important intrinsic properties which arise from their peculiar interacting structural and functional characteristics. For example, chloroplasts within leaves, driven by solar kinetic energy, possess the bioengineering capability of transforming chemical substances of low potential energy into organic compounds with high potential energy. This process furnishes substrates and chemical energy required by other living systems. Chloroplasts, however, are very much dependent upon leaf cells—at a higher structural and functional level of biological organization—to provide the compatible media for their existence. Leaf cells have the important intrinsic ability, through reproduction and development, to generate unique building blocks for the organization of the leaf. The leaf organ serves to regulate the exchange of energy and chemical materials among leaf cells, the atmosphere, and the plant conducting system which transports nutrients to various parts of the plant for utilization in the development of the whole plant organism. The plant organism (a changing body that extracts basic nutrients from its substrate and competes for incoming solar energy all the while scrubbing the atmosphere of many gaseous and particulate substances) unifies the previously mentioned biological systems allowing them to exist and operate.

Adaptations of plants (and animals) to their habitat conditions are related to the averages and ranges of the environmental circumstances which prevail. Within the range of conditions that a species can tolerate, there exists a more limited range or optimal set of conditions where it develops best. The set of optimal physical conditions for a plant organism may not coincide with its biotic optimum when growing within a vegetation mass, however, as the plant must compete with other organisms for available space, energy, and resources. The microenvironmental conditions for no two organisms in nature are exactly alike although they tend to be similar for members of the same species at the same stages in their lives. Significant differences in habitat conditions, tempered by genetically controlled plant adaptability, may heighten or lessen the effects of many environmental agents on the plants. It is well documented that the response of plants to phytotoxic air pollutants, for instance, can be markedly influenced by the microenvironmental regimes under which they grow (Brennan and Leone, 1968; Dugger and Ting, 1970; Heck *et al.*, 1965; Heck, 1968). In evaluating the impact of atmospheric phytotoxicants on the quality and quantity of vegetation it is important to be sensitive to the morphological (structural) and physiological (functional) states of the component plant organisms and plant parts. The sorptive properties of plants for air pollutants and the degree of susceptibility of the plant to the exposure dosage, although basically prescribed by their genetic constitution, are expressed through the morphological structures and physiological processes that develop in concert with the microenvironmental complex of the habitat.

A study of the mass of vegetation occupying a given area necessitates that attention be given to the organisms of which it is composed. But, important properties arising out of the coming together of the organisms and the particular interactions that occur among themselves and their continuously modified microenvironments are exhibited by the biotic community. The living, adjusting, oscillating biosystem tends toward the dynamic equilibration of its (structured) biomass with that allowed by the climatic and biotic potential of the particular ecological system. This results from homeostatic mechanisms involving checks and balances which operate in all living systems. As a result it is often found that the whole is less variable than its parts. Homeostasis allows a living system to maintain itself in a world of destructive forces.

As vegetation develops over a land area significant changes result in the physical nature of the environment as well as in the biota. The composition and energy balance of the soil and atmosphere are affected. Half of the plant biomass, and often much more, may be buried in the earth, binding the earth and changing its physical and chemical properties through a wide array of activities and processes. The roots of some species are restricted, but for most species the root systems are very extensive. Many other kinds of micro- and macrobiota and decaying organic matter also reside in or on the substratum, altering its properties. Roots not only respond to their immediate environment and change it but they depend upon the aboveground parts for organic nutrients. Likewise, the root systems vitally influence the exposed vegetation which relies upon the roots for anchorage, water, minerals, stored nutrients, and growth promoting substances. Interactions between the atmosphere and the exposed vegetation are markedly affected by the activities of the root systems. Plants grown in dry air place greater demands on water extracted from the soil by roots than the same plants grown in humid air. This is usually offset somewhat by closure of the stomata in the leaves when the roots cannot supply adequate moisture. The greater resistance to gaseous diffusion offered by epidermal tissues with closed stomata, though, inversely restrict the passage of gases from the atmosphere into the leaves. This decreases the amount of carbon dioxide available to the leaf cells for photosynthesis. It also reduces the rates at which leaf cells absorb air pollutants from the atmosphere and release oxygen back into the air. Plant leaves are specialized structures which have evolved for the extremely efficient removal of trace amounts of gases from the atmosphere. This enables leaves to remove physiologically important gases such as CO_2 from the air as well as certain potentially phytotoxic air pollutants even though the concentrations may be very low.

During the growing season plant leaves provide a major filtration and reaction surface exposed to the atmosphere—many times that of the corresponding land surface area in most parts of the United States. Moreover, cellular surface area is considerably higher than the leaf area. Since leaves are potential traps for many air pollutants, they provide a large exposed surface for removal of atmospheric pollutants. Pollutants may react with substances which are located both external and internal to plants. Superficial substances as well as components within leaves such as mesophyll cell walls, membranes, organelles, and reactants free in the inter- and intracellular aqueous media may serve as receptors and reactants. Plant cells are bathed in aqueous media into which soluble gases may dissolve and react, or perhaps be translocated to other sites in the translocation stream. The extent and site of reaction depend upon the particular pollutants, plants, and environmental conditions. Plant surfaces may also filter particulates from the air. Air flowing over and through vegetation is slowed down facilitating the deposition of particulate impurities. Because plants can be washed by rainwater, undergo growth and metabolic regenerative processes, and periodically shed replaceable leaves and other tissues, plants are self-renewing and can act as persistant sorbers of pollutants when exposed to sufficiently low concentrations and dosages. Green growing vegetation in removing unwanted gases and aerosols from the air not only cleanses the general atmosphere of these pollutants but influences most markedly their immediate microenvironment. As previously noted, however, many plants can pollute the environment also with undesirably allergenic pollen and other plant parts, noxious odors, gummy exudates, and many additional substances derived from primary and secondary metabolic processes.

Another type of "air" pollutant–vegetation interaction which is important in the microenvironment of the urban dweller is noise pollution abatement. Noise propagated through the air can be effectively attenuated by dense vegetation. Trees, shrubs, and tall grass configurations have been shown to reduce sound levels by $\frac{1}{3}$–$\frac{1}{2}$ of the measured readings recorded over equivalent distances of open surfaces (Cook and Van Haverbeke, 1971). Greater use of specific planting arrangements is likely to be utilized in the future to reduce levels in the immediate environment of man and to condition his habitat. In addition to scavenging certain airborne pollutants and abating sound, transpiring shade vegetation cools and humidifies the microenvironment while contributing to the aesthetics of the habitat.

Before progressing into detailed discussions of the interactions of gaseous air pollutants with plant canopies two terms, *interaction* and *canopy,* used in the title of this chapter should be defined in order to better understand the concepts presented and the manner in which the subject material was developed.

Interaction: Webster defines interaction as: "mutual or reciprocal action or influence." The prefix *inter* has reference to a "lacing together" of agents and processes. The suffix *action* carries the connotation of bringing about an alteration by force or through a natural agency. An action is understood when the manner and methods of performance are comprehended. Therefore, the operating mechanisms involved or models simulating the processes designed to aid in the understanding the mechanisms should be given due consideration. This chapter was formulated and developed with these concepts in mind.

Vegetation Canopy: A canopy of vegetation can be defined as a more or less regular and continuous layer of vegetative crowns or herbaceous foliage covering an area stratified fairly uniformly with respect to height. The term *canopy* refers to an overhanging shelter or shade structure, a covering. It is used in the context of a rooflike structure or awning and is therefore a specialized form of vegetative cover when applied to plant canopies.

A short discussion of the characteristic structural features of a stand of vegetation, basically related to the spatial distribution of the plant biomass, can help to clarify the concept of a canopy of vegetation. Vegetation structure can be generally classified according to three basic spatial categories: (1) vertical structure, or the vertical stratification of species and plant parts into layers or strata; (2) horizontal structure, or the spatial pattern of each species in the vegetation as a whole; and (3) quantity of vegetation based on criteria such as density of species, vegetation biomass, and ground area covered. The vegetation structural characteristics will be elaborated on briefly as they relate to plant canopies.

The layering of vegetation into strata is a common phenomenon in stands of vegetation. The profile of a north temperate zone woodland as may be found in the United States, for example, commonly exhibits a number of recognizable strata, such as an uppermost moderately regular layer of green vegetation tree crowns—the woodland canopy, an understory of discontinuous vegetation composed of shrubs and saplings, an irregular field height layer of herbaceous plants, a ground surface layer of moss, litter, and the subterranean layer of root systems, underground stems, and microflora. Tropical rain forest profiles commonly show several recognizable canopy layers in the space below the upper canopy. Treeless stands of vegetation show few layers based on stratification of species and plant parts. Dense herbaceous types, especially cultured agricultural crops, tend to have no understory vegetation and be fairly uniform throughout. The *crop canopy* layer may make up a major portion of the aboveground space occupied by these stands—excepting perhaps only the lowermost portion of the crop where light penetration is so low that foliage is thin and stems mainly prevail.

Three primary structural groupings have been proposed to assess the general pattern of vegetation on a horizontal scale (Shimwell, 1972). When the vegetation forms a continuous cover laterally, the vegetation is closed; a stratified continuous and regular layer represents a *closed canopy*. Open vegetation is applied to situations where spaces less than twice the foliar diameters of the predominant organisms exist which can be colonized. A recognizable layer of uniform vegetation having this horizontal pattern might be considered as an *open canopy*. More open vegetation than the latter case where ground space dominates the aerial view is referred to as *sparse* vegetation. The concept of a vegetation canopy becomes incongruous when considering very sparse vegetation.

IV. Air Pollutant Exchange with Plant Canopies

Exchange of air pollutants between the air and vegetation surfaces depends upon factors which affect pollutant transfer and the properties of the sources and sinks. Atmospheric pollutants with negligible settling velocities are reversibly transferred to and from surface vegetation by a combination of diffusion and flowing air movement. Except when winds are very light, the atmosphere is typically turbulent. Atmospheric turbulence is created primarily as a result of chaotic airflow over rough surfaces and rising air currents caused by solar heating of the ground layer. Gustiness, which everyone has experienced, is a manifestation of atmospheric turbulence. Turbulence is highly important in effecting air mixing. Randomly moving air parcels (turbulence elements or eddies) can transport their contents rapidly from place to place. The stability of the atmosphere—basically its tendency to suppress vertical air motion—is related to wind shear and vertical temperature structure. Vertical temperature structure, described by the atmospheric temperature lapse rate (rate of temperature *decrease* with height), has often been used as an indicator of atmospheric stability. A stable atmospheric layer over an area in which the temperature *increases* with height (called an inversion layer) suppresses convective turbulence and mixing. This condition limits dispersion and is thus of special interest in localities subject to local buildup of air pollution. Although the temperature lapse rate is commonly used as an index of stability, since chaotic winds over rough surfaces affect vertical mixing in the surface layer, the index should reflect the wind and ground roughness as well as the lapse rate. A number of useful parameters that describe the dynamics of fluid systems which have been used to assess the transport of matter and energy and to quantitatively indicate turbulence have been developed. Some parameters are given in Table I. These entities will be referred to as needed throughout this chapter in the development of the subject material.

Airflow within and immediately above vegetation couples plant and air pollutant sources, receptors, and sinks with the atmosphere. It is within this surface layer scale that interactions discussed in the following sections are concerned. Air pollutant coupling and other interactions in vegetation-atmospheric systems require interdisciplinary study.* Micrometeorological conditions influence plant and atmosphere energetics, the rates at which air pollutants and other matter and energy are exchanged in the plant-atmosphere system, and to some extent pollutant residence times on or in the plants and plant parts. The chemical and physical forms of the pollutants along with plant morphological, physiological, and biochemical states (regulated largely by the energy and chemical balance in the plant microenvironment) determine to what extent pollutants can be sorbed or emitted by the vegetation as well as the effects of a particular pollutant exposure dosage on the plants.

Inoue (1963a) has discussed airflow adjacent and within simple crop canopies and separated the air layers into three characteristic parts: (1) a logarithmic wind profile layer (boundary layer) above the canopy surface, (2) an exponential canopy–eddy layer, and (3) the lowest part of

* It has been suggested that environmental problems such as that treated here may be more satisfactorily studied from adisciplinary rather than interdisciplinary standpoints (Bodine, 1972). The interdisciplinary approach accepts the notion of separate disciplines along with the methods and bodies of knowledge commonly ascribed to them. An adisciplinary approach—one requiring free bridging and integration of the "disciplines"—more nearly characterize natural systems as they exist.

TABLE I
PARAMETERS AND RELATIONSHIPS COMMONLY USED FOR DESCRIBING TURBULENT TRANSFER OF MOMENTUM, HEAT, AND GASES[a]

	Defining expression	Expression number
Expression showing pertinent interrelationships among some important parameters listed below. (See text for descriptions of parameters)	$\tau/\rho = v_*{}^2 = (K_m + \nu)\partial\bar{u}/\partial z$ $= l^2(\partial\bar{u}/\partial z)^2$ Laminar flow $K_m = 0$ (ν, important) Turbulent flow $K_m + \nu \simeq K_m$ (ν, negligible)	T-1a

Parameter	Common units	Turbulent flow expression	Expression number
Shearing stress,[b] τ	dynes/cm²	$\tau = \rho v_*^2 = \rho K_m\, \partial\bar{u}/\partial z$	T-1b
Friction velocity, v_*	cm/sec	$v_* = (\tau/\rho)^{\frac{1}{2}} = l\, \partial\bar{u}/\partial z$	T-1c
Momentum eddy[c] Diffusivity, K_m	cm²/sec	$K_m = lv_* = l^2(\partial\bar{u}/\partial z)^2$	T-1d
Mixing length, l	cm	$l = K_m/v_*$	T-1e
Richardson number, R_i	dimensionless	$R_i = g(\partial T'/\partial z)/T(\partial\bar{u}/\partial z)$	T-1f

[a] ν, kinematic viscosity, a measure of molecular diffusivity influence; $\bar{u}$, mean wind velocity; z, height; ρ, air density; g, gravitational constant; T, temperature [T': The temperature adiabatically normalized to standard pressure conditions, known as the potential temperature, is commonly used (Munn, 1966).]

[b] Wind shear, a consequence of shearing stress, is defined as the change of wind velocity with height (or in the horizontal), i.e., $\partial\bar{u}/\partial z$.

[c] Thermal eddy diffusivities K_H and eddy diffusivities for the transfer of gases K_g have been commonly assumed to be equal to K_m when transfer is thought to occur via the same eddies. A displaced parcel of air may move in and out of an air layer, however, transferring momentum by pressure fluctuations without mixing thoroughly and exchanging its contents (Munn, 1966). Also where thermal convection (bouyancy) adds significantly to frictional turbulence, thermal diffusivities may exceed other diffusivities.

the plant–air layer in which plants and ground surface influence the wind profile (cf. Fig. 2). (A logarithmic profile probably exists very close to the ground with the wind speed decreasing to zero at ground level.) In the immediate vicinity of individual leaves gaseous pollutant transfer to leaf external and internal surfaces occurs by molecular diffusion through the leaf–air boundary layers adhering to each leaf (where a portion may react with surface substances), through the leaf epidermis (via stomata, breaks), and through the mesophyll free air spaces within the leaves. Since

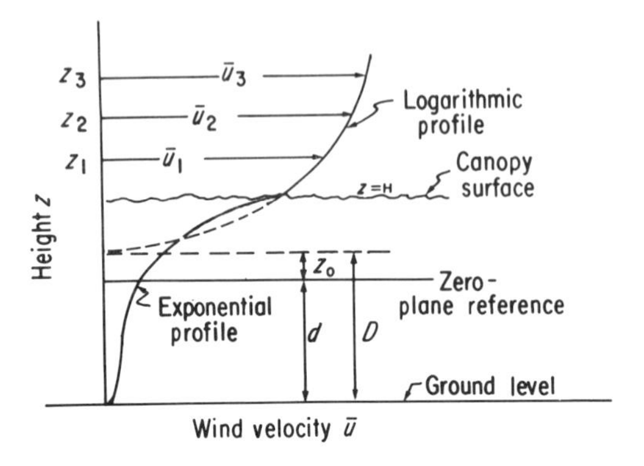

FIG. 2. Wind profile over and within simple canopy.

mesophyll cells are bathed in aqueous media and are highly structured the pollutant solubility and reactive properties, the transport of solutes within leaves, and the reaction sites influence cellular sink potentials and consequent effects on the cells.

A number of methods have been devised to assess mass and energy exchange between canopies and surrounding air in the canopy boundary layer (Barr, 1971; Barry and Chamberlain, 1963; Cionco, 1965; Evans, 1963a; Inoue, 1963; Lemon, 1965; Lemon *et al.,* 1971; Waggoner and Reifsnyder, 1968; Waggoner, 1971). Two common methods, the wind profile method and the electrical analogue simulation method, are employed here. Canopy–air–pollutant interactions described subsequently will be developed within the general framework of these model systems.

A. Wind Profile (Aerodynamic) Method

Winds largely control mass transfer processes (and to a great extent heat transfer processes) between the surface layer and vegetation. Characteristic wind profiles and turbulence properties observed in well-developed turbulent boundary layers over homogenous surfaces have motivated micrometeorologists and plant scientists to attempt to model turbulent transfer of momentum, gases, and heat between the air in the boundary layer and simple canopies.

1. Boundary Layer Airflow and Turbulent Transfer

Horizontal winds flowing over canopy surfaces are slowed by frictional drag on the vegetation. Immediately above a canopy a turbulent air boundary layer exists. Within the boundary layer the mean horizontal wind velocity is decreased logarithmically with decreasing height above the canopy surface. Outside the boundary layer wind velocity profiles are no longer logarithmic, but are characteristic of upwind conditions. Energetics (i.e., momentum and heat transfer) and canopy surface roughness influence the thickness and character of the boundary layer—hence, pollutant transfer and concentration gradients between the atmosphere and plant canopy. For a steady wind, the boundary layer thickness increases to a steady-state height with distance (fetch) traversed from the leading edge of transition from one uniform surface to another. Temperature gradients giving rise to convective bouyancy forces along with plant geometric (roughness) and elastic (plant waving) properties influence turbulence and, therefore, pollutant mixing, in the vicinity of the canopy.

Wind flow over a given surface can be laminar or turbulent. In laminar flows net vertical movement of air molecules occurs via molecular diffusion and can be represented by an equation stating Fick's first law of diffusion:

$$\frac{\partial q}{\partial t}\frac{1}{A} = D\,\frac{\partial c}{\partial x} \tag{1}$$

where $(\partial q/\partial t)$ $(1/A)$, the flux per unit cross-sectional area, denotes the one-dimensional x transfer ($x = z$ in the vertical direction) of a quantity q of substance across a cross-sectional area A per unit time t along a concentration C (or density) gradient. For turbulent flows characteristic of nature, vertical transfer of momentum, heat and gases occurs predominantly via the bulk motion of eddies, or parcels of air. As air cells leave a given air layer and move to another, in turbulent transfer, they are replaced by other cells which may have different compositions resulting in the mixing and exchange of their contents. Net vertical transfer of material and energy in the turbulent equilibrium boundary layer has been modeled by analogy to the theory of molecular diffusion. The rate of exchange between air layers can be expressed in the basic form of Eq. (1) with an "eddy diffusivity" coefficient K (a measure of the turbulent transfer rate of volumes of air) replacing the molecular diffusion coefficient D. The eddy diffusivity coefficient usually bears a subscript denoting what is being transferred in the eddies. The turbulent transfer rate may be several orders of magnitude faster than molecular diffusion.

In a fluid, such as air, retarding forces (frictional forces) on the flowing wind—originated at the surface—are distributed through the fluid producing a wind velocity gradient above the surface. The tangential force per horizontal area exerted on a flowing fluid layer causing it to move against friction is called the shearing stress. The force of retardation per unit horizontal area at the surface, given by the surface shearing stress τ_0, is related to mean reference wind velocity $\bar{u}$, the air density ρ, and an empirical (nondimensional) index of friction (a drag coefficient C_D) by $\tau_0 = C_D \rho \bar{u}^2$. For laminar flows, shearing stresses are caused by molecular agitation. Molecules moving randomly upwards from layers with light winds near the surface reduce the average momentum of stronger winds above. On the other hand, interchanged molecules moving downward increase the momentum of flowing air in the lower layers. For turbulent flow, transfer of momentum via molecular agitation is small compared to momentum transfer resulting from the bulk vertical motion of eddies. The momentum eddy diffusivity K_m representative of turbulent flow systems is a transfer coefficient quantifying the eddy transfer of momentum ($\rho\bar{u}$) downward to the surface. When considering turbulent flow systems, molecular agitation contributions are usually neglected and K_m is related to the shearing stress, mean wind velocity gradient, and a "frictional velocity" (which has units of velocity and characterizes the particular turbulence regime) as shown in expression (T-1b) in Table I.

The surface boundary layer is often defined for practical purposes as a layer in which the shearing stress is essentially constant—that is, does not vary more than 5% with height. This requires that the frictional velocity v_* also be approximately constant since the shearing stress τ varies as the square of v_* ($\tau = \rho v_*^2$) when the mean densities of the flowing layers are equal and hence do not tend to move as a layer in a direction contrary to the mean wind flow. Making the assumption that v_* is constant with height z in an equilibrium turbulent boundary layer, an equation describing the mean wind velocity gradient in the boundary layer with no bouyancy forces has been derived (Munn, 1966) and established empirically from wind tunnel studies, i.e.,

$$\frac{\partial \bar{u}}{\partial z} = \frac{v_*}{kz} \tag{2}$$

where k is the von Karman dimensionless constant which has a value of approximately 0.4. Equation (2) can be arrived at through dimensional analysis and energy balance considerations of the transfer of energy from the mean wind flow to turbulence, and the dissipation of the energy. Upon integration, Eq. (2) gives the form of the well-known boundary layer wind velocity profile equation for an homogenous surface:

$$\bar{u} = \left(\frac{v_*}{k}\right) \ln \frac{(z - d)}{z_0} \qquad z_0 < H - d < z - d \tag{3}$$

where z_0, called the "roughness length," is a constant of integration. The roughness length is used to characterize and define the relative roughness of the surface. It is assigned a value equal to the distance between a zero wind velocity extrapolation height, denoted by D in Fig. 2, and a zero-plane reference for the surface. Figure 2 gives a schematic diagram of the equilibrium wind profile for a uniform canopy surface. The theoretical zero extrapolation height D would lie within the canopy at a position $D = z_0 + d$. As vegetation grows and canopy height H increases, the boundary layer wind profile and other physical properties are displaced upward above the ground. The zero-plane displacement height d, empirically introduced into Eq. (3), characterizes this canopy-height related displacement and provides a better fit for experimental data. The two parameters z_0 and d reflect the effects of mechanical and geometrical properties of the vegetation canopy and wind velocities in the boundary layer. Sutton (1953) has reported a table of values which lists z_0 values ranging from 0.001 cm for smooth ice or mud flats to 9 cm for thick grass 50 cm tall, and d values ranging from 0 cm for short grass 1–3 cm tall to 30 cm for long grass 60–70 cm tall. Inoue (1963a) has shown that changes can occur in z_0 and d for a limber canopy with wind velocity changes which were related to the Honami (plant waving) effect caused by the wind above the canopy.

Assuming that the constraints placed upon the derivation of Eq. (3) can be tolerated and that the momentum eddy diffusivity coefficient K_{m} is equivalent to the eddy diffusivity coefficient of a gas g transferred by the eddies in the turbulent boundary layer (as predicted from the similarity theory*), Lemon (1965) has derived an integrated equation for gas exchange rates Q_g between two specified z_i levels as a function of mean wind velocities $\bar{u}_i$ and gas concentrations C_i at the two levels:

$$Q_g = \frac{C_2 - C_1}{[\ln[(z_2 - d)/(z_1 - d)]]^2/[k^2(\bar{u}_2 - \bar{u}_1)]} \qquad H - d < z - d \quad (4)$$

Since the application of this equation is restricted to neutral stability conditions, an attempt has been made to correct for nonisothermal situations that result in thermal turbulence by incorporating into the equation a correction factor containing the Richardson number (Lemon, 1965). The Richardson number is a dimensionless parameter comparing the effects of thermally induced turbulence with frictional turbulence [cf. Eq. (T-1f), Table I] and is used as an index of stability. Over transpiring (wet) vegetation, except for low wind conditions, temperature gradients are usually

* Similarity theory: A cause–effect relation theory which assumes that if experimental conditions are the same the results should be similar. The similarity principle is used extensively in micrometeorology to predict functional relationships which may be experimentally verified. According to this principle, the eddy momentum diffusivity coefficient is often substituted for the eddy transfer coefficients for gases and heat thought to be transferred in the same eddies. (Note: See footnote c, Table 1.)

relatively small owing to the release of a large part of the radiant energy absorbed by leaves as latent heat of evaporation rather than sensible heat. Corrections due to the Richardson number would be small for this condition. (Temperature gradients within canopies resulting from local heating of relatively slowly moving air around warm plant foliage exposed to radiant energy of high intensity, though, can be substantial, especially under drought conditions which cause stomatal closure in the leaves. Stomatal closure causes the leaf energy balance to shift with less of the absorbed energy lost through evapotranspiration and more going into heating of the leaves and contacting air.)

The wind profile equations and vertical exchange rate functions into which they have been incorporated assume flow over uniform, level surfaces. Even-aged, dense, agricultural crop canopies may approximate these conditions, but natural vegetation, being usually more heterogeneous and frequently located in irregular terrain, presents more complexity in the formulation of descriptive airflow and exchange models. The wind profile model, however, serves as a helpful reference guide for understanding some of the important basic principles involved in airflow over vegetation and turbulent transfer of matter and energy with simple canopies. Progress is often made in understanding more complicated systems through the extension of knowledge gained from simpler model systems.

2. Canopy Airflow and Turbulent Transfer

Turbulent airflow within simple canopies has been modeled in a fashion analogous to the aerodynamic approach taken for the surface boundary layer (allowance being made for loss of momentum in the canopy caused by the foliage) but the complexity greatly increases due to problems relating to the spatial heterogeneity of vegetation and the internal distribution of potential receptors, sinks, and sources for transferred materials and energy.

The turbulent logarithmic wind profile in the canopy boundary layer just described has been shown experimentally to change to an exponential form within simple uniform vegetation (cf. Fig. 2). Near the ground level a logarithmic form is reassumed. The turbulent exponential wind profile has been observed in a variety of homogenous vegetation types and can be represented as follows:

$$\bar{u} = \bar{u}_H \, e^{\alpha f(z,H)} \qquad z \leq H \tag{5}$$

where α is a dimensionless wind velocity attenuation constant, $\bar{u}_H$ is the mean wind velocity at the top of the canopy at height H and $f(z,H)$ defines the variable coefficient, a function, in terms of z and H. Values of α ranging from about 2–3 have been reported in the literature for several agricultural crop canopies, i.e., rice, corn, and wheat canopies (Inoue, 1963a; Lemon, 1965). An α value of 4.25 has been reported for a pine forest. The velocity attenuation parameter is reflective of the density, geometry, and elastic properties of the vegetation. A basic assumption of the canopy exponential wind profile model is that the canopy–eddy airflow structure is regulated by eddies produced at the canopy top layer.

Both the turbulent logarithmic boundary layer wind profile equation and the exponential wind profile equation have been interpreted from arguments based on the "mixing lengths" of the eddies. The mixing length characterizes the transfer properties of turbulent motion at a given level and represents the mean length eddies travel upon leaving a level before mixing at a new level. Within a number of simple experimental canopies the mixing lengths l have been reported to be nearly constant through much of the canopies (Inoue, 1963a; Lemon, 1965). When l and α are constant within an homogeneous canopy layer possessing an exponential wind profile regulated by eddies produced at the canopy top, momentum eddy diffusivities within the canopy at heights z, $K_m(z)$, may be basically related to the eddy diffusivity at the canopy top $K_m(H)$ by the equation (Lemon, 1965):

$$K_m(z) = K_m(H)\bar{u}/\bar{u}_H \qquad (6)$$

where $K_m(H)$ at the canopy top is defined by $K_m(H) = (\alpha l^2/H)\bar{u}_H$.

Difficulties exist in the evaluation of wind profiles and exchange coefficients within real canopies, but recognition of the causes of many of these difficulties provide insight into factors that should be considered when assessing interactions resulting from canopy air mass and energy exchange and notable responses of the physical and biological systems involved. To restate again, general factors that can cause significant perturbations from the model system described relate to such factors as the heterogeneity of vegetation structure and terrain, irregular wind flow patterns, and localized mass or energy sources and sinks (e.g., bouyancy forces arising from expanding air heated within the canopy–surface layer, localized sources or sinks for physiological gases and air pollutants—i.e., nonuniform absorption or generation of these substances within the canopy). The recognition that there exists spatially distributed sinks and sources within canopies and that fluxes and transfer processes in the plant–air layers can vary with height have led to the employment of several other methods for estimating canopy diffusivity coefficients. Three that have been used either independently or in conjunction with the aerodynamic method are (Inoue, 1963b): (1) the heat-budget (energy balance) method in which net radiative fluxes, sensible heat exchanged with the soil and foliage, and vertical gradients of air temperature and humidity are used to calculate exchange coefficients within plant–air layers; (2) the plant physiological method in which transpiration rate, water vapor concentration gradients, transpiring surface areas of the leaves, and "effective transpiration lengths" determined from knowledge of the plant physiology are utilized; and (3) the calculation of exchange coefficients from mean concentration gradients and eddy fluxes obtained from the statistical correlation of measured eddy velocity fluctuations.

A vegetation planting arrangement for a "sanitary protective area" that has been proposed to reduce air pollution within a protected area of man's microenvironment (e.g., around homes, playgrounds, picnic areas) can be understood by reference to Fig. 2 and the previous discussion (Hanson and Thorne, 1972). For several years tree plantings have been promoted by U.S.S.R. scientists and city planners for the purpose of reducing ground level air pollution. The sanitary protective area concept proposes that an absorptive canopy of vegetation with sufficient density to prevent the polluted atmosphere above from exchanging with vegetation cleansed air within the canopy at a rate greater than about 0.1 mph be established. As shown in Fig. 3, the area is designed to be enclosed by dense trees and shrubs (or other structures) which lifts the incoming polluted wind skyward, displacing the boundary layer above the canopy, and at the same time blocks the horizontal flow through the area at ground level. The vegetation enclosing the area serves primarily as a barrier to advective wind flow and to displace the wind upward. The exterior vegetation should

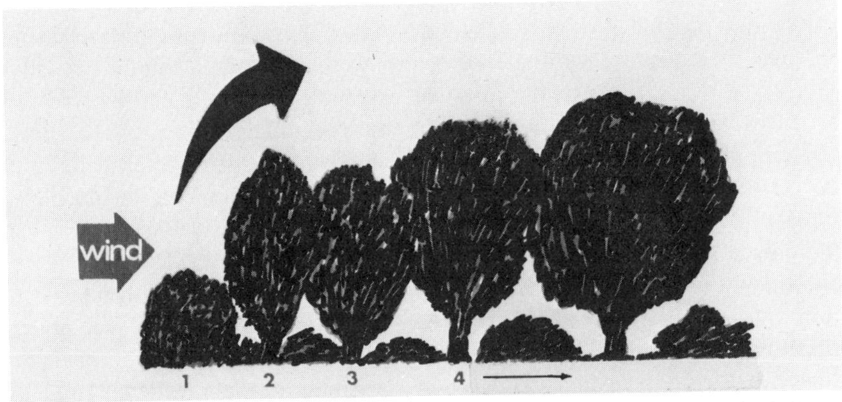

FIG. 3. Schematic drawing showing general vegetational features required for a sanitary protective area. Vegetation bordering the protected area (in zones 1–3) serve primarily to displace the incoming polluted wind above the canopy and act as a barrier to ground level wind flow. The interior vegetation (zone 4) should be chosen and strategically located to effectively lower the pollutant concentration in the air transferred into the microenvironment from above.

therefore be selected for their tolerance to phytotoxicants in the flowing air. This may require that the exposed vegetation be poor pollutant sorbers. Vegetation within the enclosed area protected from the wind, however, should be selected for their capabilities to effectively remove pollutants from the relatively stagnant air. The slow transfer rates within the protected vegetated air layers would permit the plants to more thoroughly reduce the pollutant concentrations in the air of the microenvironment than if polluted air were allowed to flow through at ground level. At the same time, because of the reduced amounts of potentially phytotoxic pollutants brought into contact with plants of the interior as a result of the restricted pollutant transfer rates, plant exposure dosages would be less than for conditions favoring more rapid ventilation providing a measure of protection to the plants. The landscaped area, in addition to bringing down pollution exposures to levels more acceptable to people within the protected area, also provides other air conditioning and aesthetic advantages previously referred to.

This section has treated general principles that regulate the turbulent transfer of mass and energy over and within simple canopies. Information presented is basic to the understanding of the transfer processes. In the following section exchange of air pollutants with vegetation will now be discussed according to a more generalized model simulating the exchange process—an electrical analogue model.

B. Electrical Analogue Simulation Method

Simple electrical analogues based on Ohm's law provide useful formats for describing and evaluating gas exchange rates with leaves and canopies. Ohm's law can be stated mathematically as

$$Q(\text{flux}) = \frac{\phi \text{ (potential difference)}}{R \text{ (Resistance to transfer)}} \tag{7}$$

For most gas exchange simulation purposes the potential difference term ϕ is equated with a concentration difference ΔC between two positions and the resistance of the pathway to transfer R is in practice visualized as the ratio of the effective distance traversed and the mean diffusivity coefficient (D or $\bar{K}$) for the process, i.e., $\Delta X/D$ or $\Delta X/\bar{K}$. [Note: instantaneous transfer rates through a very small distance leads to a differential equation which at its limit is equivalent to a rearranged form of the one-dimensional Fick's differential equation given in Eq. (1).] When applied to leaf or canopy gas exchange systems, averaged or steady-state data are usually used and interpreted as such. For convenience, the equations generally utilized in practice represent only approximate solutions for the generally highly complex exchange systems.

Figure 4 gives a simple diagrammatic representation of a vertical gas exchange system with canopies using the electrical analogue model for evaluation. The canopy–atmosphere system can be divided into strata for analysis and the model "circuits" evaluated for each strata. The central column which decreases in width within the canopy depicts the relative mean or steady-state flux magnitude as a function of height. The decreasing flux with height graphically indicates the net vertical removal of substances (air pollutants) by the canopy sink tissues. Conversely, the emission of substances by plants (e.g., CO_2 from plant respiration and H_2O vapor from transpiration) moving in the reverse direction could be represented in a similar fashion by altering the shape of the flux diagram in a representative manner. We will be concerned here only with canopy removal of gaseous air pollutants transferred into the canopy from the atmospheric surface layer above. Superimposed upon the figure is the circuitry of an electrical analogue simulator modeling the exchange process. The system is grossly analyzed according to a three-strata network. The system could, of course, be divided further into more canopy layers each with its representative circuitry.

Fluxes through the canopy boundary air layer (if no boundary layer sinks were assumed to exist) can be estimated from the vertical concentration difference between two heights above the canopy and resistance values evaluated for the air layer traversed. The isothermal boundary layer flux equation would then be equivalent to Eq. (4) described in the previous

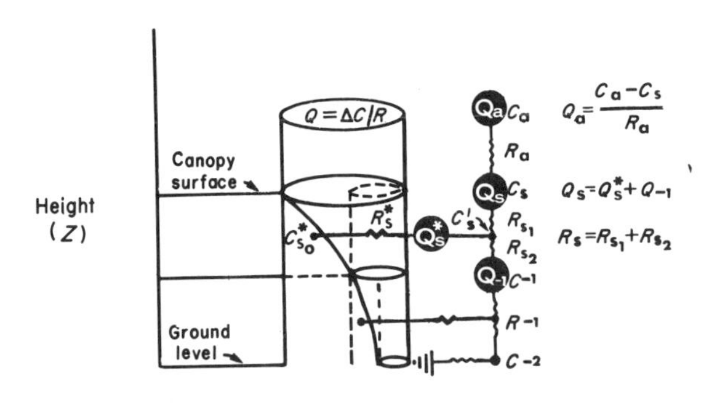

FIG. 4. Schematic representation of a vertical gaseous pollutant flux system above and through a plant canopy. Electrical analogue circuitry simulating the flux system is superimposed upon the illustration. Pollutant air concentrations are denoted by C and resistance to transfer are given by R. (See text for explanation.)

section. Within the canopy, estimations of the resistance components and fluxes become more complex. Resistances within canopies can be categorized into two general types: (1) resistances regulated by the turbulent airflow in the canopy, and (2) resistances associated with the transfer to plant sink sites on or within the plants. As polluted air moves into the canopy from above, the equilibrium influx penetrating into the uppermost canopy-air layer Q_s represents the summation of the flux penetrating through the layer to sinks below Q_{-1} plus the flux sorbed by vegetation within the layer $Q_s{}^*$, i.e., $Q_s = Q_{-1} + Q_s{}^*$. (If no boundary layer removal occurs: $Q_a = Q_s$.) Figure 4 gives the mean flux expression for the boundary layer as a function of the vertical pollutant concentration gradient and transfer resistance. Similar equations could be written for the lower canopy layers, but because of the spatially distributed sources and sinks within the canopy the analysis is more involved.

Pollutant uptake occurs throughout a canopy stratum continuously affecting the concentration gradient in the layer. However, for convenience, the circuitry representing the mean foliar uptake for the stratum can be drawn schematically through a mean level lying somewhere between the top and bottom of the layer. Using this simulation technique, the concentration gradient $C_s - C_{-1}$ is as follows:

$$C_s - C_s' = R_{s_1}(Q_{-1} + Q_s{}^*) \tag{8}$$

$$C_s' - C_{-1} = R_{s_2}Q_{-1} \tag{9}$$

$$C_s - C_{-1} = R_{s_1}Q_s{}^* + R_sQ_{-1} \tag{10}$$

where R_{s_1} characterizes the mean turbulent transfer resistance corresponding to transfer through the canopy stratum to the mean foliar sink level, R_s is the total resistance to transfer through the canopy stratum (Note: $R_s = R_{s_1} + R_{s_2}$) and C_s', which cancels out of the final expression, represents the air concentration for the average foliar sink level.

Equation (10) shows the interaction of foliar uptake within the canopy layer, uptake by sinks located below the canopy stratum, and the canopy stratum resistance to turbulent transfer on the concentration gradient established for the layer. If no sorption were to occur within the layer, the equilibrium gradient would depend only upon uptake rates by sinks below and the total resistance to turbulent transfer through the stratum. Conversely, if all uptake resulted from sorption by canopy tissues within the stratum, the gradient would depend only upon $Q_s{}^*$ and R_{s_1} (a fraction of R_s). It is important to observe that when the canopy stratum resistance to turbulent transfer is small (i.e., a condition representing rapid canopy ventilation), the concentration gradient becomes less pronounced for the same $Q_s{}^*$ and Q_{-1}. Normally rapid canopy ventilation promotes an increased uptake rate, however, by bringing the pollutant into contact with the foliage bearing sink structures at a faster rate. The concentration profile results then from a balance between the ventilation and uptake rates. It has been improperly reasoned by some investigators that unless a vertical concentration gradient can be measured within vegetation, the vegetation is not an important sink. Of course if little uptake occurred by the vegetation, distortion of the steady-state concentration profile would be minimal; but one must consider both the ventilation rates and plant uptake rates in making such an appraisal. It is possible for plants to remove pollutants rapidly from well-mixed air without a marked concentration gradient resulting. This condition is most likely to occur within sparse vegetation when air concentrations are measured in the spaces between the plants or between the foliage comprising sparse crowns of trees. Nevertheless, the uptake rates by the individual plants may be maximal under these conditions and the total uptake on large areas covered with vegetation might be substantial.

The model simulator suggests that $Q_s{}^*$ can be expressed in terms of the mean concentration difference $C_s' - C_{s_0}{}^*$ and an effective foliar resistance $R_s{}^*$ as follows:

$$Q_s{}^* = \frac{C_s' - C_{s_0}{}^*}{R_s{}^*} \tag{11}$$

where $C_{s_0}{}^*$ represents the average concentration for the vegetation sink sites and $R_s{}^*$ is an overall resistance index characterizing the foliar resistance to molecular diffusion through the leaf–air boundary layers, stomata, and mesophyll to plant sink sites. $Q_s{}^*$ can also be written as a function of the concentration gradient between the concentration at the canopy top C_s and $C_{s_0}{}^*$ by combining Eqs. (8) and (11) (eliminating C_s'):

$$Q_s{}^* = \frac{C_s - C_{s_0}{}^*}{R_{s_1} + R_s{}^*} - \left(\frac{R_{s_1}}{R_{s_1} + R_s{}^*}\right) Q_{-1} \tag{12}$$

It can be noted that if $C_{s_0}{}^*$ is considered to be negligible or zero (as has often been assumed since the net influx might irreversibly "ground" at the reaction sites), the total resistance $R(= R_{s_1} + R_s{}^*)$ between the canopy top and the mean foliar sinks within the layer is expressed by $R \simeq C_s/Q_s{}^* - R_{s_1}Q_{-1}/Q_s{}^*$. If Q_{-1} were neglible: $R \simeq C_s/Q_s{}^*$. Rearranging this latter equation to give an expression as a function of the reciprocal of the resistance (a "conductance" term), the general form of the commonly used deposition velocity V_g equation is obtained: $V_g = 1/R = Q_s{}^*/C_s$. The deposition velocity is calculated from the net rate of uptake by the "surface" and a reference air concentration determined at a specified height. The deposition velocity method assumes that the flux grounds rapidly and irreversibly at the reaction sites and the mean sink site concentration is zero.

C. Canopy Pollutant Sink Properties

Pollutant uptake studies indicate that under favorable growing conditions air pollutants (e.g., SO_2, NO_2, NO, and O_3) tend to be taken up most readily by foliage in the exposed upper portion of dense canopies (Bennett and Hill, 1973a). Uptake of metabolizable pollutants, such as certain sulfur and nitrogen oxides, is related with plant metabolism rates and is highly influenced, of course, by ventilation of the canopy and the state of the plants and plant parts (the latter being affected by environmental conditions, plant age, stomatal behavior, and other factors regulating plant structure and function). If the predominant reaction sites for a pollutant derived solute (or intermediate reaction product) were to reside within leaf cells—say at photosynthesizing chloroplasts or some other internal site—the pollutant must not only be brought into contact with the outer leaf surfaces but must diffuse through the epidermal layers, dissolve in the aqueous media of the mesophyll cells, and pass through cell walls and membrane barriers enroute to the reaction sites. The flux into leaves can then be controlled by the diffusive resistance offered by the stomata on the leaf epidermis, by the pollutants' solubility properties, and all additional physical barriers as well as chemical reaction properties and rates.

Since light plays a very important role in determining the physiological activities of leaves and in "conditioning" the reaction potentials of leaf sinks, it is not surprising that light exposure should affect plant uptake rates. Light of proper quality and intensity is required to effectively energize leaf systems and produce the chemical fuels and substrates required for metabolic activity. In addition, absorbed light energy can affect the chemical activity by altering the temperature of the leaf. Light, furthermore, markedly influences the stomatal resistance to molecular diffusion. Stomata on most mesophytic leaves characteristically open in the light, close in the dark, and are prone to be most open in actively photosynthesizing leaves (not showing water stress) located in the upper part of dense canopies. Open stomata favor the diffusion of pollutant gases into the leaves. The fate of pollutants coming into contact with leaves can be greatly determined by the position of the leaves within a canopy.

Metabolic reactions within mesophyll cells, though important in determining the uptake of certain pollutants, are not the only processes by which plants act as foliar pollutant sinks. Pollutants may react at sites on vegetation surfaces, in the aqueous solution bathing cells, with constituents of mesophyll cell walls or membranes, and with volatile substances (e.g., terpenes) emitted from the foliage. Moisture, dusts, sprays, exudates, and other reactants on leaves, stems or reproductive structures—i.e., flowers and strobili, are potential absorbants. In comparison, however, the glabrous (smooth) cutinous, epidermal surfaces of many plants are relatively ineffective pollutant sinks. (Hot surfaces may aid the decomposition of very labile gases such as O_3.) A pollutant known to be significantly deposited on plant surfaces is HF. In several HF fumigation studies in excess of 50% of the fluoride taken up was deposited externally on the foliage (Benedict *et al.,* 1965; Jacobson *et al.,* 1966; Ledbetter *et al.,* 1960). Investigations in which O_3 was made to impinge directly upon selected epidermal strippings containing extensive pubescence or saline excretions showed that O_3 can react on certain leaf surfaces (Bennett *et al.,* 1973). Ozone reacted more readily with *Kalanchoe* mesophyll tissue, however (cf. Fig. 5). Studies correlating the uptake of O_3 by bean plants with the degree of stomatal diffusivity have indicated that O_3 was taken up predominantly within the leaves of these plants (Rich *et al.,* 1970).

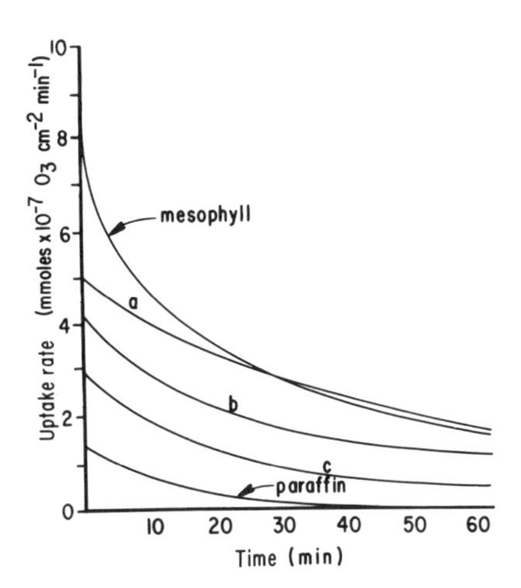

FIG. 5. Uptake of ozone as a function of time for epidermal strips taken from several plants selected for their surface characteristics. The epidermal strips were embedded in paraffin coated on Parafilm. Ozone uptake by paraffin and mesophyll tissue (*Kalanchoe tomentosa*) are also shown. Curve a represents ozone uptake by an epidermis taken from a halophyte which contained superficial saline deposits (*Atriplex confertifolia*). Curve b shows the ozone uptake by a densely pubescent epidermal surface (*Kalanchoe tomentosa*), and curve c depicts the uptake by a glabrous epidermis (*Canna hortensis*). The ozone was made to impinge directly upon a 1 cm^2 surface at a rate of 140 cc/minute in a 2 ml microchamber. The impinging ozone concentration was 20 pphm. The temperature was 23°C in the darkened chamber.

TABLE II

STEADY STATE UPTAKE RATES FOR POLLUTANTS AND CO_2, AND TRANSPIRATION WATER LOSS BY ALFALFA CANOPY. THE TABLE ALSO GIVES RELATIVE SOLUBILITIES OF THE GASES IN WATER[a]

Substance	Exchange rate/min/m² canopy	Solubility mmoles absorbed/cm³ H_2O @ P_i = 1 atm
CO	0.00 μmole/pphm	0.001
NO	0.03 μmole/pphm	0.002
O_3	0.4 μmole/pphm	0.012
NO_2	0.5 μmole/pphm	Decomposes
SO_2	0.8 μmole/pphm	1.6
HF	1.0 μmole/pphm	18
CO_2	1.1 mmoles/320 ppm	0.039
H_2O	0.72 moles	—

[a] Relative humidity in chamber atmosphere 48%. Temperature, 20°C; (For other plant and chamber conditions refer to Figs. 1 and 2.)

Ozone can readily react with constituents of cell membranes possessing sulfhydryl groups, unsaturated carbon–carbon double bonds, and many other oxidizable functional, linkage, and structural components. In green plants, O_3 probably reacts predominantly at the first membrane site encountered and may not directly interact with internal metabolic processes (Coulson and Heath, 1974). Damage to cellular membranes, however, disrupts cellular metabolism and integrity.

Results of pollutant uptake studies combined with pollutant profile investigations and observed plant physiological responses indicate that some predictable interactions occur which can be partially explained by a knowledge of the pollutants' chemical properties and probable plant reaction sites (Hill and Bennett, 1970; Hill, 1971; Bennett and Hill, 1973a,b; Bennett *et al.*, 1973). Table II gives relative uptake rates for several air pollutants of major importance by equivalent alfalfa canopies treated under the same environmental conditions. Figure 6 compares steady-state canopy pollutant concentration profiles produced within similar alfalfa canopies for five of the pollutants. The general canopy and environmental conditions are described in Fig. 7. Judging from light and CO_2 profiles, net photosynthesis occurred in the upper 20–30 cm of the canopies where light intensities, ventilation, and temperatures were greatest, and the stomata were probably most open. Net respiration exceeded CO_2 assimilation in the lowest layer.

Correlations between pollutant uptake data, solubility in aqueous media, and reaction properties should be noted. The more readily soluble a pollu-

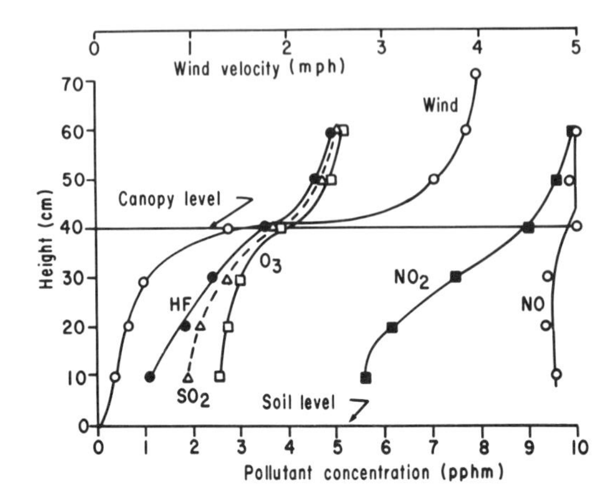

FIG. 6. Pollutant concentration and wind profiles for alfalfa canopy.

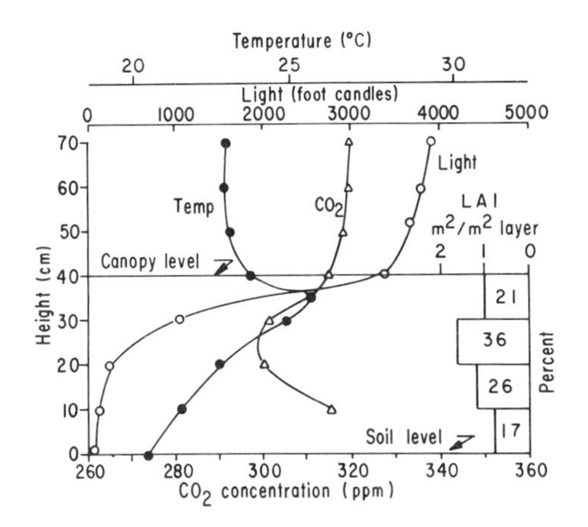

FIG. 7. Light, temperature, and CO_2 profiles for alfalfa canopy. Leaf area densities for each of four stratified layers within the canopy are plotted along the right margin. (LAI, leaf area index. LAI is given in m^2 leaf area/m^2 ground surface area for each layer.)

tant is in the cellular and extracellular solution, the more rapidly it may be expected to be absorbed under non-steady-state conditions. The steady-state absorption rate depends upon the plant's capacity to metabolize, translocate, or otherwise remove active pollutant solutes from the absorbing medium. In addition to plant uptake rates, Table II gives pollutant solubilities for the air pollutants listed. In general plant uptake rates increase as the solubility of the pollutant in water increases. This would have significance only if the pollutants were required to first dissolve into the leaf aqueous media before reacting. (Highly soluble gases (1) may react with the water, dissolved substance, tissues, and (2) by producing greater solute concentrations in the vicinity of a potential reactant results in greater chemical activity due to mass action.) Of the air pollutants compared, plant HF uptake rates and the relative canopy pollutant concentration profile gradients resulting from HF fumigations consistently were the greatest and most pronounced. Hydrogen fluoride, the most readily sorbed pollutant, is very soluble and highly reactive. Nitrogen oxide, on the other hand, is very insoluble, absorbed the slowest, and the NO concentration profiles were displaced the least (see Fig. 6). Hydrogen fluoride can be taken up effectively both by leaf mesophyll and on exposed plant surfaces. This may partially account for the greater displacement of HF concentration profiles than the other pollutant profiles at the lower canopy layers where stomatal diffusivities and metabolic rates tend to be slower. Notwithstanding the possible deposition of HF on plant surfaces, care should be taken to recognize the rapid uptake of F^- within leaves. Plant leaves are extremely effective in accumulating and concentrating fluoride from the atmosphere (Hill, 1969; MacLean and Schneider, 1973). Hydrogen fluoride exposed leaves can accumulate as much as 1,000,000 times the concentration to which they are exposed. Much of the F^- absorbed by leaves moves in the transpiration stream and becomes concentrated at the leaf margins and tips where high evaporation rates prevail due to reduced leaf boundary layers at the edges. Leaves fluctuate in fluoride content due to rain and other factors. A heavy rain can leach F^- from the leaves as well as wash it off the plant surfaces.

The oxides of sulfur and nitrogen are absorbed most efficiently by plant foliage near the canopy surface where light-induced metabolic rates and pollutant diffusivity rates are greatest. (Sulfur dioxide and NO_2 are taken up by respiring leaves in the dark, but uptake rates are much reduced.) It is known that SO_2 and NO_2 produce metabolizable substrates within mesophyll cells when exposure levels are low (Cresswell *et al.,* 1965; Bennett and Hill, 1973b; Daines, 1968; Hill and Bennett, 1970; Thomas *et al.,* 1950). Radiosulfur investigations (Thomas *et al.,* 1950) demonstrate that $^{35}SO_2$ absorbed by leaves at sublethal rates can be metabolized in essentially the same way as sulfate supplied to the plants via the roots. Sulfur dioxide is very soluble and is effectively absorbed in the cellular solution. Unless the external foliar surfaces are wet or contain reactable substances most of the gas taken up passes through the leaf stomata to the wet mesophyll cells. Sulfur dioxide is rapidly oxidized to sulfate in the cells. Metabolically reduced $^{35}SO_2$ derived sulfur can be utilized in the formation of sulfur containing amino acids and other organic compounds. The amino acids are readily incorporated into proteins—the principal sulfur bearing organic constituents of leaves. Excess sulfur in leaves, above that needed for sulfur-containing constituents, is present as sulfate. At low uptake rates SO_2 may be oxidized (with some reduction) about as rapidly as it is absorbed. Acidic oxidation components derived can be neutralized by organic bases in the leaves. High uptake rates can reduce the buffer capacities of the cells; but when SO_2 is added to leaves at very high rates the strongly phytotoxic sulfite intermediate may build up and cause injury to the cells before the buffer saturation capacities of the cells are reached. Slower uptake rates over a prolonged period, though, may lead to saturation of the cellular buffer capacities. Studies (Hill, 1971) comparing the diurnal uptake rates of NO_2 and CO_2 show that NO_2 uptake by oat canopies change with changes in solar intensity, metabolic rates, and transpiration rates (cf. Fig. 8). When gaseous NO_2 dissolves in water, it decomposes forming NO_2^- and NO_3^- ions in solution. Both of these species can be reduced to ammonia in cells in the presence of chloroplasts, ferredoxin, and light—or an appropriate reductant in the absence of light. The ammonia can be incorporated into amino acids and proteins. Chamber studies also indicate that a portion of the NO_2 absorbed by plants can be converted to NO and released back to the atmosphere (Hill and Bennett, 1970).

Figure 6 shows the NO profile to be essentially vertical throughout the canopy (except for a small deviation in the topmost 10 cm strata), indicating that foliar uptake was too slow compared with transport and mixing within the canopy to affect the steady-state NO profile substantially. The low solubility of NO limits its uptake by plants. In solution NO acts essen-

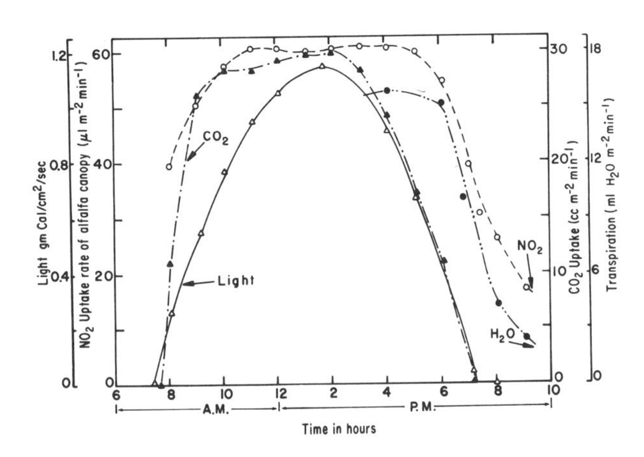

FIG. 8. Diurnal variation in the uptake rate of NO_2 by an oat canopy in sunlight. CO_2 uptake, transpiration, and light intensity are also plotted.

tially as a physically dissolved gas, obeying Henry's law (a declaration that the concentration of a gas dissolved in a medium may be expected to be proportional to the partial pressure of the external gas if the solution is ideal or nearly so). Therefore, the concentration of NO in cellular solution is tightly coupled with the concentration in the air. Nitric oxide solubility properties are stikingly evidenced in the apparent photosynthesis (CO_2 uptake) rate responses exhibited by plants exposed to NO in concentrations sufficiently high to inhibit CO_2 uptake. Apparent photosynthesis rates in plants treated with constant, inhibitory NO concentrations rapidly fall to new (depressed) equilibrium levels. Removal of NO from the atmosphere causes plant CO_2 uptake rates to quickly return to normal (see Fig. 9). This results because of the rapid loss of NO from the cellular solution in response to reduced atmospheric concentrations. In contrast to this, NO_2 which also inhibits CO_2 uptake by plants but is highly soluble, causes photosynthesis rates in fumigated plants to decrease much more slowly during fumigation with constant NO_2 air levels. Upon lowering NO_2 concentrations in the air, plants also recover more slowly (cf. Fig. 9). This indicates that NO_2 derived solutes build up in cells during fumigation and the inhibitory solutes (e.g., NO_2^-) generated are slowly depleted after termination of treatment. The slow recovery rates may be due to the necessity for the plants to decrease the cellular solutes through metabolism or translocation processes. Excess NO_2^- may depress CO_2 uptake rates by interfering with ferredoxin mediated reductive processes (Hill and Bennett, 1970). Both NO_2^- and NO exert actions at the ferredoxin site. Ferredoxin

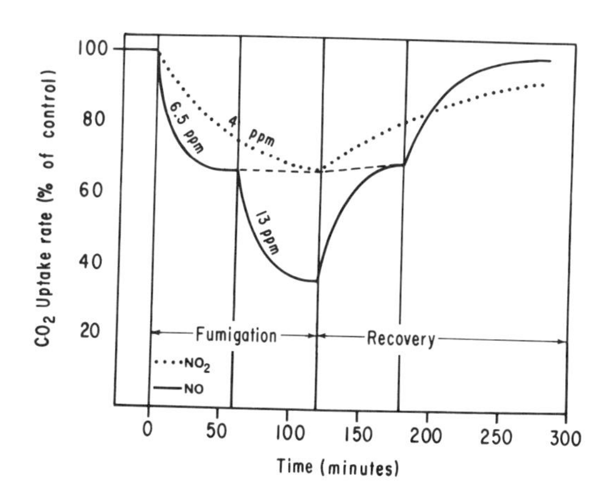

FIG. 9. Alfalfa apparent photosynthesis inhibition rates induced by fumigations with constant levels of NO and NO_2 as indicated and recovery rates after termination treatment.

is an iron-containing redox enzyme which functions in the transfer of electrons in photosynthesis and nitrogen fixation processes. Nitrite (NO_2^-) reduction and nicotinamide adenine dinucleotide phosphate (NADPH—essential reductant for photosynthetic CO_2 fixation) reduction systems converge at the ferredoxin site. The two processes may be expected to compete with each other. Nitric oxide is thought to inhibit ferredoxin activity by forming a concentration-dependent iron-NO complex (Hill and Bennett, 1970). Since NO_2^- can be reduced to NO by reducing agents commonly found in biological systems (i.e., ascorbic acid, ferrous iron, pyridine nucleotides), NO_2^- may also serve as a precursor for iron-NO complexes. (Note: Some pollutants may depress CO_2 uptake by plants by inducing stomatal closure—therefore increasing leaf resistance to CO_2 exchange; however, simultaneous transpiration and stomatal measurements made indicated this was not a significant factor for these nitrogen oxides under the conditions of the experiments.)

A potential consequence of the relatively rapid uptake of NO_2 by vegetation in comparison with NO uptake which might affect the production and concentration of another pollutant (O_3) in the air where NO_2 levels were substantially reduced in preference to NO has been suggested (Hill, 1971). Since O_3 concentration is dependent upon the NO_2/NO ratio in the air, an upset in the ratio could affect the ozone near ground level. (Although NO is removed slowly by vegetation, it can be converted to NO_2 or other substances in the atmosphere and might then be taken up more rapidly in these forms.) This potential consequence is noted here to indicate some subtle interactions which might occur near plant canopies. Similarly, volatile substances emitted into the atmosphere by vegetation may react with constituents in the air to alter pollutant levels and composition.

References

Barr, S. (1971). A modeling study of several aspects of canopy flow. *Mon. Weather Rev.* **99,** 485–493.

Barry, P. J., and Chamberlain, A. C. (1963). Deposition of iodine onto plant leaves from air. *Health Phys.* **9,** 1149–1159.

Benedict, H. M., Ross, J. M., and Wade, R. H. (1965). Some responses of vegetation to atmospheric fluorides. *J. Air Pollut. Contr. Ass.* **15,** 253–255.

Bennett, J. H., and Hill, A. C. (1973a). Absorption of gaseous air pollutants by a standardized plant canopy. *J. Air Pollut. Contr. Ass.* **23,** 203–206.

Bennett, J. H., and Hill, A. C. (1973b). Inhibition of apparent photosynthesis by air pollutants. *J. Environ. Qual.* **2,** 526–530.

Bennett, J. H., Hill, A. C., and Gates, D. M. (1973). A model for gaseous pollutant sorption by leaves. *J. Air Pollut. Contr. Ass.* **23,** 957–962.

Bodine, V. (1972). "Air Pollution." Harcourt, New York.

Brennan, E., and Leone, I. A. (1968). The response of plants to sulfur dioxide or ozone-polluted air supplied at varying flow rates. *Phytopathology* **58,** 1661–1664.

Cionco, R. M. (1965). A mathematical model for air flow in a vegetative canopy. *J. Appl. Meteorol.* **4,** 571–522.

Committee on Chemistry and Public Affairs. (1969). "Cleaning our Environment. The Chemical Basis for Action." American Chem. Soc., Washington, D.C.

Cook, D. I., and Van Haverbeke, D. F. (1971). Trees and shrubs for noise abatement. *U.S., Forest Serv., Res. Bull.* **246.**

Coulson, C., and Heath, R. L. (1974). Inhibition of the photosynthetic capacity of isolated chloroplasts by ozone. *Plant Physiol.* **53,** 32–38.

Cresswell, C. E., Hageman, R. H., Hewitt, E. J., and Hucklesby, D. P. (1965). The reduction of nitrate, nitrite and hydroxylamine to ammonia by enzymes from *Curcurbita pepo* L. in the presence of reduced benzyl viologen as electron donor. *Biochem. J.* **94,** 40–53.

Daines, R. H. (1968). Sulfur dioxide and plant response. *J. Occup. Med.* **10,** 514–524.

Dugger, W. M., and Ting, I. P. (1970). Air pollution oxidants—their effects on metabolic processes in plants. *Annu. Rev. Plant Physiol.* **21,** 215–234.

Evans, L. T., ed. (1963). "Environmental Control of Plant Growth." Academic Press, New York.

Frey, H. T., Krause, O. E., and Dickason, C. (1964). "Major Uses of Land and Water in the United States with Special Reference to Agriculture—Summary for 1964," Agr. Econ. Rep. No. 149. Econ. Res. Serv., U.S. Dept. of Agriculture, Washington, D.C.

Hanson, G. P., and Thorne, L. (1972). Vegetation to reduce air pollution. *Lasca Leaves* **20,** 60–65.

Heck, W. W. (1968). Factors influencing expression of oxidant damage to plants. *Annu. Rev. Phytopathol.* **6,** 165–188.

Heck, W. W., Dunning, J. H., and Hindawi, I. J. (1965). Interactions of environmental factors on the sensitivity of plants to air pollution. *J. Air Pollut. Contr. Ass.* **15,** 511–515.

Hill, A. C. (1969). Air quality criteria for fluoride vegetation effects. *J. Air Pollut. Contr. Ass.* **19,** 331–336.

Hill, A. C. (1971). Vegetation: A sink for atmospheric pollutants. *J. Air Pollut. Contr. Ass.* **21,** 341–346.

Hill, A. C., and Bennett, J. H. (1970). Inhibition of apparent photosynthesis by nitrogen oxides. *Atmos. Environ.* **4,** 341–348.

Hodges, L. (1973). "Environmental Pollution: A Survey Emphasizing Physical and Chemical Principles." Holt, New York.

Inoue, E. (1963a). On the turbulent structure of airflow within plant canopies. *J. Meteorol. Soc. Jap.* **41,** 317–378.

Inoue, E. (1963b). The environment of plant surfaces. *In* "Environmental Control of Plant Growth" (L. T. Evans, ed.), pp. 23–31. Academic Press, New York.

Jacobson, J. S., Weinstein, L. H., McCune, D. C., and Hitchcock, A. E. (1966). The accumulation of fluorine in plants. *J. Air Pollut. Contr. Ass.* **16,** 412–417.

Ledbetter, M. C., Mavrodineanu, R., and Weiss, A. J. (1960). Distribution studies of radioactive fluorine-18 and stable fluorine-19 in tomato plants. *Contrib. Boyce Thompson Inst.* **20,** 331–348.

Lemon, E. (1965). Micrometeorology and the physiology of plants in their natural environment. *In* "Plant Physiology: A Treatise" (F. C. Steward, ed.) Vol. IVA, pp. 203–227. Academic Press, N.Y.

Lemon, E., Stewart, D. W., and Shawcroft, R. W. (1971). The sun's work in a cornfield. *Science* **174,** 371–378.

MacLean, D. C., and Schneider, R. E. (1973). Fluoride accumulation by forage: Continuous vs. intermittent exposures to hydrogen fluoride. *J. Environ. Qual.* **2,** 501–503.

Magill, P. L., Holden, F. R., and Ackley, C., eds. (1956). "Air Pollution Handbook." McGraw-Hill, New York.

Meetham, A. R. (1950). Natural removal of pollution from the atmosphere. *Quart. J. Roy. Meteorol. Soc.* **76,** 359–371.

Meetham, A. R. (1954). Natural removal of atmospheric pollution during fog. *Quart. J. Roy. Meteorol. Soc.* **80,** 96–99.

Munn, R. E. (1966). "Descriptive Micrometeorology." Academic Press, New York.

Rich, S., Waggoner, P. E., and Tomlinson, H. (1970). Ozone uptake by bean leaves. *Science* **169,** 79–80.

Robinson, E., and Robbins, R. C. (1969). "Sources, Abundance, and Fate of Gaseous Atmospheric Pollutants Supplement," SRI Project PR-6755. Stanford Res. Inst., Menlo Park, California.

Robinson, E., and Robbins, R. C. (1971). "Emissions, Concentrations, and Fate of Particulate Atmospheric Pollutants," SRI Project SCC-8507. Stanford Res. Inst., Menlo Park, California.

Shimwell, D. W. (1972). "The Description and Classification of Vegetation." Univ. of Washington Press, Seattle.

Sutton, O. G. (1953). "Micrometeorology." McGraw-Hill, New York.

Thomas, M. D., Hendricks, R. H., and Hill, G. R. (1950). Sulfur metabolism of plants. *Ind. Eng. Chem.* **42,** 1–12.

Waggoner, P. E. (1971). Plants and polluted air. *BioScience* **21,** 455–459.

Waggoner, P. E., and Reifsnyder, W. E. (1968). Simulation of the temperature, humidity and evaporation profiles in a leaf canopy. *J. Appl. Meteorol.* **7,** 400–409.

Weinstock, B. (1969). Carbon monoxide: Residence time in the atmosphere. *Science* **166,** 224–225.

13

INTERACTION OF AIR POLLUTANTS AND PLANT DISEASES

Michael Treshow

I. Introduction

While the role of air pollutants in causing plant injury or disorders has long been documented and established (Darley and Middleton, 1966; Treshow, 1970, 1971; Jacobson and Hill, 1970), the interaction of these pollutants with the fungus, bacterial, and virus pathogens causing disease has received limited attention (Heagle, 1973). Yet, air pollutants, together with climatic and soil relations, might well provide one more environmental component influencing the development of plant disease. If air pollutants, such as sulfur dioxide, ozone, or fluoride, were directly toxic to biotic agents causing disease, they might possibly impair their growth or reproduction and thereby, partially or wholly, control the diseases they induce. On the other hand, pollutants might have the greater effect on the host, modifying its physiology, weakening it, and rendering it more susceptible to infection. Furthermore, pollutants may injure host leaf or fruit tissues causing necrotic lesions that would provide courts of entry for certain pathogens, thereby enhancing infection and disease development. Conversely, the presence of some diseases might modify the plant sensitivity to a pollutant.

As Brennan and Leone suggest (1970), two viewpoints must be considered: (1) How the disease affects the response of the plant to the pollutant, and (2) how the pollutant modifies the progress of the disease, as for instance, by modifying the external chemistry of the host, altering the quantity or quality of host exudates, inducing changes in exterior barriers to infection, or altering the chemical environment of the host surface. Another major mode of modification would be in modifying the virulence of pathogens.

It has also been suggested that air pollutants might modify disease susceptibility by impairing mycorrhizal formation, thereby affecting host nutrition and host defense against root pathogens (Parmeter and Cobb, 1972).

Most subtly, pollutant–pathogen interaction might involve neither the pathogen nor host alone, but rather a more obscure action upon their relationship. Such especially might be the case where pollutants influence such obligate parasites as virus, rust, or powdery mildew pathogens.

In the long run, air pollutants might conceivably influence plant diseases to such a degree as to alter the species composition and density in the plant community. Fungi, bacteria, and viruses are all important in the equilibrium of plant communities and the determination of their ultimate population (Treshow, 1968). Any factor influencing the ecology of such microorganisms and viruses would have a secondary impact on the vigor, stability, and composition of populations in the natural plant community. In this way, air pollutants may serve as a vital force determining the composition of forests and other natural communities (Treshow, 1968; Lighthart *et al.,* 1971).

II. Impact of Pollutants on Pathogens

A. Sulfur Dioxide

Sulfur was among the first pesticides known, and sulfur dioxide was an early fumigant used to control transit and storage diseases. As early as

1932, Tompkins reported that sulfur dioxide at a concentration of 900 ppm impaired germination and growth of *Trichoderma lignarum*. At concentrations of about 1000–2500 ppm for 30 minutes, SO_2 has been used routinely in California to control decay of vinifera table grapes caused by *Botrytis cinerea* (Nelson, 1958). The treatment kills fungi on or near the surface of the fruit, although it does not eradicate disease organisms once they have become established. SO_2 suppresses spore germination of *Botrytis* largely proportional to its concentration and the exposure time, but other environmental factors are also important. For instance, the toxicity of SO_2 increases with increasing relative humidity, being 20 times as toxic at 96% relative humidity (RH) as 75%. A concentration of 100 ppm SO_2 for 12 minutes reduced germination at 96% RH; however, lower concentrations had little effect even at the high humidity and temperature. The toxicity of SO_2 has been found to increase about 1.5 times for each 10°C rise in temperature from 1–30°C (Couey and Uota, 1961).

McCallan and Weedon (1940) studied the response of fungi and bacteria to a number of chemicals including chlorine, ammonia, and sulfur dioxide. The plant pathogens investigated were *Sclerotinia fructicola, Ceratostomella ulmi, Glomerella cingulata, Macrosporium, Pestalotia stellata, Botrytis* sp., *Rhizoctonia tuliparum,* and *Sclerotium delphinii.* The pathogens were cultured on potato dextrose agar at 21°C and exposed to the gases when approximately 60 hours old, then transferred to slants to observe growth. Exposures were for 1, 4, 15, 60, 240, and 960 minutes at concentrations of 1, 4, 16, 63, 250, and 1000 ppm.

A reciprocal relation was found between time and concentration for a given response. SO_2 and Cl_2 were the most toxic chemicals. SO_2 was toxic to *Rhizoctonia* at 4 ppm at the 960 minutes exposure. Chlorine and ammonia were only toxic above about 63 ppm at 960 minute exposures. *Sclerotium delphinii* and *Rhizoctonia tuliparum* were the most sensitive, and *Ceratostomella ulmi* was the most resistant. Spores of fungi were 5–6 times more sensitive than mycelia which were more sensitive than sclerotia.

Couey (1965) found that an SO_2 concentration of 100–400 ppm was required to impair germination of *Alternaria* sp. spores. More recently, Ham (1971) found that *Schirrhia acicola* grew normally and produced viable conidia on agar after exposure to 1 ppm SO_2 for 4 hours. Saunders (1971) discussed the selective sterilization of leaf surfaces by SO_2 as being similar to that caused by some fungicides and insecticides because it altered the composition of the microflora and microfauna by eliminating the sensitive members of the community. He found that growth of *Aspergillus niger, Alternaria brassicola,* and *Didymellina macrospora* was not affected, but a *Penicillium* sp. was slightly stimulated in nutrient solutions that contained the equivalent of 90 ppm SO_2.

B. Fluoride

The fungitoxic, preservative properties of fluoride have been recognized since at least the nineteenth century when fluoride compounds were used to impregnate greenhouse lumber to preserve it against the attack of wood decay fungi. But the concentrations of fluoride used were exceedingly high, and far greater than any foliar accumulations ever recorded (Steinherz, 1939). Consequently, data on the disease interaction, or suppression, at such concentrations have little relevance to ambient air concentrations, or concentrations of fluoride that might accumulate in host tissues. Presumably, if foliar fluoride concentrations reached a threshold above which fungus or virus development were impaired, the severity of any disease associated with the pathogen might be reduced.

The response of representative fungi to fluoride was studied by Treshow (1965) who grew the fungi in Czapek's agar into which sodium fluoride had been incorporated at concentrations ranging from 1×10^{-4} to 5×10^{-2} *M*. Fungi studied included *Alternaria oleraceae, Botrytis cinerea, Colletotrichum lindemuthianum, Cytospora rubescens, Helminthesporium sativum, Pythium debaryanum,* and *Verticillium albo-atrum*. The minimum amount of fluoride required to inhibit growth ranged from 5×10^{-4} *M* for *Pythium* to as high as 1×10^{-2} *M* for *Verticillium* and *Helminthesporium*. This would be roughly equivalent to leaf concentrations of 22–420 ppm, respectively (calculated as NaF/leaf fresh weight). Spore production of *Botrytis* was markedly inhibited, but no such effects were noted on other species. Growth of the fungi was stimulated by low fluoride concentrations so that the earliest response was one of enhanced growth above which growth was inhibited. Growth of *Botrytis* and *Colletotrichum* species was accelerated by NaF concentrations below 1×10^{-3} *M*. Growth of the later species was also stimulated significantly at 1×10^{-4} *M*. The growth inhibiting and stimulating properties of sodium fluoride have also been demonstrated by Leslie and Parberry (1972). They found that an isolate of *Verticillium lecanis* grew well until NaF concentrations in agar exceeded 0.015 *M*, above which the conidia were larger than those in the controls or agar containing less NaF.

C. Ozone

The capacity of ozone to inhibit aerial growth of fungi has been considered as a possible tool for controlling spoilage of fruits and vegetables in storage since at least 1924 (Hartman 1924), and much of the work concerning pollutant–disease interaction has dealt with this aspect.

Hartman found that ozone concentrations of 0.5–1.0 ppm were decidedly inhibitory toward nearly all forms of microorganisms although 6500 ppm were required to be germicidal. This seems extremely high however, especially in light of subsequent work that demonstrated the toxicity of far lower concentrations.

Elford and van den Ende (1942) found that 0.04 ppm (4 pphm) ozone was able to inactivate certain bacteria when they were present as unprotected singleton aerosol particles.

Bacterial inhibition by ozone was also reported by Serat *et al.* (1967) who recorded decreases in the rate of luminescence in cells of a species of luminescent bacteria exposed to ozone. Davis (1961) found that ozone was lethal to *Escherichia coli* in concentrations of 0.1–0.5 μg/ml in aqueous solution, killing 63–99.5% of the population in 1 minute. Haines (1936) found that over 200 ppm ozone in the air was needed to sterilize bacteria on nutrient agar.

While Baker (1933) reported that ozone had little effect on controlling citrus decays, Klotz (1936) found ozone at least partially inhibited germination and growth on agar of blue and green mold caused by penicillium rots of citrus fruits.

Smack and Wilson (1941) found that 0.6 ppm ozone killed spores of *Penicillium expansum* and *Sclerotinia fructicola* either wet or dry. Watson (1943) tested the effects of ozone at 0.9–3.0 ppm on *Penicillium expansum, Sclerotinia fructicola,* and *Macrosporium sarcinaeforme* by placing the spores in drops on the surface of apples and allowing them to dry. Approximately 2 hours of exposure reduced germination significantly at temperatures from 3°–34°C.

The use of ozone as a supplement and deterrent to surface molds was also studied by Schomer and McColloch (1948) who found that air containing maximum concentrations of 1.95 ppm ozone failed to reduce decay. While exposure to 3.25 ppm for 8 hours per day for 7 months of storage killed airborne spores reducing the amount of inoculum, the amount of decay was not suppressed. Rather, infection was sometimes worse due to the necrotic fleck injury ozone caused in the fruit, providing infection courts.

Spores of *Botrytis cinerea* and *Penicillium expansum* causing blue rot mold were fumigated on glass slide cultures. One hour exposures to 1.95 ppm ozone caused a definite reduction in spore germination. Four 1-hour exposures reduced germination of *Botrytis* from 98.4–1.6%. Germination of *Penicillium* was reduced from 48–0% after 4 daily 1 hour exposures.

Because of its significance as a major air pollutant, more recent attention has been directed toward the possible interaction of ozone with plant diseases. Rich and Tomlinson (1968) exposed *Alternaria solani* grown on filter paper at concentrations of 0.1 and 1.0 ppm for 4 hours and 2 hours, respectively. Both treatments stopped conidiophore elongation, caused the apical cells to swell, and often collapsed the cell wall at the tips of the conidiophores.

If sporulating cultures were exposed to ozone at 10 ppm for 30 minutes, their conidia began to germinate within 3 hours while still attached to the conidiophores. Ozone was also found to reduce the sulfhydryl content of the fungal mat significantly (Rich and Tomlinson, 1968).

The possible effects of ozone on mycelial growth, spore production, spore germination, and lipid metabolism was studied by Treshow *et al* (1969). They exposed *Helminthesporium sativum, Colletotrichum lindemuthianum,* and *Alternaria oleracea* grown on 4% potato dextrose agar in open petri dishes to ozone at 10, 40, or 60 pphm for 4 hour periods. Cultures were exposed daily until the mycelia reached the edge of the dish, then placed in an incubator held at 25°C. Spores were collected from each dish and a suspension of them germinated on 2% water agar.

Radial growth of *Colletotrichum* was reduced 6–15% even at 10 pphm, but the effect on the kind of growth was most pronounced in that during fumigation, aerial growth was inhibited, closely appressed to the substrate, and largely confined within the agar (Kormelink, 1967; Kuss, 1950). Between fumigations, the more characteristic aerial growth was resumed. Ozone at 10 pphm also inhibited development of the normal reddish pigmentation in *Colletotrichum*. If control cultures were exposed to ozone, the color disappeared within a few hours. Growth of the other species was not affected significantly. Hibben and Stotsky (1969) also found that fungus colonies maintained in an ozone atmosphere had abnormal growth characteristics. Specifically, colonies that developed from ozonated spores grew appressed to and into the agar, rather than growing aerially as was normal.

Ozonated cultures were found to have about 28% lower lipid content than controls. This was postulated to have been due to the destruction of the sulfhydryl groups vital to lipid synthesis as discussed by Rich and Tomlinson (1968).

Spore production was affected more markedly. However, spore formation was both stimulated or inhibited depending on the species and ozone concentrations. Ozone inhibited sporulation of *C. lindemuthianum* 40, 45, and 60% at concentrations of 10, 40, and 60 pphm. Sporulation of *H. sativum* was unaffected, while sporulation of *A. oleraceae* was stimulated up to 2500%. There appeared to be no affect on the viability of the spores. Richards (1949) also found that ozone stimulated fungus sporulation at concentrations below those at which inhibition occurred.

Hibben and Stotsky (1969) reported that fumigation of detached spores of 14 fungi with 10–100 pphm ozone for 1, 2, and 6 hour periods reduced the germination in relation to the spore size. Large, pigmented spores of *Chaetomium* sp., *Stemphylium sarcinaeforme, S. loti,* and *Alternaria* sp. were insensitive to 100 pphm. Germination of *Trichoderma viride, Aspergillus terreus, A. niger, Penicillium egyptiacum, Botrytis allii,* and *Rhizopus stolonifera* spores was reduced by ozone concentrations above 50 pphm. The small hyaline spores of *Fusarium oxysporum, Colletotrichum lagenarium, Verticillium albo-atrum,* and *V. dahliae* were most sensitive to ozone, and germination was reduced by 50 pphm, and occasionally doses of 25 pphm, for 4 to 6 hours. Lower doses of ozone sometimes stimulated spore germination. Ozone also delayed by several hours the emergence of germ tubes.

Kormelink (1967) found that mycelia of *Helminthosporium sativum* fumigated with 10–60 pphm ozone 4 hours on each of 3 successive days developed a marked increase in the production of globules of a highly refractive material. These globules seemed to fill the entire space of the cell leaving only a parietal band of cytoplasm. The spherical globules appearing in nonfumigated hyphae were smaller in size and number, and appeared in far fewer cells.

The number of conidiosphores produced by *Helminthosporium* and *Colletotrichum lindemuthianum* was significantly reduced with increasing ozone concentrations between 10 and 60 pphm. Kormelink also studied the percentage of spores that germinated after 24 hours incubation and found an 8–34% reduction in germination. Germination in the controls ranged from 75–95% for the three species studied while germination of fumigated spores ranged from 50–85%. The response of *Colletotrichum* was most pronounced, and even 10 pphm for 4 hours caused a 33% reduction in spore germination. This coupled with the reduced spore production could theoretically limit the reproductive success of the organism.

A further idea of the general response to ozone of a broad spectrum of fungi can be obtained from studies by Kuss (1950) who grew 30 fungus species representative of the Phycomycetes, Ascomycetes, Basidiomycetes, and Fungi Imperfecti on agar in covered petri dishes and exposed them to ozone. Sporulation was increased in 3 *Alternaria* spp., 2 *Fusarium* spp., 2 *Glomerella* spp., and a *Helminthesporium* sp. All pycnidia—or perithecia—forming fungi tested produced more fruiting bodies in air containing ozone than in normal air. Germination was decreased in most species but increased in *Diaporthe phaseolorum* and *Lenzites trabea.*

III. Impact on the Disease

The impact of pollutants on the disease itself has received some attention in recent years, but much of what is known is based on field observations in areas of intense pollution. Laboratory research has largely concerned situations where the pathogen was virtually inseparable from the host, such as with viruses, rusts, and powdery mildew fungi.

A. Sulfur Dioxide

It has been suggested that air pollutants may act as an ecological factor causing a stress that may predispose plants to infection by certain microbial pathogens (Treshow, 1968; Parmeter and Cobb, 1972). More often though, the reverse has been true where the pollutant has proved more toxic to the pathogen than its host. Such was the case where Schaeffer and Hedgecock (1955) examined several hundred trees in and outside of an area where smelter emission injury, presumably from SO_2, was severe and found marked difference in the prevalence of certain fungi between the two areas. Certain rusts *Melampsorella cerastii* and *Peridermium coloradense* causing rust witches broom, and species of *Phragmidium, Melampsora,* and *Gymnosporangium* were almost absent from the smelter zone although rather abundant in the surrounding areas. Dwarf mistletoe, usually common in Douglas fir and lodgepole pine in the Great Basin area, similarly was less prevalent near the smelter.

Wheat stem rust caused by *Puccinia graminis* was reported to be less prevalent in Sweden's industrialized Kvarntorp area than in surrounding nonindustrialized areas (Johansson, 1954).

Parasitic fungi generally appeared to be retarded or inhibited where SO_2 was present, especially where injury to forest trees and shrubs was greatest. This was chiefly noted with species of *Cronartium, Coleosporium, Melampsora, Peridermium, Pucciniastrum, Puccinia, Lophodermium, Hypoderma,* and *Hypodermella.* Two rusts, *Melampsora albertensis* on quaking aspen and *M. occidentalis* on black cottonwood, were absent in the area of greatest SO_2 injury but occurred sparsely where injury was moderate, and was abundant where injury was least. The same was true of *Pucciniastrum pustulatum* on grand and subalpine firs, *Coleosporium solidaginis* on lodgepole pine, and *Cronartium harknessii, C. comandrae,* and *Lophodermium pinastri* on ponderosa and lodgepole pines (Schaeffer and Hedgecock, 1955).

The occurrence of sporocarps, mainly of basidiomycetes, on healthy and SO_2 damaged conifers was determined in the Ore mountains of northern Czechoslovakia (Jancarik, 1961). Of the 40 species identified, 12 occurred where trees were slightly injured but were not found where trees were severely damaged. Six species occurred only on severely damaged trees. *Poria* sp., *Mycena* spp., *Schizophyllum commune,* and *Polyporus versicolor* occurred only in areas of slight SO_2 damage. Linzon (1958, 1966) reported that fewer white pine trees displayed heart rot or symptoms of blister rust (*Cronartium ribicola*) in the areas studied nearest a smelter at Sudbury, Ontario. However, he also found that the annual white pine mortality rate from unknown causes was three times higher nearest Sudbury than in the other areas. Concentrations of SO_2 in the area at the time were considered extremely high with approximately 6000 tons of SO_2 being emitted per day, and white pine was extensively damaged by fumes at distances up to 25 miles from the smelters.

Because of the possible effect of SO_2 in weakening and thereby predisposing trees to infection, one might expect a greater incidence of heart rots in areas of severe pollution injury. Schaeffer and Hedgecock (1955) determined the prevalence of heart rot in 180 lodgepole pine trees and 180 Douglas fir trees 100 or more years old growing near a Montana smelter. About 7% of the pines and 72% of the Douglas fir were infected in varying degrees by *Polyporus schweinitzii* and by *Fomes pini*. This was essentially comparable to that found in other Montana forests outside the area. On the other hand, *Armillaria mellea* on the roots and crown of the pines was most prevalent inside the area of SO_2 injury, and on weakened trees (Schaeffer and Hedgecock, 1955; Donaubauer, 1968). Other species of weakened conifers have also been reported to have been predisposed to *Armillaria* infection by SO_2 (Jancarik, 1961; Kudela and Novakova, 1962). This would be expected since the fungus is often most severe on weakened trees. Sinclair (1969) also suggested that trees may be rendered more susceptible to infection by *Armillaria*. Other wood-rotting fungi including *Glocophyllum abietinum, Trametes serialis,* and *Trametes heteromorpha* have also been found where trees were damaged by SO_2, while absent in nearby areas where trees were undamaged (Jancarik, 1961).

Aerial fungi would be particularly responsive to pollutants as found by Koch in 1935 who reported the absence of oak powdery mildew caused by *Microsphaera alni* near a paper mill in Austria. The incidence of other foliar diseases also has been diminished in areas of SO_2 pollution. These include *Hypodermella laricis, Lophodermium pinastri, Hypodermella* sp. (Schaeffer and Hedgecock, 1955), *Lophodermium juniperi, Rhytisma acerinum* (Barkman *et al.*, 1969), *Hysterium policore* (Skye, 1968), and *Venturia inaequalis* (Przybylski, 1967). Contrary to incidents of diminished infection, Jancarik (1961) found a higher incidence of *Lophodermium piceae* on spruce needles injured by SO_2. Chiba and Tanaka (1968) similarly found an increased incidence of needle blight of Japanese red pine caused by the weakly pathogenic *Rhizosphaera kalkhoffii* in an industrial area of Japan. The fungus was able only to infect injured needles. Hibben and Walker (1966) observed that lilacs grown in the polluted air of New York City and other urban areas often showed substantially less infection by the powdery mildew fungus *Microsphaera alni* than lilacs in rural areas. However, it was not ascertained whether ozone or SO_2 or both were implicated. However, they did find that SO_2 at 0.3–0.5 ppm for 72 hours decreased spore germination and disease development beyond the appressorium stage.

Rather than impair fungus development, sulfur dioxide might conversely facilitate infection of plants by aerial pathogens by its effect on increasing the stomatal aperture (Unsworth *et al.,* 1972; Williams *et al.,* 1971). Alteration in leaf acidity by acid rains might also influence predisposition to various pathogens (F. H. Borman, personal communication). The presence of sulfur dioxide resulted in more severe symptoms of *Rhizosphaera kalkhoffii* on *Pinus densiflora* (Chiba and Tanaka, 1968). *Lophodermium* on juniper and *Rhytisma* on maple may also be more severe following exposure to SO_2 (Barkman *et al.,* 1969).

B. Fluoride

Studies attempting to learn if any relationship existed between fluoride concentrations in plant foliage and the activity of viruses or other pathogens have been more limited. In one of the earlier studies of such possible relationships, Dean and Treshow (1965) placed young pinto beans in the dark for 24–48 hours after which they were removed. Primary leaves were cut off and inoculated with tobacco mosaic virus (TMV) and placed on nutrient agar in which NaF had been incorporated at concentrations ranging from 1×10^{-4} to 5×10^{-2} *M*. One-half of each leaf was placed on NaF agar; the other half on agar into which the NaF had been omitted.

Lesions were counted after 48–72 hours and leaves analyzed for fluoride content. The number of lesions was calculated as a percent of the control. Virus activity, as determined by local lesion numbers, was generally stimulated as foliar fluoride levels increased to about 300 ppm. Lesion numbers then decreased to the number in the control group at about 500 ppm above which virus activity decreased below control values.

The effect of fluoride on TMV-induced local lesions developing on intact pinto beans was also studied (Treshow *et al.,* 1967). Plants were grown in fumigation chambers in which atmospheric fluoride concentrations ranged from 0.2–2.5 $\mu g/m^3$ and foliar fluoride concentrations from 20–637 ppm. The number of lesions increased with foliar fluoride concentrations up to 500 ppm, above which the number decreased.

The tremendous variation encountered in lesion development, and the striking influence of environmental factors, such as light, temperature, and moisture, make it highly improbable that any measurable modifications of TMV activity could be detected in the field.

The only studies reported involving the interaction of bacterial or fungus disease with fluorides were those of McCune *et al.* (1973). When tender green beans were grown in controlled environment chambers during certain times of the year, plants exposed to filtered air became severely infected with powdery mildew while those exposed to HF at 7 or 10 $\mu g/m^3$ were only mildly infected. The unifoliate leaves of HF-treated plants contained 399 ppm fluoride.

Similar results were found in unpublished studies by Treshow. Chrysanthemum plants were grown in containers and half of the plants were exposed to HF at approximately 2 $\mu g/m^3$, 4 hours/day until leaf concentrations reached 350–400 ppm. After about 2 weeks, a severe mildew infection occurred on the unfumigated control plants grown in identical but filtered chambers. Plants high in fluoride had almost no mildew.

Fluoride also reduced the number of uredia per leaf of plants inoculated with bean rust (McCune *et al.,* 1973). There was a significant reduction in the severity of disease on plants exposed to HF before inoculation. It has been observed that the birch leaf rust, *Melampsoridium betulinum* was absent from birches around an aluminum plant in Norway even in 1964 when the disease was widespread over the region (Barkman *et al.,* 1969).

However, fluoride showed no significant effect on the incidence or severity of halo blight injury caused by *Pseudomonas phaseolicolus* (McCune *et al.,* 1973).

Exposing tomato plants to HF before inoculating the leaves with *Alternaria solani* causing early blight produced a significant effect. HF decreased injury when plants were exposed before inoculation, but there was no effect of exposure following inoculation. The severity of late blight, caused by *Phytophthora infestens,* was unaffected by fluoride (McCune *et al.,* 1793).

C. Ozone

The impact of ozone specifically on disease was studied by Ewell (1938) who reported ozone to be the agent most commonly used to prevent mold and bacteria in storage. The ozone concentrations used, in the range of 0.6 ppm, while higher than concentrations present in the atmosphere, were found to kill spores of *Monilinia fructicola* and *Penicillium expansum* in 3–4 hours (Smack and Wilson, 1941). However, reports of the bacteriostatic and fungistatic effectiveness of ozone have been conflicting. Spalding (1966) reported that ozone did not reduce spoilage of strawberries or peaches held at 60°F and 90% relative humidity, but did inhibit the surface growth of fungi and eliminated the typical nesting appearance.

Ridley and Sims (1966) concluded that ozone reduced disease development in strawberries and peaches held at 35°F, although their results with strawberries were not statistically significant. In studies with *Monilinia fructicola* and *Rhizopus* sp., Ridley and Sims (1967) placed inoculated peach fruits in refrigerated storage rooms into which ozone was introduced at 0, 0.25, 0.50, and 1.0 ppm. Ozone significantly reduced the overall percentage of fruit with rots, as well as the spread of the rots when exposed to greater than 0.50 ppm ozone.

Spalding (1968) fumigated fruits and vegetables in a storage room with 50 pphm ozone. Peaches were inoculated with the brown rot fungus, *Monilinia fructicola,* or spores of the rhizopus rot fungus *Rhizopus stolonifera* and then placed in the air and ozone rooms at 35°F for 7 days. They were then examined for rot, and held in air at 70°F for 2 and 4 days to determine the shelf life. Blueberries, cantaloupes, grapes, and green beans were also studied for decay responses.

Although ozone suppressed aerial growth of the fungi, it did not appreciably inhibit subsurface growth or rot. Nor did ozone significantly reduce the percentage of *Botrytis* or *Rhizopus* rot of strawberries during storage.

Cantaloupes developed the same amount of rot regardless of the presence or absence of ozone during storage. While surface growth of *Penicillium* spp. and *Alternaria tenuis* was impaired, there was no inhibition of subsurface development and rot. Gray mold rot of grape caused by *Botrytis cinerea* responded similarly. Ozone inhibited only aerial growth of the fungus. Ozone also failed to suppress *Alternaria* rot of bean, and furthermore injured 20% of the beans.

The possibility of using ozone to detoxify air in cold storage rooms for flowers, specifically to control *Botrytis* on gladiolus, also has been considered (Magie, 1960, 1963). Ozone treatment reduced stem and sheath infections to 18% of those on untreated plants. A $2\frac{1}{2}$ hour ozone exposure reduced petal infection to 12% of that on the untreated plants. There was no reduction in flower quality following 2–3 hour exposures, but no data were provided as to the concentrations used.

Manning *et al.* (1970a) exposed geranium cultivars to 2, 35, and 55 pphm ozone for 4 hour periods. Fully opened flower heads were sprayed with a spore suspension of *Botrytis cinerea* immediately before exposure to ozone. Ozone at 35 pphm did not injure the plants or prevent infection, but did inhibit pathogenesis at the lower inoculum levels, and natural infection of noninoculated flowers. Ozone at 55 pphm caused moderate injury but did not prevent infection.

Manning *et al.* (1972) later studied the effects of *Botrytis cinerea* and ozone on the bracts and flowers of poinsetta (*Euphorbia pulcherrima*). Flowers and bracts were sprayed with spores of *B. cinerea* immediately before exposure to ozone at 15, 25, 35, and 45 pphm for 4 hours. No visible increase or decrease in disease incidence occurred as a result of being exposed to ozone before incubation, and more significantly, 45 pphm ozone injured the flowers.

Since *Botrytis* typically penetrates such injured, necrotic tissue, or even senescent tissue, lesions produced by ozone toxicity would provide probable infection courts leading to increased infection. Manning *et al.* (1970b) exposed sensitive geranium plants to ozone at concentrations of 7–10 pphm for 10 hours/day. The ozone-injured leaves, as well as non-injured controls, were subsequently sprayed with suspensions of *B. cinerea* and inspected for leaf infection after 24, 48, 72, and 96 hr. Extensive infection of leaves by *B. cinerea* occurred only on those leaves with visible ozone injury.

A comparable situation was observed in the field when (Manning *et al.*, 1969) potato plants showing ozone injury were also severely blighted by *Botrytis cinerea.* When Manning inoculated ozone-injured and noninjured potato leaves with *B. cinerea,* infection was more rapid, and disease development more severe, on ozone-injured leaves. Infection was frequently observed to originate in ozone-injured leaf areas.

Cabbage seedlings planted in soil infested with *F. oxysporum* f. sp. *conglutinans* and exposed to 10 pphm ozone, 8 hours/day for 10 weeks had slightly, but not significantly, less disease (Manning *et al.*, 1971b). Perhaps the most serious aspect of pollution–disease interaction is the greater sensitivity of plants to weak pathogens caused by visible injury from the pollutant.

Ozone-induced leaf injury favors infection by weak pathogens of native plant species as demonstrated by Costonis and Sinclair (1967, 1972). *Lophodermia pinastri* and *Pullularia pullulans* were most frequently associated with needle injury of Eastern White pine in inoculation studies in which plants were fumigated with 7.0 pphm O_3 for 4.5 hours. *Lophodermia pinastri* colonized up to 6% of both fumigated and untreated attached ozone-resistant needles; 23.5% of fumigated and 10% of untreated O_3-sensitive attached needles; 32 and 40%, respectively, of detached sensitive and resistant needles naturally exposed to ozone; and 20% of fumigated O_3-sensitive, detached needles.

Rather than enhance infection, some protective mechanism also may be afforded by ozone. Yarwood and Middleton (1954) observed that bean and sunflower leaves infected with rust (*Uromyces phaseoli* or *Puccinia helianthi*) were less injured by natural or artificial smog created by combining gasoline vapors with ozone than otherwise similar, noninfected leaves. Even when heat was used to kill the rust mycelium, the rusted tissue remained resistant to smog injury. They considered that the resistance to smog injury was due to some substance which diffused beyond the limits of the rust mycelium. Similarly, Heagle and Key (1973b) found that wheat leaves inoculated with urediospores of *Puccinia graminis* f. sp. *tritici,* showed significantly less ozone injury than noninoculated leaves. The protective effect was present in inoculated areas when plants were exposed to 30 parts per hundred million (pphm) ozone 2, 3, or 4 hours after the start of urediospore incubation. Mesophyll cells beneath stomata with appressoria attached were rarely injured. They suggested that a diffusible substance was produced by germinating spores and infection structures that protect local areas of tissue from ozone injury. A similar protective agent may be produced by bacteria since chlorotic areas surrounding lesions of *Pseudomonas phaseolicola* on kidney bean leaves were not injured by ozone exposures that severely injured adjacent areas of the leaves.

The effect of mildew on ozone sensitivity was studied by C. R. Hibben (personal communication) who found that lilac mildew caused a zone of ozone resistance around the lesions. He attributed this to the higher concentration of cytokinins present around the lesions together with a greater phenol concentration, but the possible mechanisms of action of these substances were not discussed.

Heagle (1973) found that in the field, the peanut leaf spot fungus, *Cercospora arachidicola,* protected localized areas around fungus lesions from ozone injury. Similar localized protection was described around broad bean lesions caused by *B. cinerea* (Magdycz and Manning, 1973) and kidney bean infected with *Pseudomonas phaeolicola* (Kerr and Reinert, 1968), and pinto bean infected with TMV (Yarwood and Middleton, 1959).

The most prominent relationship may exist between ozone and virus infection, but not much work has been done along this line. In one such interaction study, Brennan and Leone (1969) sought to determine the effect of toxic concentrations of ozone on the development of tobacco mosaic virus. Virus-free tobacco plants and plants infected with tobacco mosaic virus were exposed to ozone fumigations at 30 pphm for 3–6 hours in an environmental chamber. During the winter months, typical symptoms of ozone toxicity consistently developed on the virus-free plants but never on those infected with TMV. However, during the summer months, both virus-free and virus-infected plants were equally damaged by ozone. During the summer months a higher ozone concentration of 40 pphm was required to injure the *Nicotiana sylvestris,* and perhaps the virus could not overcome the effects of this higher concentration. Secondly, mosaic type viruses cause less severe injury in the summer, and the resultant changes may not have been sufficient to suppress the ozone symptoms.

R. A. Reinhart (personal communication) also studied the possible impact of TMV on ozone susceptibility, but found no effect. However, he found that ozone sensitivity was substantially increased when Burley tobacco plants were infected with tobacco streak or tobacco etch viruses. Potato virus Y also increased ozone sensitivity while tobacco vein banding virus had no effect.

Biochemical data showed that ozone produced a response in healthy tobacco leaves that did not occur in virus-infected leaves. Leone *et al.* (1966) earlier found that high nitrogen content of a leaf appeared to be related to resistance to ozone. Tobacco mosaic virus-infected leaves had a higher nitrogen content than noninoculated ones. The lower carbohydrate content of infected leaves may also have provided some protection against ozone, since the concentration of sugar in the leaf is associated with the plant's response to ozone. Many other biochemical changes are also induced in the host tissue as a result of virus infection (Goodman *et al.*, 1967), including changes in phosphorus, organic acid, and phenol metabolism and elaboration of growth substances.

The effect of ozone on the severity of TMV was also studied (Brennan and Leone, 1970) on both tobacco and pinto bean. Groups of plants were first inoculated with TMV and then exposed to ozone following 0, 3, 24, or 48 hours. Treated plants were exposed to 10 pphm O_3 for 3 hours, and control plants maintained in filtered air. The number of local lesions on bean leaves was counted after 4 days. Bean leaves ozonated either immediately or 48 hours after inoculation developed the same number of lesions as the untreated plants. However, plants ozonated 3 or 24 hours after inoculation developed 19 and 16% more lesions than plants kept in filtered air.

Apparently the sensitivity of the host to ozone, the concentration and duration of ozone exposure, the specific virus, and the phase of growth at which ozonation occurs must all the considered in virus–air pollutant interactions.

Any effects of ozone on plant disease might be expected to be greatest on those fungi having the greatest exposure to the atmosphere. Such would be the case with the powdery mildews whose growth is largely external of the host.

Scheutte (1971), studying the primary infection stages of powdery mildew, found that ozone suppressed the development of infection of barley and reduced the establishment of a functional host–parasite relationship. Scheutte obtained about 93% germination in both control groups and at ozone concentrations up to 100 pphm ozone for 4 hours. The formation of mature appressoria was delayed but not suppressed by 100 pphm ozone following an 8 hour exposure.

However, sensitivity to ozone increased as the parasitic population proceeded into the penetration stage of infection. This was most clearly exhibited by the inability of the pathogen to penetrate the host and form functional secondary hyphae. The degree of response was dependent on the ozone concentration and exposure time above 25 pphm for 4 hours. Penetration normally occurs 8–13 hours following inoculation, and it is during this period that ozone exposure is most critical to infection.

Conidia developing on sporulating lesions were visibly injured following exposure to 25 or 50 pphm ozone for 8 hours. Morphological changes consisted of pre-germination and plasmolysis of the spores. Pregerminating spores began to germinate before conidia were detached from the conidiophore and failed to develop normally. Rich and Tomlinson (1968) also indicated ozone stimulated the germination of attached conidia. Plasmolyzed spores failed entirely to develop.

A significant reduction was found in the average percentage of functional secondary hyphae formed in 48 hours after exposure to 25 or 50 pphm O_3; 15 pphm had no effect. Inhibition by ozone depended on the O_3 concentration in contact with the pathogen and on the sensitivity of the infection stage occurring at the time of ozone exposure.

Heagle and Strickland (1972) also studied the interaction of ozone on powdery mildew of barley using ozone concentrations often occurring in ambient air. Plants were exposed to 0, 5, 10, or 15 pphm O_3 for 6 hours on the 7th, 8th, and 9th days after inoculation. In a second experiment, infected plants were exposed for 6 hours on 8 consecutive days after inoculation. Other plants were exposed to ozone immediately following inoculation to study its effects on spore germination.

Mildew colonies were significantly smaller on exposed plants than on nonexposed plants after two exposures. However, after subsequent exposures, the colonies were usually significantly larger on exposed plants.

Ozone did not significantly affect the number of conidiophores or the number of conidiophores bearing conidia. However, the percentage of spores that successfully infected host cells was reduced when maturing or germinating spores were exposed to ozone. Heagle and Strickland attributed this to either the easier ozone penetration through conidial papillae or immature cell walls, toxic effects unique to these stages, or a combination of these factors.

The interaction of ozone with the crown rust fungus (*Puccinia coronata* var. *avenae*) of oats was studied by Heagle (1970). Plants were exposed to 10 pphm ozone for 6 hours daily during the light period on the 12 days following inoculation. Uninoculated plants were placed in a similar, but filtered air chamber. After 10 days, pustules on plants exposed to ozone were significantly smaller than on nonexposed leaves. Ozone injury was limited to a slight flecking over less than 1% of the leaf surface over all the plants.

Urediospores on ozonated plants germinated as well, produced as many appressoria and as much infection as spores formed on nonexposed leaves. However, it appeared that only half as many spores were produced on plants exposed to ozone, thereby theoretically reducing the inoculum potential.

Heagle and Key (1973a) studied the effect of ozone on the wheat stem rust fungus. Wheat plants were inoculated, incubated, and immediately exposed to 0, 6, 12, or 24 pphm O_3 for 6 hours/day for 4 days. Ozone tended to inhibit hyphal growth after 3 and 4 exposures.

When plants were inoculated after 24 or 48 hours exposure to ozone, and when injury symptoms were present, less penetration and infection occurred on exposed than unexposed plants.

Ozone significantly reduced the number of urediospores produced at all ozone concentrations tested; however, spore germination was not affected. But by reducing spore formation, ozone could conceivably reduce the inoculum potential of cereal rusts in areas of elevated ozone concentrations.

D. Particulates

In one of the few studies concerning particulate pollution, Schönbeck (1960) dusted a field planting of sugar beets biweekly at the rate of 2.5 g/M^2 and found that infection by the leaf-spotting fungus, *Cercospora beticola* was significantly greater than in nondusted plots. He postulated that the lime dust altered the physiological balance and increased the plant's susceptibility to infection.

Sassafras and wild grape plants continuously exposed to emissions of limestone dust in southwestern Virginia were more susceptible to infection by *Guignardia bidwellii* and *Gleosporium* sp., the causal agents of leaf spot disease of grape and many other species (Manning, 1971). Dusty grape and sassafrass leaves also had far more, but not kinds, of fungi and bacteria when compared to clean leaves. Bacteria were greatly reduced in number on dusty hemlock leaves while fungi were increased. Streptomyces were absent and bacteria and fungi drastically reduced in kind and number on grape leaves heavily encrusted with dust.

Sharp (1967) found that high ion concentrations in the atmosphere reduced germination of *Puccinia striiformis* urediospores, but the significance of this was not determined.

IV. Modes of Action

What mechanisms might account for the effects air pollutants exert on the disease process? Most obvious are the direct effects on the pathogen.

Such is certainly the case when high concentrations of sulfur dioxide or ozone impair the development of a parasite. The toxicity of SO_2 has been demonstrated with oak mildew and rose black leaf spot in the field. The mode of action of sulfur toxicity is presumed to be related to the effect of SO_2 in upsetting the balance of oxidized and reduced sulfur in the cell.

Saunders (1970) suggests that the marked sensitivity of lichens and other epiphytes to atmospheric pollution, notably sulfur, may be due partly to the possession of highly effective mechanisms for accumulating and concentrating such substances from very dilute solutions, and partly to the reliance of lichens upon nutrients derived from substances precipitated by rainwater. The acidification of rainwater and the nutrient solution on the lichen may also be important to the toxicity.

Since fungi vary markedly in their sensitivity to SO_2 (Saunders, 1966) and other pollutants, considerable variation similarly would be expected in the response of the disease, and disease–pollutant interaction.

The potential impact of ozone is more conjectural and can only be implied from laboratory research. But such work provides a basis for speculating on how ozone might affect a parasite. E. Brennan (personal communication) relates the interaction of ozone and TMV to the effect of the pollutant on RNA production and the permeability of the host cell.

Rich and Tomlinson (1968) concluded that the injury of conidiophores by 10 pphm ozone, if repeated as is common in Connecticut, may have a decided effect on development of the fungus. Ozone might stimulate spore germination either by destroying a germination inhibitor or altering the permeability of the cell wall to gases. The sulfhydryl changes induced by ozone may oxidize the sulfhydryl compounds vital to normal extension of the conidiophores. Nickerson and van Rij (1949) have demonstrated the importance of sulfhydryl groups in cell division of a yeast.

Weakening and strengthening of the cell wall in yeast is related to the activity of a protein disulfide reductase, which controls the $SH \rightleftharpoons S—S$ linkage in the polysaccharide-protein cell wall components (Nickerson and Falcone, 1956). Scheutte (1971) suggests that such enzymatically controlled changes may be necessary for the emergence of the penetration peg in infection by powdery mildew fungi. Sulfhydryl functional groups are susceptible to oxidation by ozone in fungi (Rich and Tomlinson, 1968), and may be oxidized to disulfide linkages thus inhibiting the formation of a penetration peg. Since sulfhydryl groups are active sites of many enzymes, as well as potential cross-linking structural groups, irreversible oxidation may inactivate enzymes or other reduced components in the parasite. Ozone caused no significant reduction in the percentage of secondary hyphae formed (Scheutte, 1971), and once the penetration was completed and the penetration peg was established in the host cell, the fungus became less sensitive to ozone.

Ozone may also act on the exposed surface of the conidia, or may penetrate the cell wall and membrane and react with vital endogenous constituents, but this appears to be questionable. Rather, the increased plasmolysis of ozonated mycelia indicated that the molecular architecture of the cell wall and/or membrane was affected. Changes in permeability may have been caused by the direct oxidation of membrane lipids such as the unsaturated fatty acid components, oxidation of membrane SH proteins, or through an inhibition of lipid synthesis. Sulfhydryl enzymes essential for fatty acid and lipid synthesis may be inhibited or destroyed by ozone (Treshow *et al.,* 1969; Tomlinson and Rich, 1969). Inhibition of lipid synthesis may affect membrane integrity and reduce lipid components of the conidial wall. Observations by Scheutte (1971) that presumed viable conidia could form mature appressoria, but were unable to infect a host, indicated that ozone may have affected endogenous constituents necessary to complete infection.

Regarding the pregermination of fungus spores exposed to ozone, Rich and Tomlinson (1968) suggested this may be caused by ozone destroying a germination inhibitor or by altering the permeability of the cell wall membrane to gases.

Ozone can also modify the chemical nature of the host as in causing an increase in the phenolic substances (Howell, 1970; Menser and Chaplin, 1969) or peroxidase activity (Curtis and Howell, 1971; Dass and Weaver, 1968). The decreased infection and invasion by rust fungi found in leaves exposed to ozone may be related to increases in phenol concentration and/or the activation of oxidative enzymes by ozone causing accumulation of quinones.

Another mode of action might be indirect where the pathogen was affected by structural, physiological or metabolic changes in the host tissues. Such is especially the case with weaker parasites, or facultative saprophytes, which favor weakened host tissues. Heagle and Key (1973a) believe that ozone affects the rust fungi primarily through its effect on the host mesophyll tissue.

Perhaps the most significant mechanism of air pollutant–disease interaction, is where the pollutant actually injures the tissues producing foliar lesions that facilitate entry by certain fungi or bacteria. *Botrytis cinerea* especially has been noted to infect such tissues under field conditions.

Conversely, certain pollutants at low concentrations appear to stimulate the metabolic activity of the host and thereby favor such parasites as develop most vigorously in active tissue. This would include especially viruses and obligate parasites. We have seen that tobacco mosaic virus development is thus enhanced by low but elevated fluoride concentrations. Also, the inducement of pregermination of certain fungi by ozone might increase the inoculum potential. However, the negative effects of ozone in impairing appressoria formation would tend to counter this, and on a field basis, any effects would be unlikely.

The effect of fluoride, where the development of powdery mildew is impaired, might exemplify direct inhibition of the parasite. Treshow (1965) demonstrated the tremendous variation existing in sensitivity of fungi to fluoride, and it is conceivable that some species might be more sensitive than the flowering plants. Thus, fungal growth might be inhibited at foliar fluoride concentrations entirely harmless to the host plant. In any host–parasite interaction where the pathogen is more sensitive than the host, some suppression of the disease would be expected. In one study attempting to explain the action of fluoride directly on fungi, Lal and Bhargava (1962) found than NaF in nutrient agar decreased sugar utilization and formation of itaconic acid by *Aspergillus terreus*. This effect was reversed when pyruvate was added to the substrate, leading to the theory that NaF inhibits the normal glycolytic pathway as a step in the conversion of pyruvate to itaconic acid. This is similar to some of the effects of fluoride on higher plants, and it is likely that the mode of action of fluoride generally would be much the same as in the phanerogams.

Treshow *et al.* (1967) postulated that the modification of lesion numbers caused by fluorides was due largely to the effects of fluoride on the physiological activity of the host, since fluoride is known to stimulate host metabolism and growth at the same general concentrations that were found to stimulate virus activity.

V. Conclusions

While fungal responses to air pollutants can be demonstrated under the controlled regimen of the laboratory, we should most significantly ask whether such air pollutant–disease interactions actually occur in the field. Where pollutant concentrations are sufficiently high to impair development of a fungus, yet remain below the pollutant injury threshold of the host plant, some degree of disease control might be expected. Such was the case where oak mildew (*Micosphaera alni* var. *quercina*) was completely absent in the area downwind from a paper mill (Koch, 1935). The fungus was widespread in the surrounding areas where the sulfurous gases were absent. This was among the first reports of an air pollutant modifying disease severity.

In another instance, James (1965) observed that blackspot of roses was absent in areas of high atmospheric pollution and attributed this to the presence of SO_2. Later, Saunders observed that the disease was checked or completely eliminated in areas where the SO_2 concentrations exceeded 100 $\mu g/m^3$ (Saunders, 1966). An inverse relation existed between SO_2 concentrations in the same concentration range and lower, and infection near urban pollution sources. Saunders found that germination of conidia of *Diplocaryon rosae* was inhibited in a solution made by dissolving 35 ppm SO_2 in water. Inhibition was permanent after 3 days. Gaseous SO_2 in air at 100 $\mu g/m^3$ markedly reduced infection of rose leaves after exposure for 2 days. Lower concentrations, however, tended to stimulate infection. Sulfur dioxide suppression of disease apparently occurs only at concentrations that are rarely found today. With the improved control of SO_2 emissions, such incidents are likely to be still less important in the future.

Fluoride emissions also are no longer sufficiently high to expect much if any interaction with disease. The limited studies available suggest that while fluoride limits the development of certain diseases, it does so only at concentrations over 10 times as high as would be expected in the field even near major fluoride pollution sources. Also the effects under field conditions are likely to be obscured by the many more prominent and significant factors influencing disease development.

The potential impact of ozone might be greater since prospects for controlling this pollutant are less promising. However, ozone is not nearly as fungitoxic as SO_2 so more is required to influence either fungi or flowering plants. Furthermore, only certain stages of infection appear to be at all sensitive to ozone. With ozone–powdery mildew interaction, for instance, the developmental sequence of infection is continual, and under field conditions, various developmental stages of the parasite are occurring at different times for the parasitic population. Since only specific stages of infection are suppressed by ozone, only a small segment of the parasitic population would be at a susceptible stage during an acute exposure period. Regardless of the ozone concentration applied, a major percentage of the population would be in a tolerant stage and remain functional. However, with the reduction in numbers of viable conidia, some reduction in the inoculum potential might theoretically be possible.

One of the most significant ramifications of pollutant–disease interaction may occur where ozone injury is enhanced by previous infection by certain viruses. In the long run though, no true comparisons should be made between pollutant–disease interaction in the laboratory and field. Pollutant uptake as well as the response may be quite different in a field situation.

Studies of air pollutant–disease interaction are few. The limited field observations and research only indicate what might happen in the case of a few diseases and pollutants. Interaction is likely to be greatest where high concentrations of a pollutant accumulate in the host tissue, as with fluoride, or where a major amount of the pathogen is exposed to the atmosphere, as with the powdery mildews; and interaction has been demonstrated in these instances when pollutant concentrations were sufficiently high.

The limited interaction even under such conditions though suggests that effects at lower pollutant concentrations, or on internal pathogens are unlikely. Realistically, the only probable pollutant–disease interaction is where the pollutant physically injures the host tissue to where significant infection courts are provided. The physiological changes induced by pollutants that might influence disease development occur only at rather prolonged exposures to pollutant concentrations now infrequent or unlikely, and then might be expected only on the most pollutant-sensitive plant species.

Secondary effects of pollutants such as SO_2 through their impact on the soil and soil flora might be more significant. Sobotka (1964) for instance, found a more diverse array of soil fungi in polluted areas. Conversely though, Mrkva and Grunda (1969) found fewer fungi. They also found that the numbers of bacteria in the soil were reduced, although not the types, near a smelter where the soil pH and soil potassium and magnesium were decreased. The potential effects on plant pathogenic bacteria, or the numbers of species present, were not discussed. Sobotka also reported that mycorrhizal associations in spruce were abnormal in polluted areas with the Hartig net sometimes hypertrophic and the mantle often thinner, tuft-like, or even absent where trees were injured by SO_2. Thus the secondary effects of SO_2 might be considerable.

Studies by Manning and Papia (1972) have shown that ozone can stimulate mycoflora on leaves and roots either directly or indirectly. Long exposure to low ozone doses of 6 pphm for 8 hours each of 28 days injured leaves of pinto bean plants, and more fungus propagules developed. Ozone caused similar stimulatory effects on fungi of the uninjured root surface (Manning *et al.,* 1971a). About 20% more colonies of fungi were isolated from roots of plants with ozone injured leaves than on uninjured plants.

The effect of ozone on the root microflora was also demonstrated by Reinert *et al.* (1971). Exposure of soybean plants to 12–15 pphm ozone for 15 days over a 3-week period decreased the number of *Rhizobium* root nodules. Lower doses of 6 pphm, 8 hours/day for 20–60 days decreased the number, size and weight of *Rhizobium* nodules on pinto bean roots (Manning *et al.,* 1972). Tingey and Blum (1973) found that a single, 1 hour exposure to 75 pphm ozone decreased the number and size of nodules on soybean plants. They also reported that ozone decreased the leghemoglobin content in plant roots and postulated that if ozone so drastically affected the relationship between legume roots and *Rhizobium*, other organisms might probably be affected similarly.

Nothing is yet known of possible synergistic mechanisms where two or more pollutants are present in the atmosphere. Combined effects of ozone and sulfur dioxide have been demonstrated (Menser and Heggestad, 1966; Applegate and Durrant, 1969; Dochinger and Bender, 1971), and SO_2 and NO_x (Tingey *et al.*, 1971; White, 1973) on a few species of agronomic plants, but no work has been conducted on the response of fungi, bacteria or viruses, or pollutant disease interaction involving such combinations. However, field observations of diseases in areas where such combinations exist have failed to reveal any changes in the incidence of endemic diseases.

References

Applegate, H. A., and Durrant, L. C. (1969). Synergistic action of ozone–sulfur dioxide on peanuts. *Environ. Sci. Technol.* **3,** 759–760.

Baker, C. E. (1933). Effect of ozone upon apples in cold storage. *Ice Refrig.* **84,** 402–404.

Barkman, J. J., Rose, F., and Westhoff, V. (1969). Discussion in Section 5: The effects of air pollution on non-vascular plants. *In Air Pollut., Proc. Eur. Congr. Influence Air Pollut. Plants & 1st, 1968,* pp. 237–241.

Brennan, E., and Leone, I. A. (1969). Suppression of ozone toxicity symptoms in virus-infected tobacco. *Phytopathology* **59,** 263–264.

Brennan, E., and Leone, I. A. (1970). Interaction of tobacco mosaic virus and ozone in *Nicotiana sylvestris*. *J. Air Pollut. Contr. Ass.* **20,** 470.

Chiba, O., and Tanaka, K. (1968). The effect of sulfur dioxide on the development of pine needle blight caused by *Rhizosphaera kalkhoffii* Bubak (L.) *J. Jap. Forest. Soc.* **50,** 135–139.

Costonis, A. C., and Sinclair, W. A. (1967). Effects of *Lophodermium pinastri* and *Pullularia pullulans* on healthy and ozone injured needles of *Pinus strobus*. *Phytopathology* **57,** 807 (abstr.).

Costonis, A. C., and Sinclair, W. A. (1972). Susceptibility of healthy and ozone-injured needles of *Pinus strobus* to invasion by *Lophodermium pinastri* and *Aureobasidium pullulans*. *Eur. J. Forest Pathol.* **2,** 65–73.

Couey, H. M. (1965). Inhibition of germination of Alternaria spores by sulfur dioxide under various moisture conditions. *Phytopathology* **55,** 525–527.

Couey, H. M., and Uota, M. (1961). Effect of concentration, exposure time, temperature, and relative humidity on the toxicity of sulfur dioxide to the spores of *Botrytis cinerea*. *Phytopathology* **51,** 739–814.

Curtis, C. R., and Howell, R. K. (1971). Increases in peroxidase isoenzyme activity in bean leaves exposed to low doses of ozone. *Phytopathology* **61,** 1306–1307.

Darley, E. F., and Middleton, J. T. (1966). Problems of air pollution in plant pathology. *Annu. Rev. Phytopathol.* **4,** 103–118.

Dass, H. C., and Weaver, G. M. (1968). Modification of ozone damage to *Phaseolus vulgaris* by antioxidants, thiols and sulfhydryl reagents. *Can. J. Plant Sci.* **48,** 569–574.

Davis, I. (1961). Microscopic studies with ozone-quantitative lethality of ozone for *Escherichia coli. U.S. Air Force Aerosp. Med. Cent., Rep.* **61-54.**

Dean, G., and Treshow, M. (1965). Effects of fluoride on the virulence of tobacco mosaic virus *in vitro. Utah Acad. Sci., Arts Lett., Proc.* **42,** 236–239.

Dochinger, L. S., and Bender, F. W. (1971). Chlorotic dwarf of Eastern white pine caused by ozone and sulfur dioxide interaction. *Nature (London)* **225,** 476.

Donaubauer, E. (1968). Sek und ävschäden in Osterreichischen Rauch schadensgebieten. Schwierigkeiter Der diagnose und Bezuertung. *Niedzynarodowei Konf. Wpylw Zanieczyszczen Powietrza na Lasy, 6th,* Katowice, Poland.

Elford, W. J., and van den Ende, J. (1942). An investigation of the merits of ozone as an aerial disinfectant. *J. Hyg.* **42,** 240–265.

Ewell, A. W. (1938). Present use and future prospects of ozone in food storage. *Food Res.* **3,** 101–108.

Goodman, R. N., Kiroly, Z., and Zaitlin, M. (1967). "The Biochemistry and Physiology of Infections of Plant Disease." Van Nostrand-Reinhold, Princeton, New Jersey.

Haines, R. B. (1936). The effect of pure ozone on bacteria. *Gt. Brit., Dep. Sci. Ind. Res., Food Invest. Bd., Spec. Rep.* No. 1935, pp. 30–31.

Ham, D. L. (1971). The biological interactions of sulfur dioxide and *Schirrhia acicola* on loblolly pine. Ph.D. Thesis, Duke University, Durham, North Carolina.

Hartman, F. E. (1924). The industrial application of ozone. *J. Amer. Soc. Heat. Vent. Eng.* **30,** 711–727.

Heagle, A. S. (1970). Effect of low-level ozone fumigations on crown rust of oats. *Phytopathology* **60,** 252–254.

Heagle, A. S. (1973). Interactions between air pollutants and plant parasities. *Annu. Rev. Phytopathol.* **11,** 365–388.

Heagle, A. S., and Key, L. (1973a). Effect of ozone on the wheat stem rust fungus. *Phytopathology* **62,** 397–400.

Heagle, A. S., and Key, L. (1973b). Effect of *Puccinia graminis* f. sp. *tritici* on ozone injury in wheat. *Phytopathology* **63,** 609–613.

Heagle, A. S., and Strickland, A. (1972). Reaction of *Erisiphe graminis* f. sp. *hordei* to lower levels of ozone. *Phytopathology* **62,** 1144–1148.

Hibben, C. R. (1966). Sensitivity of fungal spores to sulfur dioxide and ozone. *Phytopathology* **56,** 880 (abstr.).

Hibben, C. R., and Stotsky, G. (1969). Effects of ozone on the germination of fungus spores. *Can. J. Microbiol.* **15,** 1187–1196.

Hibben, C. R., and Walker, J. T. (1966). A leaf roll-necrosis complex of lilacs in an urban environment. *Proc. Amer. Soc. Hort. Sci.* **89,** 636–642.

Howell, R. K. (1970). Influence of air pollution on quantities of caffeic acid isolated from leaves of *Phaseolus vulgaris. Phytopathology* **60,** 1626–1629.

Jacobson, J. S., and Hill, A. C. (1970). "Recognition of Air Pollution Injury to Vegetation: A Pictorial Atlas," Tech. Rep. No. 1. Air Pollut. Contr. Ass., Pittsburgh, Philadelphia.

James, R. (1965). Prevention of rose diseases. *Gardeners Chron.* **157,** 452.

Jancarik, V. (1961). Výskyt drevokaznych hub v kourem poskozovani oblasti Krusných hor. *Lesnictvi* **7,** 667–692.

Johansson, O. (1954). "Rapport över ett studium av luft och nederbörd omkring Svenska Klifferolje Aktie bologets anlaggningar vid Kvarhtorp Med Specioll hänsyn till tistributionen av svavaloch des inverkan pä växtarna Lic. dvh vid K." Lantbruk Shögskolar, Uppsala.

Kerr, E. O., and Reinert, R. A. (1968). The responses of bean to ozone as related to infection by *Pseudomonas phaseolicola*. *Phytopathology* **58,** 1055 (abstr.).

Klotz, L. J. (1936). Nitrogen trichloride and other gases as fungicides. *Hilgardia* **10,** 27–52.

Koch, G. (1935). Eichenmehltau und Rauchgasschaden. *Z. Pflanzenkr.* **14**, 44–45.

Kormelink, J. R. (1967). Effects of ozone on fungi. M.S. Thesis, University of Utah, Salt Lake City.

Kudela, M., and Novakova, E. (1962). Lesní skudci a skody zveri v lesich poskozovaných Kourem. *Lesnictvi* **6,** 493–502.

Kuss, F. R. (1950). The effect of ozone on fungus sporulation. M.Sc. Thesis, University of New Hampshire,

Lal, M., and Bhargava, P. M. (1962). Reversal by pyruvate of fluoride inhibition in *Aspergillus terreus*. *Biochim. Biophys. Acta* **68,** 628–630.

Leone, I. A., Brennan, E., and Daines, R. H. (1966). Effect of nitrogen nutrition on the response of tobacco to ozone in the atmosphere. *J. Air Pollut. Contr. Ass.* **16,** 191–196.

Leslie, R., and Parberry, D. G. (1972). Growth of *Verticillium lecanii* on medium containing sodium fluoride. *Trans. Brit. Mycol. Soc.* **58,** 351–352.

Lighthart, B., Hiatt, V. E., and Rossano, A. T., Jr. (1971). The survival of airborne *Serratia marcescens* in urban concentration of sulfur dioxide. *J. Air Pollut. Contr. Ass.* **21,** 639–642.

Linzon, S. N. (1958). The influence of smelter fumes on the growth of white pine in the Sudbury region. *Ont., Dep. Lands Forests, Can. Dep. Agr. Pub.* pp. 1–45.

Linzon, S. N. (1966). Damage to eastern white pine by sulfur dioxide, semimature tissue needle blight and ozone. *J. Air Pollut. Contr. Ass.* **16,** 140–144.

McCallan, S. E. A., Weedon, F. R. (1940) Toxcicity of ammonia, chlorine, hydrogen cyanide, hydrogen sulphide, and sulphur dioxide gases. II. Fungi and bacteria. *Contrib. Boyce Thomp. Inst.* **11,** 331–342.

McCune, D. C., Weinstein, L. H., Mancini, J. F., and van Leuken, P. (1973). Effects of hydrogen fluoride on plant-pathogen interactions. *Proc. Int. Clean Air Congr., Dusseldorf, 1973* (in press).

Magdycz, W. P., and Manning, W. J. (1973). *Botrytis cinerea* protects broad beans against visible ozone injury. *Phytopathology* **63** (in press).

Magie, R. O. (1960). Controlling gladiolus botrytis bud rot with ozone gas. *Proc. Fla. State Hort. Soc.* **73,** 373–375.

Magie, R. O. (1963). Botrytis disease control on gladiolus, carnations, and chrysanthemums. *Proc. Fla. State Hort. Soc.* **76,** 458–461.

Manning, W. J. (1971). Effects of limestone dust on leaf condition, foliar disease incidence, and leaf surface microflora of natural plants. *Environ. Pollut.* **2,** 69–76.

Manning, W. J., and Papia, P. M. (1972). Influence of long-term low levels of ozone on the leaf surface mycoflora of pinto bean plants. *Phytopathology* **62,** 497 (abstr.).

Manning, W. J., Feder, W. A., Perkins, I., and Glickman, M. (1969). Ozone injury and infection of potato leaves by *Botrytis cinerea*. *Plant Dis. Rep.* **53,** 691–693.

Manning, W. J., Feder, W. A., and Perkins, I. (1970a). Ozone and infection of geranium flowers by *Botrytis cinerea*. *Phytopathology* **60,** 1302 (abstr.).

Manning, W. J., Feder, W. A., and Perkins, I. (1970b). Ozone injury increases infection of geranium leaves by *Botrytis cinerea*. *Phytopathology* **60,** 669–670.

Manning, W. J., Feder, W. A., Papia, P. M., and Perkins, I. (1971a). Influence of foliar ozone injury on root development and root surface fungi of pinto bean plants. *Environ. Pollut.* **1,** 305–312.

Manning, W. J., Feder, W. A., Papia, P. M., and Perkins, I. (1971b). Effect of low levels of ozone on growth and susceptibility of cabbage plants to *Fusarium oxysporum* f. sp. *conglutinans*. *Plant Dis. Rep.* **55,** 47–49.

Manning, W. J., Feder, W. A., and Perkins, I. (1972). Effects of *Botrytis* and ozone on bracts and flowers of Poinsetta cultivars. *Plant Dis. Rep.* **56,** 814–816.

Menser, H. A., and Chaplin, S. F. (1969). Air pollution: Effects on the phenol and alkaloid content of cured tobacco leaves. *Tob. Sci.* **13,** 169–170.

Menser, H. A., and Heggestad, H. E. (1966). Ozone and sulfur dioxide synergism: Injury to tobacco plants. *Science* **156,** 424–425.

Mrkva, R., and Grunda, B. (1969). Einfluss von immisionen auf due walkböden und ihre mikroflora im gebiet von südmahren. *Acta Univ. Agr., Brno, Fac. Silvicult.* **38,** 247–270.

Nelson, K. E. (1958). Some studies of the action of sulfur dioxide in the control of Botrytis rot of Tokay grapes. *Proc. Amer. Soc. Hort. Sci.* **71,** 183–189.

Nickerson, W. J., and Falcone, G. (1956). Enzymatic reduction of disulfide bonds in cell wall protein of baker's yeast. *Science* **142,** 315–319.

Nickerson, W. J., and van Rij, N. J. W. (1949). The effect of sulfhydryl compounds, penicillin and cobalt on the cell division mechanisms of yeasts. *Biochem. Biophys. Acta* **3,** 461–475.

Parmeter, J. R., and Cobb, F. W., Jr. (1972). Long-term impingement of aerobiology systems on plant production systems. *In* "US/IBP Aerobiology Program Handbook" No. 2, pp. 61–68. Univ. of Michigan Press, Ann Arbor.

Przybylski, Z. (1967). Results of observation of the effect of SO_2, SO_3 and H_2SO_4 on fruit trees, and some harmful insects near the sulfur mine and sulfur processing plant at Machow near Tarnobrzeg. *Postepy Nauk Roln* **2,** 111–118.

Reinert, R. A., Tingey, D. T., and Coons, C. E. (1971). The early growth of soybean as influenced by ozone stress. *Agronomy* **63,** 148 (abstr.).

Rich, S., and Tomlinson, H. (1968). Effects of ozone on conidiophores and conidia of *Alternaria solani*. *Phytopathology* **58,** 444–446.

Richards, M. C. (1949). Ozone as a stimulant for fungus sporulation. *Phytopathology* **39,** 20 (abstr.).

Ridley, J. D., and Sims, E. T., Jr. (1966). Preliminary investigations on the use of ozone to extend the shelf life and maintain the market quality of peaches and strawberries. *S. C., Agr. Exp. Sta., Res. Ser.* **70,** 1–22.

Ridley, J. D., and Sims, E. T., Jr. (1967). The response of peaches to ozone during storage. *S. C., Agr. Exp. Sta., Tech. Bull.* **1027,** 1–24.

Saunders, P. J. W. (1966). The toxicity of sulfur dioxide to *Diplocaryon rosae* Wolf causing blackspot of roses. *Ann. Appl. Biol.* **58,** 103–114.

Saunders, P. J. W. (1970). Air pollution in relation to lichens and fungi. *Lichenologist* **4,** 337–349.

Saunders, P. J. W. (1971). Modification of the leaf surface microorganisms. *In* "Ecology of Leaf Surface Microorganisms" (T. F. Preece, and C. H. Dickinson, eds.), pp. 81–89. Academic Press, New York.

Schaeffer, T. C., and Hedgecock, G. G. (1955). Injury to Northwestern forest trees by sulfur dioxide from smelters. *U.S., Dep. Agr., Tech. Bull.* **1117,** 1–49.

Scheutte, L. R. (1971). Response of the primary infection process of *Erysiphe graminis* f. sp. *hordei* to ozone. Ph.D. Thesis, University of Utah, Salt Lake City.

Schomer, H. A., and McColloch, L. P. (1948). Ozone in relation to storage of apples. *U.S., Dep. Agr., Circ.* **765.**

Schönbeck, H. (1960). Beobachtungen zur frage des Einflusses von industriellen immissionen auf die Krankheitsbereitchaft der Pflanze. *Ber. Landesanst. Bodennutzungsschutz* **1,** 89–98.

Serat, W. F., Budinger, F. E., Jr., Mueller, P. K. (1966). Toxicity evaluation of air pollutants by use of luminescent bacteria. *Atmos. Environ.* **1,** 21–32.

Sharp, E. L. (1967). Atmospheric ions and germination of uredospores of *Puccinia struformis. Science* **156,** 1359–1360.

Sinclair, W. A. (1969). Polluted air: Potent new selective force in forests. *J. Forest.* **67,** 305–309.

Skye, E. (1968). Lichens and air pollution. *Acta Phytogeogr. Suec.* **52,** 1–23.

Smack, R. M., and Wilson, R. D. (1941). Ozone in apple storage. *Refrig. Eng.* **42,** 97–101.

Sobotka, A. (1964). Vliv prumyslovych exhalátu na pudni Zivenu Smrkovjch porostu Krusnych hor. *Les. Cas.* **10,** 987–1002.

Spalding, D. H. (1966). Appearance and decay of strawberries, peaches, and lettuce treated with ozone. *U.S., Dep. Agr., Mktg. Res. Rep.* **756.**

Spalding, D. H. (1968). Effects of ozone atmospheres on spoilage of fruits and vegetables. *U.S., Dep. Agr., Mktg. Res. Rep.* **801.**

Steinherz, D. (1939). Fluoride compounds as wood preservatives. A review of methods and application. *Can. Chem Process Ind.* **23,** 601.

Tingey, D. T., and Blum U. (1973). Effects of ozone on soybean nodules. *J. Environ. Qual.* (in press).

Tingey, D. T., Reinert, R. A., Dunning, J. A., and Heck, W. W. (1971). Vegetation injury from the interaction of nitrogen dioxide and sulfur dioxide. *Phytopathology* **61,** 1506–1511.

Tomkins, R. G. (1932). The action of certain volatile substances and gases on the growth of mould fungi. *Proc. Roy. Soc., Ser. B* **111,** 210–226.

Tomlinson, H. and Rich, S. 1969. Relating lipid content and fatty acid synthesis to ozone injury of tobacco leaves. *Phytopathol.* **59,** 1284–1286.

Treshow, M. (1965). Response of some pathogenic fungi to sodium fluoride. *Mycologia* **57,** 216–221.

Treshow, M. (1968). Impact of air pollutants on plant populations. *Phytopathology* **58,** 1103–1113.

Treshow, M., Dean, G., and Harner, F. M. (1967). Stimulation of tobacco mosaic virus-induced lesions on bean by fluoride. *Phytopathology* **57,** 756–758.

Treshow, M. (1970). "Environment and Plant Response." McGraw-Hill, New York.

Treshow, M. (1971). Fluorides as air pollutants affecting plants. *Annu. Rev. Phytopathol.* **9,** 21–44.

Treshow, M., Harner, F. M., Price, H. E., and Kormelink, J. R. (1969). Effects of ozone on growth, lipid metabolism and sporulation of fungi. *Phytopathology* **59,** 1223–1225.

Unsworth, M. H., Biscae, P. V., and Pinckey, H. R. (1972). Stomatal response to sulfur dioxide. *Nature* (*London*) **239,** 458–459.

Watson, R. O. (1943). Some factors influencing the toxicity of ozone to fungi in cold storage. *Refrig. Eng.* **46,** 103–106.

White, K. (1973). Synergistic inhibition of apparent photosynthetic rate of alfalfa by combinations of SO_2 and NO_2. M.S. Thesis, University of Utah, Salt Lake City.

Williams, R. J., Lloyd, M. M., and Ricks, G. R. (1971). Effects of atmospheric pollution on deciduous woodland. I. Some effects on leaves of *Quercus petraea* (Mattuschba) Leibl. *Environ. Pollut.* **3,** 57–68.

Yarwood, C. E., and Middleton, J. T. (1954). Smog injury and rust infection. *Plant Physiol.* **29,** 393–395.

Yarwood, C. E., and Middleton, J. T. (1959). Virus infection and heating reduce smog damage. *Plant Dis. Rep.* **43,** 129–130.

14

INTERACTIONS OF AIR POLLUTION AND AGRICULTURAL PRACTICES

Saul Rich

I. Introduction

Agricultural practices can interact with air pollution effects on plants either by changing the susceptibility of plants to the pollutants, or by generating airborne compounds or materials that can reduce the value of affected crops. The effect of air pollution on plants is not necessarily limited to crop injury. For example, volatile nitrogenous compounds can enrich lakes and reservoirs and increase rates of eutrophication. The discussion of agricultural practices can be restricted to the immediate production of crops or it can include the production of forests and forest products, cotton linters as well as cotton bolls, greenhouse crops, food processing, and finally the production of agricultural chemicals.

Cultural practices that can change crop susceptibility are fertilizer application, pesticide application, irrigation, kind and cultivars of crops planted, and even time of planting. The effect of fertilizers may be influenced by their composition and time of application. Growth regulators, fungicides, and nematicides can change the sensitivity of crops to ozone. The effect of irrigation on susceptibility of plants to air pollution can be modified by proper timing. The choice of crops or cultivars to be planted could depend on the frequency with which a particular air pollutant is present in a particular area. Crops sensitive to the air pollutant would either not be grown in the area or the growers would be forced to use resistant cultivars if they were available. Certain crops could be planted either early or late in order to avoid periods with the greatest frequency of air-pollution episodes.

Air pollutants that can affect plant populations may be produced by a number of agricultural practices. Airborne pesticides such as certain formulations of 2,4-dichlorophenoxyacetic acid (2,4-D) can injure crops far from the points of application. Drifting insecticides can kill honeybees and eliminate apple crops by preventing cross-pollination. If not used with proper care, fumigants used to control soil-borne pathogens can cause injury to adjacent crops. Fertilizers applied to neutral soil can produce ammonia that will volatilize and eventually add to the nitrogenous burden of eutrophying lakes. The manufacture of superphosphate for fertilizer mixes is a well-known source of fluorides that damage adjacent crops. Deliberate burning of agricultural wastes or crop residues can produce combustion products that affect plants. Dusts and particulate matter in sufficient quantity to reduce photosynthesis can be produced by land preparation or crop harvesting. Finally, there is the persistent observation that secondary pollutants such as ozone can be produced photochemically from volatile hydrocarbons that are present in the air above forests.

II. Cultural Practices

A. Plant Nutrition

Farmers may be able to control injury from air pollutants by carefully controlling the kinds and amounts of fertilizers they apply to plants. The ability to do this can result only from specific information about the reaction of specific crops to particular air pollutants.

Heck *et al.* (1965) concluded that plants grown at low fertility levels are more sensitive to air pollutants than plants grown at higher fertility rates. However, they also observed that high nitrogen levels increase sensitivity. This general observation appears to be contradicted by Menser and Street (1962) who found that tobacco (*Nicotiana tabacum* L.) in the field is more sensitive to ozone when the nitrogen supply is restricted. Leone *et al.* (1966) found that optimum rates of nitrogen increased ozone injury to *Nicotiana rustica* L. Dunning *et al.* (1972) reported that low rates of potassium increased ozone injury to both beans (*Phaseolus vulgaris* L.) and soybeans (*Glycine max* L.). Leone and Brennan (1970) made tomatoes (*Lycopersicon esculentum*) more sensitive to ozone by increasing application of phosphate. Adedipe *et al.* (1972) found that increasing sulfur nutrition decreased the ozone sensitivity of beans. Brewer *et al.* (1961) reported that spinach becomes more sensitive to ozone at high nitrogen levels. However, this effect could be largely prevented by concomitant increases in both potash and phosphorus. That application of fertilizers can be used practically to protect plants from ozone has been demonstrated by Will and Skelly (1974). They found that eastern white pine (*Pinus strobus* L.) Christmas trees can be made resistant to ozone by autumn application of 25–9–9 fertilizer.

Leone and Brennan (1972) studied the effect of nitrogen and sulfur nutrition on the injury of tobacco and tomatoes by sulfur dioxide. They reported that optimum rates of these nutrients increased the sensitivity of these plants to sulfur dioxide. Excessive nitrogen decreased sensitivity, while excessive sulfur increased sensitivity.

Brennan *et al.* (1950) grew tomato plants at different rates of nitrogen, calcium, and phosphate and subjected them to soil applications of NaF and to HF fumigation. They reported that optimum levels of the 3 nutrients increased plant injury by both NaF and HF. Plants deficient in the 3 nutrients accumulated less fluorine. Excessive rates of nitrogen and calcium made the plants less sensitive to fluoride injury. Within the concentration range used, phosphorus applications had no effect on sensitivity of tomatoes to fluorides. MacLean *et al.* (1969) reported that foliar accumulation of fluoride in HF-fumigated tomatoes was suppressed by magnesium deficiency. Fluroide accumulation was unaffected by changes in calcium nutrition, but fluoride accumulation was increased by potassium deficiency. MacLean (1970) continued this study and found that foliar accumulation of fluoride was inhibited in plants given the lowest and highest rates of magnesium. However, magnesium-deficient tomatoes were more sensitive, and plants given excess magnesium were more resistant to HF injury than were plants growing at optimum magnesium nutrition. Pack (1966) grew tomatoes at both low and optimum rates of calcium nutrition and fumigated both with HF continuously through flowering and fruit development. He found that HF fumigation inhibited the mechanism by which calcium aids the fertilization of tomato blossoms. He also reported that fluoride injury to tomato foliage was most severe at the lower rate of calcium. Adams and Sulzbach (1961) noted that nitrogen-deficient beans were the only ones to show foliar injury in their HF fumigation tests, even though the foliage of plants deficient in potassium, calcium, nitrogen, and phosphorus accumulated higher concentrations of fluoride. McCune *et al.* (1966) grew 2 different cultivars of gladiolus (*Gladiolus hortulanus*) on solutions deficient in various nutrients. After subjecting the plants to weekly fumigations of HF, they found no effect of nutrient differences on fluoride accumulation within either cultivar. However, tipburn was most severe in the potassium- and phosphorus-deficient plants of one cultivar and in the magnesium-deficient plants of the other cultivar.

These examples illustrate that fertilizer composition, rate, and perhaps time of application can influence the sensitivity of crops to varius air pollutants.

B. Irrigation

Controlled application of water to a crop can influence its sensitivity to air pollutants. Irrigation does so in at least 3 ways: (1) by controlling water stress, (2) by changing relative humidity around leaves, and (3) by changing the soil–oxygen diffusion rates. By now it is well established that the first two factors influence sensitivity to air pollutants by affecting stomatal aperture.

A supply of soil moisture sufficient to minimize water stress increases the sensitivity of plants to air pollutants. This has been reported for ozone sensitivity of tobacco by Taylor *et al.* (1960), Walker and Vickery (1961), Menser and Street (1962), Dean and Davis (1967). This is also true for other plants as well (Seidman, 1963). Zimmerman and Hitchcock (1956) reported that wilting makes leaves resistant to both sulfur dioxide and to hydrogen fluoride. Wiebe and Poovaiah (1974) found that moisture stress before or during hydrogen fluoride fumigation of soybeans closed stomata, reduced fluoride uptake, and reduced injury. However, moisture stress immediately following hydrogen fluoride fumigation increased subsequent injury.

A relation between relative humidity and ozone sensitivity of plants has been reported by Leone and Brennan (1969). The drier the air, the more likely it is that stomata will close (Otto and Daines, 1969) resulting in increased resistance to ozone. Wilhour (1970) also found that high relative humidity increases the ozone sensitivity of white ash (*Fraxinus americana* L.). Daines *et al.* (1952) reported that plants absorbed more fluoride and were damaged more at high humidities than in drier air. MacLean *et al.* (1973) found that gladioli fumigated with hydrogen fluoride were more severely injured at 85% relative humidity (RH) than at either 65 or 50% RH, even though fluoride accumulation did not follow this exact pattern.

Rich and Turner (1972) demonstrated that stomatal behavior in the presence of ozone depends on moisture stress, relative humidity, and plant species or cultivar. The stomata of water-stressed beans closed faster in the presence of ozone than did the stomata of beans that were not water stressed. Also bean stomata were more sensitive to ozone in dry air than in moist air. A 30-minute exposure to ozone [25 parts per hundred million (pphm)] decreased stomatal aperture in dry air but not in moist air. The rate at which tobacco stomata closed, however, depended not only on plant–water relations but also on the cultivar. Stomata of both tobacco cultivars began to close at the same rate when ozone was present in moist air. In ozone-containing dry air, however, stomata of the ozone-resistant tobacco cultivar closed much faster than did those of the ozone-susceptible cultivar. Mansfield and Majernik (1970) observed yet a different effect. They found that at 42% RH the stomata of broad bean (*Vicia faba* L.) opened wider in the presence of sulfur dioxide than they did in its absence, but at 32% RH they opened less widely in the presence of sulfur dioxide than they did in its absence.

MacDowall (1966) examined the relation between free moisture on tobacco leaves and their sensitivity to ozone. He concluded that free moisture on the leaf surface tended to increase resistance to ozone by impeding the movement of gases into the leaves.

Water application can also change the diffusion of gases through the soil. Stolzy *et al.* (1961) varied the soil–diffusion rate of oxygen in the soil of containers in which tomato plants were growing. They found that both transpiration rate and ozone sensitivity of these plants varied directly with diffusion rate and partial pressure of oxygen. The faster the soil–diffusion rate, the faster the transpiration, and the greater the sensitivity of foliage to ozone in the air.

C. Resistant Crops

Growers can reduce losses from air pollutants by using resistant species or cultivars. Unless this is done, air pollution can change agriculture in certain areas. It is difficult to raise good crops of peaches or citrus, for example, in the vicinity of industrial plants that manufacture aluminum or superphosphate. In both cases sufficient fluorides are emitted to damage these sensitive crops (Benson, 1959); Brewer *et al.,* 1960). It is also difficult to raise greenhouse orchids in southern California and spinach in Connecticut. The ethylene coming from automobile exhausts in smoggy Los Angeles causes the dry sepal injury of orchids (Davidson, 1949). Because it is difficult to remove ethylene from the air entering greenhouses, orchid growers have had to move elsewhere. Spinach growing in Connecticut has also declined sharply, in part because spinach is so susceptible to ozone (Daines *et al.,* 1960).

Selection, breeding, and production of tobacco cultivars resistant to ozone have saved the important shade-tobacco industry of Connecticut. The tobacco grown under shade tents is used as the wrapper, or outside leaf of the more expensive cigars. To be of commerical value, these leaves must be unblemished. Ozone injury causes a flecking and spotting that persists through the processing and reduces the value of the final product. Before it was even known what caused such flecking and spotting, plant breeders in Connecticut had begun the slow process of producing plants that would resist this injury. Now all tobacco plants under the shade tents are genetically resistant to ozone injury.

Ozone-resistant cultivars have, of necessity, been sought in alfalfa (*Medicago sativa* L.) (Howell *et al.,* 1971), beans (*Phaseolus vulgaris* L.) (Davis and Kress, 1974), onions (*Allium cepa* L.) (Engle and Gabelman, 1966), safflower (*Carthamus tinctorius* L.) (Howell and Thomas, 1972), and other crops. Lists of plant species that differ in sensitivity to various air pollutants are presented in the publication edited by Jacobson and Hill (1970) and in Ryder (1973).

Growers can limit losses caused by air pollution by either growing crops that are less sensitive to air pollutants in their area, or by choosing cultivars that are known to resist injury by these pollutants.

III. Pesticides

Pesticides are used in agriculture as herbicides, insecticides, fungicides, and fumigants. Formulations in each class have caused crop losses as air pollutants.

Herbicides are used to kill weeds. In 1971, 428 million pounds of synthetic organic herbicides were produced in the United States nearly 38% of the national production of synthetic organic pesticides (Fowler and Mahan, 1973). The most widely used herbicides are 2,4-dichlorophenoxyacetic acid (2,4-D) and 2,4,5-trichlorophenoxyacetic acid (2,4,5-T).

Insecticides are used to control insects that reduce the yield or quality of crops. In recent years, ecological restrictions have been placed on the use of insecticides, and new and stricter federal and state regulations are now being formulated. Even though production of certain insecticides, such as DDT, aldrin, and dieldrin, is being terminated, the present discussion will include research on these compounds as illustrations of the aerial movement of insecticides. In 1971, 558 million pounds of synthetic organic insecticides were produced in the United States, about 49% of the total United States production of synthetic organic pesticides (Fowler and Mahan, 1973).

Fungicides are applied to plants or plant parts to protect them from molds, rots, blights, wilts, and other problems caused by fungi. Of the major classes of pesticides, they have caused the least injury as air pollutants. In 1971, 149 million pounds of synthetic organic fungicides were produced in the United States, about 13% of the total synthetic organic pesticide production in the nation (Fowler and Mahan, 1973).

Fumigants are volatile pesticides injected into soil, or into containers to rid them of insects, fungi, or nematodes. Because they are volatile, they can easily act as air pollutants. Consequently, they are used with great care to ensure that no live plants are nearby, or that treated soil is not planted before the fumigants have disappeared. The most important single use of fumigants is for control of nematodes in soil. Methyl bromide is the only fumigant for which production figures are available. In 1971, United States production of methyl bromide was about 24 million pounds (Fowler and Mahan, 1973).

Pesticides appear in the air as pollutants by drift of aerial or ground applications of spray or dust formulations, by volatilization, by the wind-borne movement of pesticide-treated soil particles, and by codistillation with water.

As air pollutants, pesticides can directly affect agriculture in at least 3 ways: (1) by phytotoxicity, (2) by preventing pollination, and (3) by changing the metabolism or chemical composition of plants so that they cannot be used safely.

A. Herbicides and Growth Regulators

Herbicides offer the most serious threat as phytotoxic pesticides. Because they are designed to inhibit the growth of some plants—weeds—they can produce significant economic losses when they are permitted to escape off-target. Akesson and Yates (1964) describe the legal difficulties produced by drifting 2,4-D after it came into widespread use in the early 1950's. They reported that 2,4-D drifting from a sprayed area injured sensitive crops 15 miles away, and concluded that such hazards led to the Miller Amendment of 1954.

The ubiquity of herbicides can be realized from the research of Adams *et al.* (1964), who were able to detect and identify 2,4-D and esters in the air near Pullman, Washington on more than 80% of 106 days sampled between May 4 and August 14, 1963. Sherwood *et al.* (1970) are convinced that there is sufficient 2,4-D in Iowa air to affect seriously the yield of sensitive crops such as grapes.

Day *et al.* (1959) examined the relative volatility of herbicides in the field. They carefully sprayed herbicides on enclosed cotton plots during the summer in southern California and observed the appearance of herbicides symptoms on adjacent plots after the enclosures were removed. Esters of 2,4-D injured plants over areas of adjacent plots roughly proportional to the amount of herbicide applied to the target plots. The acid formulation of 2,4-D was less volatile and 2,4-D alkanolamine and isopropylamine salts showed the least volatility. The phytotoxic vapors appeared to be generated during the heat of the day and not at night. There was no injury on plots adjacent to those sprayed with aminotriazole.

Flint *et al.* (1963) were able to separate 2,4-D esters into 2 classes of volatility. The less volatile esters are the butyoxyethanol, isooctyl, and the propylene glycol butyl ether, which are 10–20 times less volatile than the lower alkyl esters, such as isopropyl and *N*-butyl ester.

Other herbicides have also shown volatility. Gentner (1964) found that vapors of 4-amino-3,5,6-trichloropicolinic acid were more toxic to beans than were vapors of some esters and salts of 2,4-D. Kearney *et al.* (1964) measured the loss by volatility of seven different *S*-triazine herbicides from soil. The group included ametryne, atrazine, prometone, prometryne, propazine, and simazine. They found that relative herbicide volatility was affected by soil type and moisture content of the soil, and that not all of the compounds responded similarly. For example, prometone losses from the 5 soils that they studied were directly related to percent sand and inversely to percent clay and organic matter. Losses of atrazine and simazine were less influenced by these properties. Simazine was more quickly lost from dry soil than from wet soil, whereas prometryne and ametryne were more quickly lost from wet soil than from dry soil. The rate of propazine loss was not greatly influenced by soil moisture content. In a later study, Talbert *et al.* (1971) demonstrated that prometryne adsorbed more strongly to clay soils and hence was lost more slowly from heavy soils than from light soils. As one would expect, volatility increased with temperature. They also found that the kind of formulation also influenced loss by volatility. For example, prometryne was lost faster from wettable powders than from emulsifiable concentrates.

Loss of herbicides from the treated soil to the air above it is also affected by air movement. Danielson and Gentner (1964) studied the effect of air velocities of $\frac{3}{4}$ to 4 mph on the persistence of EPTC (ethyl *N,N*-di-*n*-propylthiolcarbamate) applied to the soil surface. They found that persistence was inversely related to air velocity. EPTC loss was more rapid from sand than from soil, and the technical material was lost more quickly than the commerical formulation.

Off-target movement of herbicides can cause direct losses by reducing crop yields. This can also cause agricultural losses by changing the metabolism of the affected crops so that they may become poisonous to livestock. Stahler and Whitehead (1950) reported that 2,4-D on sugar beets increased the potassium nitrate level of the foliage sufficiently to be dangerous to ruminants. Ruminant bacteria change nitrates to toxic nitrites. Swanson and Shaw (1954) found that 2,4-D increased the hydrocyanic acid and nitrate content of treated sudangrass to a level that could be dangerous to livestock. Frank and Grigsby (1957) tested the effect of 6 commonly used herbicides on nitrate accumulation by 14 different weed species. They reported significant increases of nitrate in *Eupatorium maculatum* L. and in *Impatiens biflora* treated with 4 of the herbicides. The danger is obvious.

Herbicides can also interact with other air pollutants. Hodgson (1970) reported that ozonation of corn (*Zea mays* L.) and sorghum (*Sorghum bicolor* L.) could alter the metabolism of atrazine in the fumigated leaves. Hodgson *et al.* (1973) also found that metabolism of diphenamid was altered in tomato plants exposed to ozone. Carney *et al.* (1973) studied phytotoxic interactions in plants treated with commonly used herbicides and exposed to ozone. The responses were additive, antagonistic, or synergistic. The last two responses are interesting. They found less injury on tobacco cultivars treated with benefin and ozone than with either treatment alone. However, the tobacco cultivars were injured more severely than expected by ozone fumigation when they were first treated with either pebulate or chloramben.

In this section on the interaction of herbicides and ozone, it is appropriate to include a discussion of the interaction of growth regulators and ozone. The most commonly used growth regulators are compounds that inhibit stem elongation of ornamental plants. They are used to make shorter, sturdier, bushier plants that appear to be a mass of flowers because their blossoms are closer together. Cathey and Heggestad (1972) tested a number of such compounds for their ability to make petunias more resistant to ozone. The two compounds, 2,4-dichlorobenzyltributylphosphorium chloride (CBBP) and succinic acid-2,2-dimethyl hydrazide (SADH) and certain of its analogs that did retard growth of petunias, also made them less sensitive to ozone. These compounds were more effective as foliar sprays than as soil drenches, and only leaves that developed after treatment were resistant to ozone.

B. Insecticides

The presence of synthetic organic insecticides in the air as pollutants has been reported many times. Newsom (1967) stated that samples of air collected from 4 California cities in 1963 contained detectable amounts of DDT, and that samples taken on flights detected DDT in the air over remote areas of Canada. Abbott *et al.* (1965) detected DDT, dieldrin, and isomers of benzene hexachloride in rainwater collected on rooftops at 2 locations in London, England. Later Abbott *et al.* (1966) also detected these organochlorine insecticides in air samples. Peterle (1969) reported DDT concentrations of 40 parts per trillion in meltwater from Antarctic ice.

Drift is a form of air pollution by pesticides that has been of particular concern to applicators of insecticides. The main factors that influence drift are the equipment and method used for application, the physical form or formulation of the spray material, and microclimate (Akesson and Yates, 1964).

Because large scale drift is most likely to be a problem of aerial application of pesticides, most drift investigations have focused on this aspect of pesticide use. The actual patterns developed by various types of aircraft used to apply pesticides have been described by Chamberlin *et al.* (1955) and Yates (1962). The physical form of the material being applied has a profound influence on the distance the pesticide will drift. The relation is best described by the data in Table I (Brooks, 1947).

Most insecticide formulations are used in the form of dusts, sprays, or granules. The danger of drift is greatest for dusts and least for granules. Akesson and Yates (1964) stated that in the 1940's dusts were used in over 60% of the aircraft applications in the United States, but that their use was so hazardous that by the 1960's, dust applications from the air had

TABLE I

THE RELATION OF PARTICLE SIZE TO DISTANCE THAT PARTICLE WILL DRIFT HORIZONTALLY IN A 3 MPH WIND WHILE FALLING 10 FT VERTICALLY

Particle size (μm)	Particle type	Horizontal drift (ft)
400	Coarse sprays, aircraft	8
150	Medium sprays, aircraft	22
100	Fine sprays, aircraft	48
50	Mist blowers	178
20	Fine sprays and dusts	1,109
10	Commonly used dusts	4,435
2	Aerosols	110,880

decreased to about 20%. Sprays are much more convenient to use because their droplet size can be largely controlled by the type of spray nozzle used; by additives such as surfactants; and by materials such as amine stearates that restrict reduction of droplet size by evaporation between nozzle and target. Granules, because of their large particle size, 200–600 μm, have little tendency to drift. Their use, however, is restricted to application of pesticides in or on the soil.

The meteorological conditions that restrict or limit danger from air pollution by drift are actually those that might favor danger from other forms of air pollution. These include minimum air movement and conditions that lead to temperature inversions. For example, it is common practice for orchardists to do their spraying very early in the morning or in the early evening when convection currents have died down. The most favorable conditions for restricting drift—a clear, summer evening—also favor formation of an inversion layer. The lower the inversion layer, the lower the mixing depth, and the less the possibility for drifting of pesticides sprayed from the ground. This is not always true, however. Akesson and Yates (1964) found that an unusually low inversion layer at 10–20 ft allowed no upward diffusion of the spray material and consequently the applied insecticide drifted further downwind. Argauer *et al.* (1968) examined drift of malathion and azinphosmethyl sprayed from the air during periods when cross winds were 4–10 mph. In some cases they were able to detect the insecticides as far as $\frac{1}{2}$ mile from the line of application. From 18–96% of the amount applied was found on plots adjacent to the treated swaths. When the application was made from a height of about 8 ft the swath was about 50 ft wide. Applied from a height of about 30 ft, the spray made a swath 95 ft wide.

Insecticides also enter the air by volatilization. Lichtenstein (1972) claimed that "volatilization is chiefly responsible for the loss of pesticidal residues from a given substance to which they had originally been applied." Lichtenstein and Schulz (1961) investigated the effect of soil cultivation, soil surface, and water on persistence of insecticides in the soil. They found that aldrin is displaced from soil particles by water and that the insecticide was then lost from soil by volatilization. Aldrin was also lost from water surfaces, but such loss was not affected by water evaporation. In the field, daily disking of a loam soil treated with aldrin or DDT reduced the aldrin soil residues by 38% and DDT residues by 25% over a 3 month period. The experimental plots were examined again 10 years later, even though they had not been disked in the intervening years. Lichtenstein *et al.* (1971a) found that the nondisked particles of the plots still contained 44% of the applied DDT as its isomers and degradation products, and 11% of the applied aldrin as dieldrin, its breakdown product. The portions that had been disked contained 31% of applied DDT as its isomers and degradation product and 5% of the applied aldrin as dieldrin.

Lichtenstein *et al.* (1971b) compared the effect of cropping versus repeated cultivations in field plots that had been treated once with aldrin and heptachlor in 1960 and then examined 10 years later. Compared to the residues in the soil of the plots cropped continuously with alfalfa, the cultivated plots had 76–82% less insecticidal residues.

Lichtenstein and Schulz (1970) also studied volatilization of various ^{14}C-labeled insecticides from various substrates in the laboratory. The rate of volatilization proved to be determined by the physical properties of the particular insecticide and of the substrate to which it was applied. Of the compounds tested, aldrin was most volatile and azinphosmethyl the least. The highest rates of volatilization were from water, and the lowest from soil. Volatility of a specific insecticidal application from a substrate could be slowed by adding soil or a detergent.

Cliath and Spencer (1971) compared relative rates of loss of dieldrin and lindane from field plots. Lindane was lost more rapidly than dieldrin from the soil. They concluded that one reason for this difference was the 10-fold higher vapor pressure of lindane. Cliath and Spencer suggested that volatilization of these insecticides could be reduced by keeping the soil surface dry, or by mixing the insecticide with the soil rather than merely applying it to the soil surface. Caro and Taylor (1971) concluded that volatilization accounted for most of the dieldrin lost from treated field plots. Mackay and Wolkoff (1973) measured rates of evaporation of hydrocarbons and chlorinated hydrocarbons from water surfaces. The insecticides tested included DDT, lindane, dieldrin, and aldrin. Based on water solubility and vapor pressure, Mackay and Wolkoff calculated the respective half-lives in days of these compounds in a stream of water as 3.7, 289, 723, and 10.1. They concluded that transfer of chlorinated hydrocarbons from water to air may be much faster than is generally appreciated.

The escape of synthetic organic compounds from treated soil by codistillation with water has been suggested by Bowman *et al.* (1959), Weidhaas *et al.* (1960), and Acree *et al.* (1963). However, Spencer and Cliath (1969) could find no evidence that codistillation occurred when dieldrin was lost from soil or from mixtures with water.

Long distance transport of insecticides on dust particles was proposed by Cohen and Pinkerton (1966). They analyzed rainwater and dust for organochlorines and detected them in all the samples they examined. Some of the compounds could be identified as originating from insecticides. Using meteorological evidence, Cohen and Pinkerton concluded that some of the contaminated dust came from great distances. Risebrough *et al.* (1968) detected and identified chlorinated hydrocarbons in dust carried by trade winds to Barbados from the European–African land mass. They calculated that the amount of pesticides contributed to the tropical areas of the Atlantic by trade winds was about the same as the amount carried to the sea by major river systems.

Improper disposal of pesticide containers can also contaminate the air with potentially phytotoxic compounds. Kennedy *et al.* (1972) incinerated various synthetic organic pesticides at 900°C and detected the following gases in the fumes: chlorine, hydrogen chloride, hydrogen sulfide, and nitric acid.

Air pollution by insecticides can affect crops (1) by directly injuring sensitive plants, (2) by killing pollinating insects, (3) or by destroying insect parasites that control populations of economic pests. Potentially phytotoxic insecticides, such as arsenicals or dinitriphenols, can cause plant injury if they drift onto sensitive crops. Many synthetic organic insecticides are highly toxic to honeybees (Anderson and Atkins, 1968) and other pollinating insects. If improperly applied insecticides drift at pollination time across an apple orchard or an alfalfa crop being raised for seed, the fruit or seed yield generally is drastically reduced. The increase in the population of an economic pest by accidental destruction of its parasites has been reported by Hart *et al.* They found (1966b) that methyl parathion can produce an increase in the population of brown soft scale (*Coccus hesperidum* L.) in citrus. Later (1966a) they observed that the scale population increased in an adjacent grove that had been contaminated by drift of methyl parathion.

C. Fungicides and Nematicides

Air pollution by fungicides has in rare instances caused crop losses. However, most investigations of interactions between fungicides and air pollution in agriculture have dealt with the effect of fungicides on the sensitivity of plants to air pollutants such as ozone and sulfur dioxide.

Dimond and Stoddard (1955) presented a well-documented case of injury by a fungicide as an air pollutant. They examined the poisoning of roses in a greenhouse by mercury vapors emanating from fungicides incorporated in the sashbar paint used to protect the paint film against fungi. Zimmerman and Crocker (1934) had earlier described mercury injury to a number of rose cultivars as well as other plants. In 1954, Butterfield reported that many greenhouse growers in the northeastern United States had lost a great number of Better Times, Briarcliff and related rose cultivars that had apparently been damaged by mercury injury. The mercury vapors came from paints that contained the fungicide di(phenylmercuric) dodecenyl succinate (DPMDS).

The problem that Dimond and Stoddard investigated was brought to them by greenhouse growers who had painted their sashbars in July and had not seen injury until the ventilators were closed in late fall and early winter. Dimond and Stoddard were able to detect and identify mercury vapors physically, chemically, and biologically (Dimond and Waggoner, 1955). They found that DPMDS is decomposed by a number of other components in the paint and is released as mercury vapor. They pointed out that a stable formulation of paint containing DPMDS may start producing mercury vapors if another paint with destabilizing components is applied in the same confined space. There was evidence that the mercury vapors could not only enter the roses directly but could also enter the plants through the soil. To correct this pollution, the contaminated soil was removed from the greenhouse before new, uninjured roses were planted. To prevent mercury vapors from escaping from the paint film, Dimond and Stoddard covered the paint with a coating of lime sulfur in a paste of wheat flour. A single application of the paste brushed onto the offending sashbars was effective for at least 5 months.

Fungicides applied to plants or to soil in which plants are growing can make them either more or less sensitive to air pollutants. Fortunately, there are few instances of fungicides causing increased sensitivity to air pollutants. Matsushima and Harada (1965) observed that orange trees (*Citrus nobilis*) sprayed with Bordeaux mixture were more sensitive to injury by sulfur dioxide than were trees that had not been sprayed with Bordeaux.

Fungicides, if they have an effect, usually make plants more resistant to air pollutants. Middleton (1956) reported that plants sprayed with zineb (zinc ethylenebisdithiocarbamate) were protected from ozone. Walker (1961) and Silber (1964) protected tobacco from ozone injury in field plots by treating them with dichlone (2,3-dichloro-1,4-naphthoquinone). Kendrick *et al.* (1954) protected bean plants from injury by exposures to ozonated gasoline or hexene-1 by first dusting or spraying the leaves with dithiocarbamate fungicides. Ordin *et al.* (1962) tested a number of compounds, including zineb, as dusts to protect plants against photochemical air pollutants. They found that these compounds protected more effectively against injury by ozone than by peroxyacetyl nitrate (PAN). Kendrick *et al.* (1962) tested fungicides as protectants against both ozone and ozonated hexene. Of the materials tested the dithiocarbamate fungicides gave the best protection against ozone. The effective fungicide gave protection only when applied to the lower surface of leaves, and the protection was restricted to the portion of the leaf covered.

Development of synthetic organic system fungicides has produced a number of compounds that can protect plants against ozone. Taylor (1970) first reported that sprays or dusts of the systemic fungicide benomyl [methyl (1-butylcarbamoyl)benzimidazol-2-yl-carbamate] protected field plots of tobacco against ozone. Since then Pellissier *et al.* (1972) observed that both benomyl and a related systemic fungicide, thiabendazole [2-(4′-thiazolyl)benzimidazole] can protect beans against ozone. Taylor and Rich (1973) used soil treatments of benomyl and another systemic fungicide, carboxin (2,3-dihydro-6-methyl-5-phenylcarbamoyl-1,4-oxathiin) to protect field plots of tobacco against ozone. Manning and Vardaro (1973) reduced ozone injury to beans in the field by spraying them with benomyl, and prevented ozone injury completely by applying carboxin over bean seeds at planting. Moyer *et al.* (1974a) protected annual bluegrass (*Poa annua* L.) from ozone by applying soil treatments of benomyl and two other systemic fungicides, thiophanate [1,2-bis(3-ethoxycarbonyl-2-thioureido)benzene] and thiophanate-methyl [1,2-bis(3-methoxycarbonyl-2-thioureido)benzene]. Both thiophanate and thiophanate-methyl are related chemically and in biological activity to benomyl. Rich *et al.* (1974) reported the relative activity of various 1,4-oxathiin derivatives as protectants against ozone. Moyer *et al.* (1974b) suppressed naturally occurring oxidant injury on azaleas with soil drenches and foliar sprays of benzimidazole and oxathiin compounds.

Not much is known about the effect of nematicides on sensitivity to plants to air pollutants. Taylor and Miller (1970) reported that tobacco plants growing in field plots treated with a number of contact and fumigant nematicides were made more sensitive to ozone injury.

IV. Burning

Burning in agriculture is used (1) to remove agricultural wastes left after harvesting, logging, or pruning, (2) to destroy plant pathogens, insects, or weeds that cause problems in succeeding crops, (3) to burn prescribed forest areas as a method for controlling wildfires, and (4) in greenhouses for space heating and for increasing the CO_2 content of the greenhouse air.

In 1970, burning was used to dispose of 3.6 million tons of agricultural waste on 1.1 million acres of land (Shaw *et al.*, 1970). In the same year 3.5 million acres of forest subjected to prescribed burning (Dieterich, 1971) emitted an estimated 6.5 million tons of particulates.

Prescribed burning in forests and the burning of huge piles of agricultural wastes are often the only means of removing unwanted residues. There are also many instances in agriculture where burning may be perhaps the only way to control or reduce the incidence of plant diseases or pests. For example, burning the pruned limbs in apple orchards can reduce the amount of overwintering *Physalospora obtusa* the fungus that causes apple black rot (Miller and Anagnostakis, 1973). The incidence of *Verticillium albo-atrum,* a wilting fungus, has been reduced in potato fields by burning the vines at the end of the season (Powelson and Gross, 1962). Horner and Dooley (1965) flamed peppermint (*Mentha piperata* L.) fields in order to destroy overwintering sclerotia of *Verticillium dahliae.*

Perhaps the most widely known example of burning to control plant pests is the annual burning of grass seed fields in the Willamette Valley of western Oregon. Over a quarter million acres of this cropland is burned each summer, releasing about 700,000 tons of material into the air and contributing over 25% of the total annual quantity of suspended particulates (Shum and Loveland, 1974) in that area.

Grass seed fields in Oregon are burned primarily (Hardison, 1963) to control 3 problems: blind-seed disease caused by the fungus *Gloeotinia temulenta;* ergot caused by the fungus *Claviceps purpurea;* and grass seed nematodes *Anguina* sp. Hardison (1964) stated that burning was the only known method for controlling these pests, and had these additional benefits: it eliminated weeds, returned potash and other crop minerals to the soil, thinned the sod, destroyed insects and rodents, avoided the use of chemical pesticides, and accomplished the necessary removal of straw. Chilcote (1967) reported that burning increased seed yield of various grasses by about 50 to 350%.

Boubel *et al.* (1969) found that for each ton of fuel burned, the Oregon grass fields produced 101 lb of carbon monoxide, 15.6 lb of particulates, 2.8 lb of olefin, 1.8 lb of saturates plus acetylene, and 1.7 lb of ethylene. At their hottest, these fires produced 29 ppm of NO_x.

Air pollution from burning or combustion in agriculture can affect plants in at least 3 ways: (1) Smoke or flue gasses can contain sufficient ethylene to hasten fruiting or ripening of sensitive crops such as pineapple (Rodríguez, 1932), or cause crop damage in greenhouses (Hanan, 1973). (2) Hydrocarbons in the smoke can generate ozone when exposed to sunlight (Evans *et al.,* 1974). (3) The vast quantities of particulates from forest fires may perhaps effect rainfall, but Cooper (1971) does not think this is likely except under very special conditions.

V. Fertilizers and Feedlots

Air pollution can occur from production and use of phosphate fertilizers, nitrogenous fertilizers, and limestone.

The processing of phosphate rock for fertilizer produces gaseous and particulate fluorides that can reduce crop yields wherever such facilities exist. Phosphate fertilizers are produced from a fluorapatite [3 $Ca_3(PO_4)_2CaF_2$]. This mineral is treated with heat or acid to produce superphosphates, and in the process can emit HF, SiF_4 and H_2SiF_6 (Hendrickson, 1970).

Symptoms of fluoride injury to plants and their relative sensitivity are described in Treshow and Pack (1970). The biochemical effects of fluorides on vegetation are discussed by Weinstein and McCune (1970).

Plants in the vicinity of factories producing nitrogenous fertilizers can also be affected adversely. Sierpinski (1971) reported that about 130 ha of pines were killed by factory fumes that contained ammonium sulfate, ammonia, nitrogen oxides, and sulfur dioxide.

Limestone processing can throw enough dust on a nearby forest so that the accumulation can inhibit tree growth. Brandt and Rhoades (1973) measured the annual rings of trees covered with dust from limestone processing factories. They found that the lateral growth of *Acer rubrum* L., *Quercus prinus* L., and *Q. rubra* L. were reduced by at least 18% because of the accumulated dust. However, the lateral growth of dust-laden *Liriodendron tulipifera* L. increased by 76% (Brandt and Rhoades, 1973).

The estimated total annual production of livestock wastes in the United States is 1.1 billion tons of solids and 435 million tons of liquids (Wadleigh, 1968). There has been vigorous debate about the amount and fate of ammonia and other gaseous nitrogenous products from animal feedlots (Viets, 1971). Not only is ammonia evolved but other nitrogenous compounds as well. Mosier *et al.* (1973) trapped and identified a series of aliphatic amines emanating from cattle feedlots. Hutchinson and Viets (1969) estimated that a lake about $\frac{1}{4}$ mile from a large feedlot absorbed volatilized ammonia at the rate of 65 lb of nitrogen per acre per year, and that another lake about 1 mile from the feedlot absorbed about half that amount. They concluded that this amount of nitrogen can speed eutrophication. Gaseous ammonia can also be absorbed directly through the leaves of crops. Porter *et al.* (1972) demonstrated that corn seedlings exposed to 1 ppm of ammonia for 24 hours can absorb 43%, of the total ammonia in the air, and those exposed to 10 and 20 ppm of ammonia can absorb 30%.

VI. Photochemical Oxidants from Forest Volatiles

Forests, part of agriculture, make a contribution to air pollution (Went 1955). He noticed that blue summer haze is most pronounced over the forested areas of the world. Because it looked like the urban smog, Went called it natural smog. He proposed that it was generated photochemically by the action of sunlight on terpenes and other hydrocarbons that are volatilized from the large masses of vegetation. Rasmussen and Went (1965) identified α-pinene, β-pinene, myrcene, and isoprene in the air over plants. They found the highest concentrations of terpenes in the air when leaves were dying in the autumn, and when meadows were mowed. Their estimate of the annual world production of plant volatiles was 438×10^6 tons. Rasmussen (1972) measured the release of volatile hydrocarbons from different tree species and concluded that 70% of the forested areas of the United States emit terpenes to the atmosphere. He estimated that the world's forests annually emitted 175×10^6 tons of reactive hydrocarbons, 6.2 times greater than the emission of such compounds from man's activities. Ripperton *et al.* (1971) obtained experimental evidence that α-pinene can be active in a photochemical system that produces ozone. They suggested that these reactions produce nonurban, tropospheric ozone in nature. Their suggestion is reinforced by data of Stasiuk and Coffey (1974). By making simultaneous measurements, Stasiuk and Coffey found that ozone concentrations in urban areas of upstate New York were as high and often higher than the ozone concentrations in New York City and other densely populated urban areas. In addition, ozone levels in the rural areas remained high even during the night, while that of the cities dropped to low concentrations after sunset.

This rural source of ozone may well be the cause of such diseases of eastern white pine (*Pinus strobus* L.) as emergence tipburn (Berry and Ripperton, 1963), chlorotic dwarf (Dochinger, 1968) and semimature-tissue needle blight (Linzon, 1967). Hepting (1968) suggested that all of these symptoms, known at one time or another as white pine blight, may be oxidant damage. In 1966, he noted that emergence tipburn has been known for at least 70 years.

VII. Other Air Pollutants from Agriculture

Production of airborne dusts in agriculture is merely mentioned in this chapter, not because they are generated in small amounts, but because their effects on plants have not been properly defined. Wadleigh (1968) stated that each year 30 million tons of natural dusts enter the atmosphere. Industries related to agriculture, such as cotton ginning, alfalfa mills, and lime kilns, plus the extracting industries, such as cement factories, smelters, and mines, annually generate 17 million tons of dust. Wadleigh noted that the dust can accumulate on plants and impair the growth or quality of crops.

The effect of limestone dust on plants has already been discussed

(Brandt and Rhoades, 1973). Darley (1966) examined the effect of cement-kiln dust on vegetation. He found that in the presence of free moisture the cement dust encrusted on leaves could cause injury by excluding light and because of the alkalinity of the solution. He also compared the carbon dioxide exchange rates of dusted and nondusted bean leaves. Only the finer particles of cement-kiln dust interfered with carbon dioxide exchange rates. Darley concluded that the toxicity of this dust depends on its chemical composition, particle size, and deposition rate. This is probably true for dusts from other sources as well.

Dusts may also influence the effect of other air pollutants on plants. Vasiloff and Drummond (1974) were able to demonstrate that road dust applied to the upper surface of leaves partially protected buckwheat (*Fagopyron esculentum*) from sulfur dioxide and beans from ozone.

Finally, there is one more source of air pollution in agriculture that does not fit in the other categories. Wills and Patterson (1970) reported that fluorescent lights, such as those used in plant growth chambers or greenhouses, can produce enough ethylene to affect the growth of peas (*Pisum sativum* L.). They found that the ethylene originated in the ballast chokes, and that at operating temperature a single ballast generated 1.5 μg/hour of ethylene. Ethylene was also obtained from lacquered copper wire such as that used in the ballasts. Other hydrocarbons given off by the ballasts in lesser amounts were methane, ethane, propane, propene, 2-methylpropane, butane, but-1-ene, 2-methylpropene, and pentane.

References

Abbott, D. C., Harrison, R. B., Tatton, J. O., and Thomson, J. (1965). Organochlorine pesticides in the atmospheric environment. *Nature* (*London*) **208,** 1317–1318.

Abbott, D. C., Harrison, R. B., Tatton, J. O., and Thomson, J. (1966). Organochlorine pesticides in the atmosphere. *Nature* (*London*) **211,** 259–261.

Acree, F., Jr., Beroza, M., and Bowman, M. C. (1963). Co-distillation of DDT with water. *J. Agr. Food Chem.* **11,** 278–280.

Adams, D. F., and Sulzbach, C. W. (1961). Nitrogen deficiency and fluoride susceptibility of bean seedlings. *Science* **133,** 1425–1426.

Adams, D. F., Jackson, C. M., and Bamesberger, W. L. (1964). Quantitative studies of 2,4-D esters in the air. *Weeds* **12,** 280–283.

Adedipe, N. O., Hofstra, G., and Ormrod, D. P. (1972). Effects of sulfur nutrition on phytotoxicity and growth responses of bean plants to ozone. *Can. J. Bot.* **50,** 1789–1793.

Akesson, N. B., and Yates, W. E. (1964). Problems relating to application of agricultural chemicals and resulting drift residues. *Annu. Rev. Entomol.* **9,** 285–318.

Anderson, L. D., and Atkins, E. L., Jr. (1968). Pesticide usage in relation to beekeeping. *Annu. Rev. Entomol.* **13,** 213–238.

Argauer, R. J., Mason, H. C., Corley, C., Higgins, A. H., Sauls, J. N., and Liljedahl, L. A. (1968). Drift of water-diluted and undiluted formulations of malathion and azinphosphmethyl applied by airplane. *J. Econ. Entomol.* **61,** 1015–1020.

Benson, N. R. (1959). Fluoride injury on soft suture and splitting of peaches. *Proc. Amer. Soc. Hort. Sci.* **74,** 184–198.

Berry, C. R., and Ripperton, L. A. (1963). Ozone, a possible cause of white pine emergence tipburn. *Phytopathology* **53,** 552–557.

Boubel, R. W., Darley, E. F., and Schuck, E. A. (1969). Emissions from burning grass, stubble, and straw. *J. Air Pollut. Contr. Ass.* **19,** 497–500.

Bowman, M. C., Acree, F., Jr., Schmidt, C. H., and Beroza, M. (1959). Fate of DDT in larvacide suspensions. *J. Econ. Entomol.* **52,** 1038–1042.

Brandt, C. J., and Rhoades, R. W. (1973). Effects of limestone dust accumulation on lateral growth of forest trees. *Environ. Pollut.* **4,** 207–213.

Brennan, E. G., Leone, I. A., and Daines, R. H. (1950). Fluorine toxicity in tomato as modified by alterations in the nitrogen, calcium and nutrition of the plant. *Plant Physiol.* **25,** 736–747.

Brewer, R. F., Creveling, R. K., and Guillemet, F. B. (1960). The effects of hydrogen fluoride gas on seven citrus varieties. *Proc. Amer. Soc. Hort. Sci.* **75,** 236–243.

Brewer, R. F., Guillemet, F. B., and Creveling, R. K. (1961). Influence of N-P-K fertilization on incidence and severity of oxidant damage to mangels and spinach. *Soil Sci.* **92,** 298–301.

Brooks, F. A. (1947). The drifting of poisonous dusts applied by airplanes and land rigs. *J. Agr. Eng.* **28,** 233–239.

Butterfield, N. W. (1954). Roses affected by mercury compounds in greenhouse paints. *Florists Exch. Hort. Trade World* Oct. 30, 14.

Carney, A. W., Stephenson, G. R., Ormrod, D. P., and Ashton, G. C. (1973). Ozone-herbicide interactions in crop plants. *Weed Sci.* **21,** 508–511.

Caro, J. H., and Taylor, A. W. (1971). Pathways of loss of dieldrin from soils under field conditions. *J. Agr. Food Chem.* **19,** 379–384.

Cathey, H. M., and Heggestad, H. E. (1972). Reduction of ozone damage to *Petunia hybrida* Vilm. by use of growth regulating chemicals and tolerant cultivars. *J. Amer. Soc. Hort. Sci.* **97,** 695–700.

Chamberlin, J. C., Getzendaner, C. W., Hessig, H. H., and Young, V. D. (1955). Studies of airplane spray-deposit patterns at low flight levels. *U.S., Dep. Agr., Tech. Bull.* **1110,** 1–45.

Chilcote, D. O. (1967). Grass seed field burning report. *Proc. 27th Annu. Meet. Oreg. Seed Growers League* pp. 67–70.

Cliath, M. M., and Spencer, W. F. (1971). Movement and persistence of dieldrin and lindane in soil as influenced by placement and irrigation. *Soil Sci. Soc. Amer. Proc.* **35,** 791–795.

Cohen, J. M., and Pinkerton, C. (1966). Widespread translocation of pesticides by air transport and rain-out. *Advan. Chem. Ser.* **60,** 163–176.

Cooper, C. F. (1971). Effects of prescribed burning on the ecosystem. *Proc. Prescribed Burning Symp. (Charleston, S. C.), U.S. Dep. Agr. Forest Serv., Asheville, N. C.* pp. 152–160.

Daines, R. H., Leone, I. A., and Brennan, E. (1952). The effect of fluorine on plants as determined by soil nutrition and fumigant studies. *In* "Air Pollution" (L. C. McCabe, ed.), pp. 97–105. McGraw-Hill, New York.

Daines, R. H., Leone, I. A., and Brennan, E. (1960). Air pollution as it affects agriculture in New Jersey. *N.J. Agr. Exp. Sta., Bull.* **794,** 1–14.

Danielson, L. L., and Gentner, W. A. (1964). Influence of air movement on persistence of EPTC on soil. *Weeds* **12,** 92–94.

Darley, E. F. (1966). Studies on the effect of cement-kiln dust on vegetation. *J. Air Pollut. Contr. Ass.* **16,** 145–150.

Davidson, O. W. (1949). Effects of ethylene on orchid flowers. *Proc. Amer. Soc. Hort. Sci.* **53,** 444–446.

Davis, D. D., and Kress, L. (1974). The relative susceptibility of ten bean varieties to ozone. *Plant Dis. Rep.* **58,** 14–16.

Day, B. E., Johnson, E., and Dewlen, J. L. (1959). Volatility of herbicides under field conditions. *Hilgardia* **28,** 255–267.

Dean, C. E., and Davis, D. R. (1967). Ozone and soil moisture in relation to the occurrence of weather fleck on Florida cigar-wrapper tobacco in 1966. *Plant Dis. Rep.* **51,** 72–75.

Dieterich, J. H. (1971). Air-quality aspects of prescribed burning. *Proc. Prescribed Burning Symp. (Charleston, S.C.), U.S. Dep. Agr. Forest Serv., Asheville, N.C.* pp. 139–151.

Dimond, A. E., and Stoddard, E. M. (1955). Toxicity to greenhouse roses from paints containing mercury fungicides. *Conn., Agr. Exp. Sta., New Haven, Bull.* **559,** 1–19.

Dimond, A. E., and Waggoner, P. E. (1955). Bioassay of mercury vapor arising from a phenyl mercury compound. *Plant Physiol.* **30,** 374–376.

Dochinger, L. S. (1968). The impact of air pollution on eastern white pine: The chlorotic dwarf disease. *J. Air Pollut Contr. Ass.* **18,** 814–816.

Dunning, J. A., Tingey, D. T., and Heck, W. W. (1972). Foliar sensitivity of legumes to ozone as affected by temperature, potassium nutrition, and dose. *N.C. State Univ., Agr. Exp. Sta., J. Ser., Pap.* **3931,** 1–22.

Engle, R. L., and Gabelman, W. H. (1966). Inheritance and mechanism for resistance to ozone damage in onion *Allium cepa* L. *Proc. Amer. Soc. Hort. Sci.* **89,** 423–430.

Evans, L. F., King, N. K., Packman, D. R., and Stephens, E. T. (1974). Ozone measurements in smoke from forest fires. *Environ. Sci. Technol.* **8,** 75–76.

Flint, G. W., Alexander, J. J., and Funderburk, H. P. (1963). Vapor pressures of low volatile esters of 2,4-D *Weed Sci.* **16,** 541–543.

Fowler, D. L., and Mahan, J. N. (1973). "The Pesticide Review, 1972." U.S. Dept of Agr., Washington, D.C.

Frank, P. A., and Grigsby, B. H. (1957). Effects of herbicidal sprays on nitrate accumulation in certain weed species. *Weeds* **5,** 206–217.

Gentner, W. A. (1964). Herbicidal activity of vapors of 4-amino-3,5,6-trichloropicolonic acid. *Weeds* **12,** 239–240.

Hanan, J. J. (1973). Ethylene pollution from combustion in greenhouses. *HortScience* **8,** 23–24.

Hardison, J. R. (1963). Control of Gloeotinia temulenta in seed fields in *Lolium perenne* by cultural methods. *Phytopathology* **53,** 460–464.

Hardison, J. R. (1964). Justification for burning grass fields. *Proc. 24th Annu. Meet. Oreg. Seed Growers League* pp. 93–96.

Hart, W. G., Ingle, S., Garza, M., and Mata, M. (1966a). The response of soft scale and its parasites to repeated insecticide pressure. *J. Rio Grande Val. Hort. Soc.* **20,** 64–68.

Hart, W. G., Balock, J. W., and Ingle, S. (1966b). The brown soft scale, *Coccus hesperidum* L. (Hemiptera: Coccidae), in citrus groves in Rio Grande Valley. *J. Rio Grande Val. Hort. Soc.* **20,** 69–73.

Heck, W. W., Dunning, J. A., and Hindawi, I. J. (1965). Interactions of environmental factors on the sensitivity of plants to air pollution. *J. Air Pollut. Contr. Ass.* **15,** 511–515.

Hendrickson, E. R. (1970). The fluoride problem. *In* "Impact of Air Pollution on Vegetation" (S. N. Linzon, ed.), pp. 29–36. Air Pollut. Contr. Ass., Pittsburgh, Pennsylvania.

Hepting, G. H. (1966). Air pollution impacts to some important species of pine. *J. Air Pollut. Contr. Ass.* **16,** 63–65.

Hepting, G. H. (1968). Diseases of forest and tree crops caused by air pollutants. *Phytopathology* **58,** 1098–1101.

Hodgson, R. H. (1970). Alteration of triazine metabolism by ozone. *Abstr., Weed Sci. Soc. Amer.* p. 28.

Hodgson, R. H., Frear, D. S., Swanson, H. R., and Regan, L. A. (1973). Alteration of diphenamid metabolism in tomato by ozone. *Weed Sci.* **21,** 542–549.

Horner, C. E., and Dooley, H. L. (1965). Propane flaming kills *Verticillium dahliae* in peppermint stubble. *Plant Dis. Rep.* **49,** 581–582.

Howell, R. K., and Kremer, D. F. (1972). Ozone injury to soybean cotyledonary leaves. *J. Environ. Qual.* **1,** 94–97.

Howell, R. K., and Thomas, C. A. (1972). Relative tolerance of twelve safflower cultivars to ozone. *Plant Dis. Rep.* **56,** 195–197.

Howell, R. K., Devine, T. E., and Hanson, C. H. (1971). Resistance of selected alfalfa strains to ozone. *Crop Sci.* **11,** 114–115.

Hutchinson, G. L., and Viets, F. G., Jr. (1969). Nitrogen enrichment of surface water by absorption of ammonia volatilized from cattle feedlots. *Science* **166,** 514–515.

Jacobson, J. S., and Hill, A. C., eds. (1970). "Recognition of Air Pollution Injury to Vegetation: A Pictorial Atlas," Inform. Rep. No. 1, TR-7 Agr. Comm. Air Pollut. Contr. Ass., Pittsburgh, Pennsylvania.

Kearney, P. C., Sheets, T. J., and Smith, J. W. (1964). Volatility of seven *s*-triazines. *Weeds* **12,** 83–87.

Kendrick, J. B., Jr., Middleton, J. T., and Darley, E. F. (1954). Chemical protection of plants from ozonated olefin (smog). *Phytopathology* **44,** 494–495.

Kendrick, J. B., Jr., Darley, E. F., and Middleton, J. T. (1962). Chemotherapy for oxidant and ozone induced plant damage. *Int. J. Air Water Pollut.* **6,** 391–402.

Kennedy, M. V., Stojanovic, B. J., and Shuman, F. L., Jr. (1972). Chemical and thermal aspects of pesticide disposal. *J. Environ. Qual.* **1,** 63–65.

Leone, I. A., and Brennan, E. (1969). The importance of moisture in ozone phytotoxicity. *Atmos. Environ.* **3,** 399–406.

Leone, I. A., and Brennan, E. (1970). Ozone toxicity in tomato as modified by phosphorous deficiency. *Phytopathology* **60,** 1521–1524.

Leone, I. A., and Brennan, E. (1972). Modification of sulfur dioxide injury to tobacco and tomato by varying nitrogen and sulfur nutrition. *J. Air Pollut. Contr. Ass.* **22,** 544–550.

Leone, I. A., Brennan, E., and Daines, R. H. (1966). Effect of nitrogen nutrition on the response of tobacco to ozone in the atmosphere. *J. Air Pollut. Contr. Ass.* **16,** 191–196.

Lichtenstein, E. P. (1972). Environmental factors affecting fate of pesticides. *In* "Degradation of Synthetic Organic Molecules in the Biosphere," pp. 190–205. Nat. Acad. Sci., Washington, D.C.

Lichtenstein, E. P., and Schulz, K. R. (1961). Effect of soil cultivation, soil surface, and water on the persistence of insecticidal residue in soils. *J. Econ. Entomol.* **54,** 517–522.

Lichtenstein, E. P., and Schulz, K. R. (1970). Volatilization of insecticides from various substrates. *J. Agr. Food Chem.* **18,** 814–818.

Lichtenstein, E. P., Fuhremann, T. W., and Schulz, K. R. (1971a). Persistence and vertical distribution of DDT, lindane, and aldrin residues, 10 and 15 years after a single soil application. *J. Agr. Food Chem.* **19,** 718–721.

Lichtenstein, E. P., Schulz, K. R., and Fuhremann, T. W. (1971b). Effects of a cover crop and soil cultivation on the fate and vertical distribution of insecticide residues in soil, 7 to 11 years after soil treatment. *Pestic. Monit. J.* **5,** 218–222.

Linzon, S. N. (1967). Ozone damage and semimature-tissue needle blight of eastern white pine. *Can. J. Bot.* **45,** 2047–2061.

McCune, D. C., Hitchcock, A. E., and Weinstein, L. H. (1966). Effect of mineral nutrition on the growth and sensitivity of gladiolus to hydrogen fluoride. *Contrib. Boyce Thompson Inst.* **23,** 295–299.

MacDowall, F. D. H. (1966). The relation between dew and tobacco weather fleck. *Can. J. Plant Sci.* **46,** 349–353.

Mackay, D., and Wolkoff, A. W. (1973). Rate of evaporation of low-solubility contaminants from water bodies to atmosphere. *Environ. Sci. Technol.* **7,** 611–614.

MacLean, D. C. (1970). Influence of magnesium nutrition on the sensitivity of tomato plants to air pollution. *HortScience* **5,** 333 (abstr.).

MacLean, D. C., Roark, O. F., Folkerts, G., and Schneider, R. E. (1969). Influence of mineral nutrition on the sensitivity of tomato plants to hydrogen fluoride. *Environ. Sci. Technol.* **3,** 1201–1204.

MacLean, D. C., Schneider, R. E., and McCune, D. C. (1973). Fluoride toxicity as affected by relative humidity. *Proc. Int. Clean Air Congr., 3rd,* 1973 A143-A145.

Manning, W. J., and Vardaro, P. M. (1973). Suppression of oxidant injury on beans by systemic fungicides. *Phytopathology* **63,** 1415–1416.

Mansfield, T. A., and Majernik, O. (1970). Can stomata play a part in protecting plants against air pollutants? *Environ. Pollut.* **1,** 149–154.

Matsushima, J., and Harada, M. (1965). Sulfur dioxide gas injury to fruit trees. III. Leaf fall of citrus trees by Bordeaux spray before sulfur dioxide fumigation. *J. Jap. Soc. Hort. Sci.* **34,** 272–276.

Menser, H. A., and Street, O. E. (1962). Effects of air pollution, nitrogen levels, supplemental irrigation, and plant spacing on weather fleck and leaf losses of Mayland tobacco. *Tob. Sci.* **6,** 165–169.

Middleton, J. T. (1956). Response of plants to air pollution. *J. Air Pollut. Contr. Ass.* **6,** 1–4.

Miller, P. M., and Anagnostakis, S. L. (1973). Piles of apple prunings as sources of conidia of *Physalospora obtusa. Phytopathology* **63,** 1080.

Mosier, A. R., Andre, C. E., and Viets, F. G., Jr. (1973). Identification of aliphatic amines volatilized from cattle feedyards. *Environ. Sci. Technol.* **7,** 642–644.

Moyer, J. H., Cole, H., Jr., and Lacasse, N. L. (1974a). Reduction of ozone injury on *Poa annua* by benomyl and thiophanate. *Plant Dis. Rep.* **58,** 41–44.

Moyer, J. W., Cole, H., Jr., and Lacasse, N. L. (1974b). Suppression of naturally occurring oxidant injury on azalea plants by drench or foliar spray treatment with benzimidazole or oxathiin compounds. *Plant Dis. Rep.* **58,** 136–138.

Newsom, L. D. (1967). Consequences of inseticide use on nontarget organisms. *Annu. Rev. Entomol.* **12,** 257–286.

Ordin, L., Taylor, O. C., Propst, B. E., and Cardiff, E. A. (1962). Use of antioxidants to protect plants from oxidant type air pollutants. *Int. J. Air Water Pollut.* **6,** 223–227.

Otto, H. W., and Daines, R. H. (1969). Plant injury by air pollutants: Influence of humidity on stomatal apertures and plant response to ozone. *Science* **163,** 1209–1210.

Pack, M. R. (1966). Response of tomato fruiting to hydrogen fluoride as influenced by calcium nutrition. *J. Air Pollut. Contr. Ass.* **16,** 541–544.

Pellissier, M., Lacasse, N. L., and Cole, H., Jr. (1972). Effectiveness of benzimidazole, benomyl, and thiabendazole in reducing ozone injury to pinto beans. *Phytopathology* **62,** 580–582.

Peterle, T. J. (1969). DDT in Antarctic snow. *Nature (London)* **224,** 620.

Porter, L. K., Viets, F. G., Jr., and Hutchinson, G. L. (1972). Air containing nitrogen-15 ammonia:Foliar absorption by corn seedlings. *Science* **175,** 759–761.

Powelson, R. L., and Gross, A. E. (1962). Thermal inactivation of *Verticillium albo-atrum* in diseased potato vines. *Phytopathology* **52,** 364.

Rasmussen, R. A. (1972). What do hydrocarbons from trees contribute to air pollution? *J. Air Pollut. Contr. Ass.* **22,** 537–543.

Rasmussen, R. A., and Went, F. W. (1965). Volatile organic material of plant origin in the atmosphere. *Proc. Nat. Acad. Sci. U.S.* **53,** 215–220.

Rich, S., and Turner, N. C. (1972). Importance of moisture on stomatal behavior of plants subjected to ozone. *J. Air Pollut. Contr. Ass.* **22,** 718–721.

Rich, S., Ames, R., and Zuckel, J. W. (1974). 1,4-oxathiin derivatives protect plants against ozone. *Plant Dis. Rep.* **58,** 162–164.

Ripperton, L. A., Jeffries, H., and Worth, J. J. B. (1971). Natural synthesis of ozone in the troposphere. *Environ. Sci. Technol.* **5,** 246–248.

Risebrough, R. W., Haggett, R. J., Griffin, J. J., and Goldberg, E. D. (1968). Pesticides: Transatlantic movements in Northeast trades. *Science* **159,** 1233–1236.

Rodríguez, A. G. (1932). Influence of smoke and ethylene on the fruiting of the pineapple (*Ananas sativus* Shult). *J. Dep. Agr. P. R.* **16,** 5–18.

Ryder, E. J. (1973). Selecting and breeding plants for increased resistance to air pollutants. *Advan. Chem. Ser.* **122,** 78–84.

Seidman, G. (1963). Water availability and sensitivity of plants to photochemical air pollution. *Plant Physiol.* **38,** Suppl., xxxvi.

Shaw, W. C., Heck, W. W., and Heggestad, H. E. (1970). The role of agriculture in air pollution and its control. *Proc. 19th Annu. Meet. Agr. Res. Inst.* pp. 30–43.

Sherwood, C. H., Weigle, J. L., and Denisen, E. L. (1970). 2,4-D as an air pollutant: Effects on growth of representative horticultural plants. *HortScience* **5,** 211–213.

Shum, Y. S., and Loveland, W. D. (1974). Atmospheric trace element concentrations associated with agricultural field burning in the Willamette Valley of Oregon. *Atmos. Environ.* **8,** 645–655.

Sierpinski, Z. (1971). Secondary noxious insects of pine in stands growing in areas with industrial air pollution containing nitrogen compounds. *Sylwan* **115,** 11–18.

Silber, G. (1964). Effectiveness of some chemicals for the prevention of weather fleck of field-grown tobacco. *Tob. Sci.* **8,** 93–95.

Spencer, W. F., and Cliath, M. M. (1969). Vapor density of dieldrin. *Environ. Sci. Technol.* **3,** 670–674.

Stahler, L. M., and Whitehead, E. I. (1950). The effect of 2,4-D on potassium nitrate level in leaves of sugar beets. *Science* **112,** 749–751.

Stasiuk, W. N., Jr., and Coffey, P. E. (1974). Rural and urban ozone relationships in New York State. *J. Air Pollut. Contr. Ass.* **24,** 564–568.

Stolzy, L. H., Taylor, O. C., Letey, J., and Szuszkiewicz, T. E. (1961). Influence of soil-oxygen diffusion rates on susceptibility of tomato plants to air-borne oxidants. *Soil Sci.* **91,** 151–155.

Swanson, C. R., and Shaw, W. C. (1954). The effect of 2,4-dichlorophenoxyacetic acid on the hydrocyanic acid and nitrate content of sudangrass. *Agron. J.* **46,** 418–421.

Talbert, R. E., Smith, D. R., and Frans, R. E. (1971). Volatilization, leaching, and absorption of prometryne in relation to selectivity in cotton. *Weed Sci.* **19,** 6–10.

Taylor, G. S. (1970). Tobacco protected against fleck by benomyl and other fungicides. *Phytopathology* **60,** 578.

Taylor, G. S., and Miller, P. M. (1970). Weather fleck on tobacco made worse by certain nematicides. *Phytopathology* **60,** 578–579.

Taylor, G. S., and Rich, S. (1973). Ozone fleck on tobacco reduced by benomyl and carboxin in soil. *Phytopathology* **63,** 208.

Taylor, G. S., DeRoo, H. G., and Waggoner, P. E. (1960). Moisture and fleck of tobacco. *Tob. Sci.* **4,** 62–68.

Treshow, M., and Pack, M. R. (1970). Fluoride. *In* "Recognition of Air Pollution Injury to Vegetation: A Pictorial Atlas" (J. S. Jacobson and A. C. Hill, eds.), pp. D1-D17. Air Pollut. Contr. Ass., Pittsburgh, Pennsylvania.

Vasiloff, G. N., and Drummond, D. B. (1974). The effectiveness of road dust as a protective agent on buckwheat and pinto bean against sulfur dioxide and ozone. *Phytopathology* **64,** 588.

Viets, F. G., Jr. (1971). The mounting problem of cattle feedlot pollution. *Agr. Sci. Rev.* **9,** 1–7.

Wadleigh, C. H. (1968). Wastes in relation to agriculture and forestry. *U.S., Dep. Agr., Misc. Publ.* **1065,** 1–112.

Walker, E. K. (1961). Chemical control of weather fleck in flue-cured tobacco. *Plant Dis. Rep.* **45,** 583–586.

Walker, E. K., and Vickery, L. S. (1961). Influence of sprinkler irrigation on the incidence of weather fleck on flue-cured tobacco in Ontario. *Can. J. Plant Sci.* **41,** 281–287.

Weidhaas, D. E., Schmidt, C. H., and Bowman, M. C. (1960). Effect of heterogeneous distribution and codistillation on the results of tests with DDT against mosquito larvae. *J. Econ. Entomol.* **53,** 121–125.

Weinstein, L. H., and McCune, D. C. (1970). Effects of fluorides on vegetation. *In* "Impact of Air Pollution on Vegetation" (S. N. Linzon, ed.), pp. 81–86. Air Pollut. Contr. Ass., Pittsburgh, Pennsylvania.

Went, F. W. (1955). Air pollution. *Sci. Amer.* **192,** 62–72.

Wiebe, H. H., and Poovaiah, B. W. (1974). Influence of moisture, heat, and light stress on hydrogen fluoride injury to soybeans. *Plant Physiol.* **52,** 542–545.

Wilhour, R. G. (1970). The influence of temperature and relative humidity on the response of white ash to ozone. *Phytopathology* **60,** 579.

Will, J. B., and Skelly, J. M. (1974). The use of fertilizer to alleviate air pollution damage to white pines (*Pinus strobus*) Christmas trees. *Plant Dis. Rep.* **58,** 150–154.

Wills, R. B. H., and Patterson, B. D., (1970). Ethylene, a plant hormone from fluorescent lighting. *Nature* (*London*) **225,** 199.

Yates, W. E. (1962). Spray pattern analysis and evaluation of deposits from agricultural aircraft. *Trans. ASAE* (*Amer. Soc. Agri. Eng.*) **5,** 49–53.

Zimmerman, P. W., and Crocker, W. (1934). Plant injury caused by vapors of mercury and compounds containing mercury. *Contrib. Boyce Thompson Inst.* **6,** 167–187.

Zimmerman, P. W., and Hitchcock, A. E. (1956). Susceptibility of plants to hydrofluoric acid and sulfur dioxide gases. *Contrib. Boyce Thompson Inst.* **18,** 263–279.

SUBJECT INDEX

B

C

D

E

K

L

P

Q

R

S

T

U

B
C
D 8
E 9
F 0
G 1
H 2
I 3
J 4